Oxford Series in Ecology and Evolution
Edited by Robert M. May, H. Charles J. Godfray, Jennifer A. Dunne, and Ben Sheldon

The Comparative Method in Evolutionary Biology
Paul H. Harvey and Mark D. Pagel

The Cause of Molecular Evolution
John H. Gillespie

Dunnock Behaviour and Social Evolution
N. B. Davies

Natural Selection: Domains, Levels, and Challenges
George C. Williams

Behaviour and Social Evolution of Wasps: The Communal Aggregation Hypothesis
Yosiaki Itô

Life History Invariants: Some Explorations of Symmetry in Evolutionary Ecology
Eric L. Charnov

Quantitative Ecology and the Brown Trout
J. M. Elliott

Sexual Selection and the Barn Swallow
Anders Pape Møller

Ecology and Evolution in Anoxic Worlds
Tom Fenchel and Bland J. Finlay

Anolis Lizards of the Caribbean: Ecology, Evolution, and Plate Tectonics
Jonathan Roughgarden

From Individual Behaviour to Population Ecology
William J. Sutherland

Evolution of Social Insect Colonies: Sex Allocation and Kin Selection
Ross H. Crozier and Pekka Pamilo

Biological Invasions: Theory and Practice
Nanako Shigesada and Kohkichi Kawasaki

Cooperation Among Animals: An Evolutionary Perspective
Lee Alan Dugatkin

Natural Hybridization and Evolution
Michael L. Arnold

The Evolution of Sibling Rivalry
Douglas W. Mock and Geoffrey A. Parker

Asymmetry, Developmental Stability, and Evolution
Anders Pape Møller and John P. Swaddle

Metapopulation Ecology
Ilkka Hanski

Dynamic State Variable Models in Ecology: Methods and Applications
Colin W. Clark and Marc Mangel

The Origin, Expansion, and Demise of Plant Species
Donald A. Levin

The Spatial and Temporal Dynamics of Host-Parasitoid Interactions
Michael P. Hassell

The Ecology of Adaptive Radiation
Dolph Schluter

Parasites and the Behavior of Animals
Janice Moore

Evolutionary Ecology of Birds
Peter Bennett and Ian Owens

The Role of Chromosomal Change in Plant Evolution
Donald A. Levin

Living in Groups
Jens Krause and Graeme D. Ruxton

Stochastic Population Dynamics in Ecology and Conservation
Russell Lande, Steiner Engen, and Bernt-Erik Sæther

The Structure and Dynamics of Geographic Ranges
Kevin J. Gaston

Animal Signals
John Maynard Smith and David Harper

Evolutionary Ecology: The Trinidadian Guppy
Anne E. Magurran

Infectious Diseases in Primates: Behavior, Ecology, and Evolution
Charles L. Nunn and Sonia Altizer

Computational Molecular Evolution
Ziheng Yang

The Evolution and Emergence of RNA Viruses
Edward C. Holmes

Aboveground–Belowground Linkages: Biotic Interactions, Ecosystem Processes, and Global Change
Richard D. Bardgett and David A. Wardle

Principles of Social Evolution
Andrew F. G. Bourke

Maximum Entropy and Ecology: A Theory of Abundance, Distribution, and Energetics
John Harte

Ecological Speciation
Patrik Nosil

Energetic Food Webs: An Analysis of Real and Model Ecosystems
John C. Moore and Peter C. de Ruiter

Evolutionary Biomechanics: Selection, Phylogeny, and Constraint
Graham K. Taylor and Adrian L. R. Thomas

Quantitative Ecology and Evolutionary Biology: Integrating Models with Data
Otso Ovaskainen, Henrik Johan de Knegt, and Maria del Mar Delgado

Quantitative Ecology and Evolutionary Biology

Integrating Models with Data

OTSO OVASKAINEN
Department of Biosciences, University of Helsinki, Finland

HENRIK JOHAN DE KNEGT
Department of Environmental Sciences, Wageningen University, The Netherlands

MARIA DEL MAR DELGADO
Research Unit of Biodiversity (UMIB, UO-CSIC-PA), Oviedo University – Campus Mieres, Mieres, Spain

OXFORD
UNIVERSITY PRESS

Great Clarendon Street, Oxford, OX2 6DP,
United Kingdom

Oxford University Press is a department of the University of Oxford.
It furthers the University's objective of excellence in research, scholarship,
and education by publishing worldwide. Oxford is a registered trade mark of
Oxford University Press in the UK and in certain other countries

© Otso Ovaskainen, Henrik Johan de Knegt & Maria del Mar Delgado 2016

The moral rights of the authors have been asserted

First Edition published in 2016

Impression: 1

All rights reserved. No part of this publication may be reproduced, stored in
a retrieval system, or transmitted, in any form or by any means, without the
prior permission in writing of Oxford University Press, or as expressly permitted
by law, by licence or under terms agreed with the appropriate reprographics
rights organization. Enquiries concerning reproduction outside the scope of the
above should be sent to the Rights Department, Oxford University Press, at the
address above

You must not circulate this work in any other form
and you must impose this same condition on any acquirer

Published in the United States of America by Oxford University Press
198 Madison Avenue, New York, NY 10016, United States of America

British Library Cataloguing in Publication Data
Data available

Library of Congress Control Number: 2015957822

ISBN 978–0–19–871486–6 (hbk.)
ISBN 978–0–19–871487–3 (pbk.)

Preface

We have written this book primarily for two kinds of target audiences. The first one consists of those ecologists and evolutionary biologists who may not be very experienced with mathematical and statistical modelling yet, but who are interested in learning more about modelling approaches in order to integrate them with data. The second audience consists of those mathematicians, statisticians, and computer scientists who are interested in ecological and evolutionary questions, and thus wish to see how the mathematical, statistical, and computational tools that they already master can be applied in these fields of sciences. We have intended the book for researchers and graduate students, as well as advanced master students. We also hope it will be useful as a textbook for graduate-level courses in ecological modelling, statistical ecology, theoretical ecology, and spatial ecology.

After an introductory chapter, we devote one chapter to movement ecology, one to population ecology, one to community ecology, and one to genetics and evolutionary ecology. We have followed the same structure in each chapter. We start with a conceptual section, which provides the necessary biological background and motivates the modelling work. The next three sections present mathematical modelling approaches, followed by one section devoted to statistical approaches. We end each chapter with a perspectives section, where we summarize some of the key messages of the chapter, discuss some limitations of the approaches we have considered, and mention some alternative approaches.

We assume that the reader is familiar with the basics of calculus and probability. These are covered in good depth in the context of ecological modelling, e.g., in the books of Otto and Day (2007), Mangel (2006), and Hastings (1996). From the statistical side, we expect the reader to have basic knowledge of linear models (e.g., ANOVA and regression), covered, e.g., by the book by Grafen and Hails (2002) or that of Zuur et al. (2007). Some parts of this book utilize rather advanced mathematical and statistical methods compared with the background typical for graduate students in ecology and evolution. Therefore, we finish the book with two appendices, one related to mathematical and the other to statistical methods. In these appendices, we summarize a wide range of methods that are utilized in this book and that we expect to be useful for the reader more widely.

In a single book, it is not possible to present a comprehensive review of all the mathematical and statistical approaches developed for movement ecology, population ecology, community ecology, population genetics, and evolutionary biology.

Therefore, the selection of approaches and models we present is necessary limited, and biased by our own research interests. The reason why we cover such a broad range of topics rather than focusing on a more limited scope is that one key aim of this book is to link modelling approaches used in different fields of ecology and evolutionary biology to each other. With this aim, we have constructed the models in a stepwise manner, starting from simple baseline models, and then adding gradually new components. As one example, in Chapter 2 we start from a random walk model in homogeneous space. We then extend the model to heterogeneous space by considering how moving individuals may respond to spatial variation in their environments, after which we model movements in a highly fragmented landscape as a particular case study. In Chapter 3, we add to the movement model the births and deaths of individuals, resulting in a model that we will call the butterfly metapopulation model. As another example, in Chapter 2 we link dispersal kernels to movement models. In Chapter 3, we utilize dispersal kernels to construct a model for the ecological dynamics of sessile organisms, which we call the plant population model. In Chapter 4, we add to the plant population model competition among heterospecifics, resulting in what we call the plant community model. Finally, in Chapter 5 we extend the plant population model into an eco-evolutionary model by letting the individuals carry genes that influence the dispersal of their propagules.

While our aim is to integrate models with data, we discuss real data only in the context of few motivating examples. All the data analysed in the statistical sections have been generated by the mathematical models of the same chapter. We have decided to do so because with simulated data the reader has the full knowledge of the process that has generated the data, which would be never the case with real data. Therefore, knowing the underlying process enables one to pinpoint the potentials and pitfalls of different kinds of statistical methods more effectively than would be the case with real data.

As our starting point is movement ecology, much (though not all) of our emphasis is on the spatial aspects of ecology and evolution, especially on the interplay between environmental heterogeneity and ecological processes. In particular, each chapter has one section that applies the mathematical modelling approaches to study the consequences of habitat loss and fragmentation. In addition to presenting perspectives to ecological modelling, we thus hope to provide some insights on how habitat loss and fragmentation influence the movements of individuals, the dynamics of populations and communities, and evolutionary dynamics.

Writing this book would not have been possible without the support we have obtained from many people. First of all, working with the publisher has been as smooth as one could hope for. We would therefore like to thank Charles Godfray for taking the initiative of inviting us to write this book, Ian Sherman for guiding us with the initial planning, and Jennifer Dunne for providing feedback. Lucy Nash has worked as an excellent editor; her help has been invaluable through the writing process. For permission to reproduce copyrighted material, we would like to thank the National Academy of Sciences USA (Figures 2.2A–C and 4.2A–C), the Ecological

Society of America (Figure 2.2D), Oxford University Press (Figure 3.2CD), Nature Publishing Group (Figure 3.2E), and John Wiley and Sons (Fig. 5.2A,B,C).

Most importantly, the book has benefited greatly from comments that we have received from several colleagues. We would thus like to thank Nerea Abrego, Florian Hartig, Arild Husby, Jussi Jousimo, Etsuko Nonaka, Iñaki Odriozola, Bart Peeters, Johannes Signer, Tord Snäll and Eugenia Soroka for providing a large number of suggestions and corrections that greatly improved the contents. We thank Kaisa Torppa for helping us to compile the index, Sami Ojanen for producing the maps of Figures 2.2 and 3.2., and Juha Merilä for permission to use the photo of Fig. 5.2D. Our warmest thanks go to Ilkka Hanski, the leader of the Metapopulation Research Centre, where this book has been written. This is not only for the feedback and encouragement that he has provided during the writing process, but also and especially for sharing with us his passion for science.

The research that resulted in the development of the modelling approaches presented in this book, as well as the writing of this book, has been supported by grants from the European Research Council (Grant 205905), Academy of Finland (Grants 12424, 129636, and 25044), the Kone Foundation (Grant 44-6977), The Netherlands Organisation for Scientific Research (Rubicon Grant 825.11.009), and the Research Council of Norway (CoE Programme Grant 223257).

Contents

1	**Approaches to ecological modelling**	**1**
1.1	Forward and inverse approaches	2
1.2	The interplay between models and data	3
1.3	The many choices with mathematical and statistical models and methods	6
1.4	What a biologist should learn about modelling	8
2	**Movement ecology**	**10**
2.1	Why, where, when, and how do individual organisms move	10
	2.1.1 Internal state: why to move	11
	2.1.2 Motion capacity: how to move	12
	2.1.3 Navigation capacity: when and where to move	13
	2.1.4 Different types of movement	13
	2.1.5 Approaches to movement research	14
	2.1.6 Outline of this chapter	15
2.2	Movement models in homogeneous environments	17
	2.2.1 The Lagrangian approach	18
	2.2.2 Translating the Lagrangian model into an Eulerian model	20
	2.2.3 Dispersal kernels	21
	2.2.4 Adding directional persistence: correlated random walk models	23
	2.2.5 Adding directional bias: home-range models	25
2.3	Movement models in heterogeneous environments	27
	2.3.1 Random walk simulations in heterogeneous space	27
	2.3.2 Diffusion models with continuous spatial variation in movement parameters	29
	2.3.3 Diffusion models with discrete spatial variation in movement parameters	34
	2.3.4 Using movement models to define and predict functional connectivity	35
	2.3.5 The influence of a movement corridor	40
2.4	Movements in a highly fragmented landscape	43
	2.4.1 The case of a single habitat patch	44
	2.4.2 The case of a patch network	47

2.5 Statistical approaches to analysing movement data 50
    2.5.1 Exploratory data analysis of GPS data 51
    2.5.2 Fitting a diffusion model to capture-mark-recapture data 60
2.6 Perspectives 63
    2.6.1 Limitations and extensions of random walk and diffusion models 64
    2.6.2 The many approaches of analysing movement data 66

3 Population ecology 68

3.1 Scaling up from the individual level to population dynamics 68
    3.1.1 Factors influencing population growth through birth and death rates 69
    3.1.2 How movements influence population dynamics 70
    3.1.3 How population structure influences population dynamics 71
    3.1.4 The outline of this chapter 72
3.2 Population models in homogeneous environments 73
    3.2.1 Individual-based stochastic and spatial model 76
    3.2.2 Simplifying the model: stochasticity without space 77
    3.2.3 Simplifying the model further: without stochasticity and space 80
    3.2.4 Another way of simplifying the model: space without stochasticity 81
3.3 Population models in heterogeneous environments 83
    3.3.1 Environmental stochasticity 84
    3.3.2 Spatial heterogeneity in continuous space: the plant population model 86
    3.3.3 Spatial heterogeneity in discrete space: the butterfly metapopulation model 89
    3.3.4 The Levins metapopulation model and its spatially realistic versions 93
3.4 The persistence of populations under habitat loss and fragmentation 97
    3.4.1 Habitat loss and fragmentation in the plant population model 98
    3.4.2 Habitat loss and fragmentation in the butterfly metapopulation model 100
    3.4.3 Habitat loss and fragmentation in the Levins metapopulation model 100
3.5 Statistical approaches to analysing population ecological data 102
    3.5.1 Time-series analyses of population abundance 102
    3.5.2 Fitting Bayesian state-space models to time-series data 105
    3.5.3 Species distribution models 112
    3.5.4 Metapopulation models 115
3.6 Perspectives 117
    3.6.1 The invisible choices made during a modelling process 118
    3.6.2 Some key insights derived from population models 119
    3.6.3 The many approaches to analysing population data 120

4 Community ecology — 122

- 4.1 Community assembly shaped by environmental filtering and biotic interactions — 122
  - 4.1.1 Ecological interactions — 124
  - 4.1.2 Fundamental and realized niches and environmental filtering — 125
  - 4.1.3 Organizational frameworks for metacommunity ecology — 126
  - 4.1.4 The outline of this chapter — 127
- 4.2 Community models in homogeneous environments — 129
  - 4.2.1 Competitive interactions — 129
  - 4.2.2 Resource–consumer interactions — 135
  - 4.2.3 Predator–prey interactions — 138
- 4.3 Community models in heterogeneous environments — 140
  - 4.3.1 The case of two competing species — 141
  - 4.3.2 The case of many competing species — 142
- 4.4 The response of communities to habitat loss and fragmentation — 145
  - 4.4.1 Endemics-area and species-area relationships generated by the plant community model — 145
- 4.5 Statistical approaches to analysing species communities — 150
  - 4.5.1 Time-series analyses of population size in species communities — 150
  - 4.5.2 Joint species distribution models — 155
  - 4.5.3 Ordination methods — 161
  - 4.5.4 Point-pattern analyses of distribution of individuals — 162
- 4.6 Perspectives — 164
  - 4.6.1 Back to the metacommunity paradigms — 164
  - 4.6.2 Some insights derived from community models — 165
  - 4.6.3 The many approaches to modelling community data — 167

5 Genetics and evolutionary ecology — 168

- 5.1 Inheritance mechanisms and evolutionary processes — 168
  - 5.1.1 Genetic building blocks and heritability — 168
  - 5.1.2 Selection, drift, mutation, and gene flow — 170
  - 5.1.3 Connections between ecological and evolutionary dynamics — 172
  - 5.1.4 The outline of this chapter — 173
- 5.2 The evolution of quantitative traits under neutrality — 175
  - 5.2.1 An additive model for the map from genotype to phenotype — 175
  - 5.2.2 Coancestry and the additive genetic relationship matrix — 177
  - 5.2.3 Why related individuals resemble each other? — 179
  - 5.2.4 The animal model — 180
  - 5.2.5 Why related populations resemble each other? — 181
- 5.3 The evolution of quantitative traits under selection — 184
  - 5.3.1 Evolution by drift, selection, mutation, recombination, and gene flow — 185
  - 5.3.2 Selection differential and the breeder's equation — 186
  - 5.3.3 Population divergence due to drift and selection — 190

5.4 Evolutionary dynamics under habitat loss and fragmentation ... 194
    5.4.1 Evolution of dispersal in the Hamilton–May model under adaptive dynamics ... 194
    5.4.2 Evolution of dispersal in the plant population model under quantitative genetics ... 198
5.5 Statistical approaches to genetics and evolutionary ecology ... 201
    5.5.1 Inferring population structure from neutral markers ... 201
    5.5.2 Estimating additive genetic variance and heritability ... 203
    5.5.3 Using association analysis to detect loci behind quantitative traits ... 204
    5.5.4 Detecting loci under selection from genotypic data ... 206
    5.5.5 Detecting traits under selection from genotypic and phenotypic data ... 207
5.6 Perspectives ... 209
    5.6.1 Mathematical approaches to modelling genetics and evolution ... 209
    5.6.2 Some insights derived from evolutionary models on dispersal evolution ... 211
    5.6.3 The many uses of genetic data ... 212

Appendix A: Mathematical methods ... 215

A.1 A very brief tutorial to linear algebra ... 215
A.2 A very brief tutorial to calculus ... 217
    A.2.1 Derivatives, integrals, and convolutions ... 217
    A.2.2 Differential equations ... 218
    A.2.3 Systems of differential equations ... 220
    A.2.4 Partial differential equations ... 221
    A.2.5 Difference equations ... 222
A.3 A very brief tutorial to random variables ... 222
    A.3.1 Discrete valued random variables ... 222
    A.3.2 Continuous valued random variables ... 223
    A.3.3 Joint distribution of two or more random variables ... 225
    A.3.4 Sums of random variables ... 226
    A.3.5 An application of random variables to quantitative genetics ... 228
A.4 A very brief tutorial to stochastic processes ... 229
    A.4.1 Markov chains ... 229
    A.4.2 Markov processes ... 231

Appendix B: Statistical methods ... 233

B.1 Generalized linear mixed models ... 233
    B.1.1 Linear models ... 233
    B.1.2 Link functions and error distributions ... 234
    B.1.3 Relaxing the assumption of independent residuals ... 236
    B.1.4 Random effects ... 238

B.1.5 Multivariate models	241
B.1.6 Hierarchical models	244
B.2 Model fitting with Bayesian inference	245
B.2.1 The concepts of likelihood, maximum likelihood, and parameter uncertainty	246
B.2.2 Prior and posterior distributions, and the Bayes theorem	246
B.2.3 Methods for sampling the posterior distribution	249
References	253
Index	277

1

Approaches to ecological modelling

The motivation for conducting any kind of research in ecology and evolutionary biology, including modelling, is to advance knowledge about ecological and evolutionary processes. This happens mostly with small steps that refine the existing views by bringing new pieces of knowledge, but sometimes through revolutionary ideas that challenge existing paradigms. For an excellent introduction to the philosophy of science in the context of ecological research, we refer to the book by Hilborn and Mangel (1997). Briefly, the process of doing science involves the interplay between theories, hypotheses, models, and data. Theories summarize our present understanding on how we consider nature to work. They are the highest form of scientific knowledge, as they have resisted repeated testing and scrutiny through experiments and observations. Hypotheses are proposed explanations of how nature might work, requiring new data to test the validity of the proposed explanation. Theories and hypotheses can be expressed verbally, but they can also be made more specific by expressing them with the help of mathematical models. Different models of the same phenomena can be viewed as alternative hypotheses, and alternative parameter values of those models can be seen as even more specific hypotheses. Data are essential for examining the extent to which hypotheses are supported. This can be done by examining whether a single hypothesis is supported by data, or by quantifying the degree of support for alternative hypotheses simultaneously. Observing a contradiction between a hypothesis and data is a key step for making scientific progress, as it makes one think why nature does not work as was hypothesized. Such thinking can lead to new hypotheses, which can refine the existing theories or even eventually replace them entirely by new ones.

The word 'modelling' is used in ecological research in many different ways, ranging from mathematical models used to develop theoretical ecology to statistical models used to analyse empirical data. Mathematical modelling helps to understand the causal pathways from ecological mechanisms to the resulting phenomena, whereas statistical models are needed to confront theories and models with data. As a simple example, a theoretically minded ecologist might model resource–consumer dynamics with a system of differential equations (as we will do in Chapter 4). Insights from the model may be derived using just pen and paper, e.g. to show that a consumer population will persist in the long term if and only if the rate of resource production exceeds some threshold value, or to understand more generally how the dynamics of the consumer depends on the underlying ecological processes and their

Quantitative Ecology and Evolutionary Biology. Otso Ovaskainen, Henrik Johan de Knegt & Maria del Mar Delgado. © Otso Ovaskainen, Henrik Johan de Knegt & Maria del Mar Delgado 2016. Published 2016 by Oxford University Press. DOI 10.1093/acprof:oso/ 9780198714866.001.0001

parameter values. An empirically minded researcher could study the same system by acquiring experimental data on consumer abundance for different levels of resource availability. Statistical models may then be applied to show that in the experiment the growth rate of the consumer increases with the availability of resources, as suggested by the mathematical models constructed by the theoretical ecologist.

1.1 Forward and inverse approaches

One way of classifying different approaches to ecological modelling is that of 'forward' and 'inverse' approaches. Forward approaches are those where assumptions are made about the underlying mechanisms, and mathematical or simulation tools are used to study the consequences of those assumptions. With an inverse approach, the aim is to uncover the mechanisms that might have generated the patterns observable in empirical data. Both the forward (moving from mechanisms to patterns) and the inverse (moving from patterns to mechanisms) approaches involve their own challenges. To illustrate, let us consider a researcher who is interested in the movement ecology of butterflies, specifically on how the characteristics of the species and the structure of the environment influence the displacements that the individuals make over their lifetimes (as we will do in Chapter 2). If taking the forward approach, the researcher could track the detailed movement tracks of individuals to learn about the mechanisms that influence lifetime displacements, such as flight speed and directional persistence under different environmental conditions. Such data could be used to construct a movement model, which could be used to simulate lifetime movements in heterogeneous landscapes. While this can be relatively straightforward to do, the critical question is how much the model can be trusted. This is because the researcher is likely to obtain detailed movement data only for a limited number of individuals, only for a limited set of environmental conditions, and only over small temporal and spatial scales. This can be problematic, as the key features of movement behaviour may differ among individuals, environmental conditions, and temporal and spatial scales. More generally, a major challenge in the forward approach is that of extrapolation.

Conversely, if taking the inverse approach, the researcher could start by acquiring capture-mark-recapture data, i.e. marking and releasing butterflies when encountering them, and then recapturing some of the marked individuals later at another location. Such data may be acquired at the scale of lifetime displacements, and thus they would avoid the problem of extrapolation. However, it can be highly challenging to use such data for inference about the mechanisms that might have generated the observed displacements. This is because the data are simultaneously influenced by a myriad of processes, such as the intrinsic behaviour of the species, the spatial distribution of habitat characteristic, the prevailing weather conditions, and where and when the captures were attempted. Inverse problems are generally ill posed in the mathematical terminology, meaning that they do not have a unique solution. For example, assume that only few butterflies were recaptured. This may be because the death rate was high or alternatively because the emigration rate out of the study area was high, or because the capture rate was low.

Much of the current research in ecology and evolutionary biology uses both forward and inverse approaches in an integrated manner. This is partly due to the increasing interest in conducting ecological research at the interface between theory and data, and partly due to developments in hardware and software that allow for computationally demanding analyses. In this book, our emphasis is to combine the two. To do so, we will use mathematical models to generate various kinds of data sets. We will then apply statistical methods to those data, with the aim of asking how much of the assumed mechanisms we can recover.

1.2 The interplay between models and data

Figure 1.1 illustrates the interplay between models and data in the context of the forward approach. Often one wishes to understand how some phenomenon at a higher level of biological organization emerges from the processes at a lower level of biological organization. In the example of Figure 1.1, we are interested in how the dynamics of a population emerge from the deaths and births of individuals. In the process of model construction, knowledge about the ecological context is required. This could involve e.g. whether the organism has discrete or overlapping generations, and whether it inhabits a seasonal or unseasonal environment. These assumptions translate into the model structure. In addition to deciding about the model structure, one needs to decide about the parameter values. In scenario simulations, the parameter values are assumed rather than estimated, which is a perfectly valid approach if the aim is to ask how the parameter values influence the model's behaviour. To confront the model with data, the parameter values need to be estimated. In the most pure form of the forward approach, the model parameters are simply measured one by one. For example, one might monitor the lifetimes of individuals at different population densities to measure the mean death rate and its dependency on population density. Once the parameters have been measured, the model can be used to generate predictions. To validate the model, the predictions are to be contrasted with independent data. For example, one may compare the dynamics generated by the model with data acquired directly at the population level. If the match between model prediction and data is good, one may conclude that the model is compatible with the data. This is, of course, encouraging, yet it does not mean that the model is correct, as different kinds of models may end up with the same prediction. To refine the model or to select between alternative types of models, new data, ideally from a different context, are needed. Testing whether the model can predict such new data is called cross-validation. If the model does not match well with the data, that is actually not such a bad result. In this case, the conclusion is that the hypothesis that corresponds to the model is not compatible with the data. Thus, something is wrong either with the structural model assumptions, or with the parameter values. Through analysing the mismatch between the model's prediction and the data in more detail, one may

4 • *Approaches to ecological modelling*

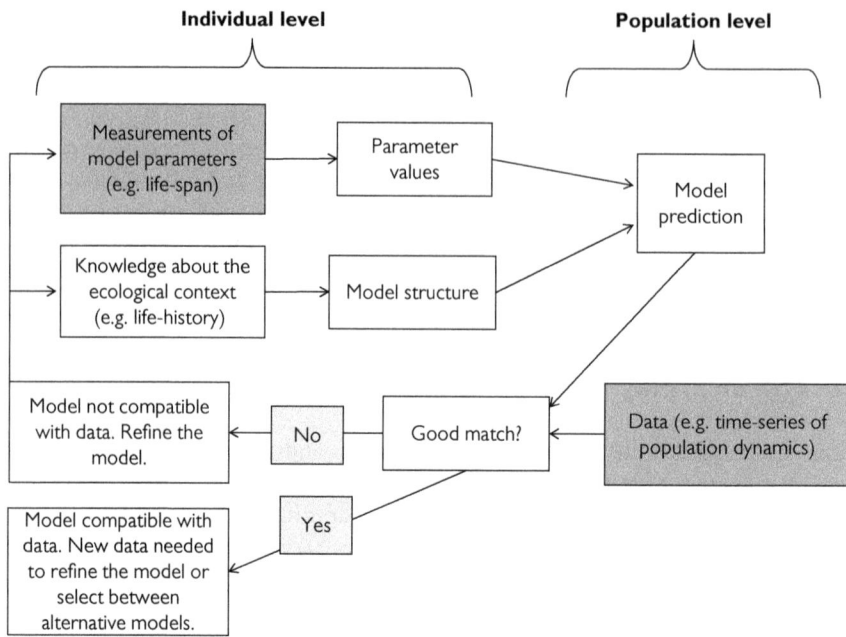

Figure 1.1 A flow diagram illustrating the interplay between models and data by the forward approach. We have assumed here that the model is constructed at the individual level, with the aim of predicting dynamics at the population level.

be able to infer reasons for the failed predictions, and thus learn more about the ecological context or the parameter values.

Applying the pure forward approach of Figure 1.1 can be challenging due to three problems. The first problem is that ecological processes are complex and influenced by manifold factors; thus, it is very difficult to derive the structural model assumptions from first principles. The second problem is that it is usually very difficult to perform direct measurements of all the model parameters. Even if this would be possible, the third problem is that the parameter estimates usually involve large amounts of uncertainty, and when a large number of uncertain estimates are put together through uncertain and nonlinear model structures, the amount of uncertainty becomes amplified.

Figure 1.2 presents an overview of the processes of model construction, parameterization, and inference, now performed in a manner that combines elements from both the forward and the inverse approaches. Like with the forward approach, the researcher uses prior knowledge about the ecological context to formulate a model, or a family of models, that describe the structural assumptions about the assumed underlying biology. While in the pure forward approach we assumed that all model parameters were measured directly, here we assume that direct information is possibly available only for some of the parameters, and that this information

1.2 The interplay between models and data

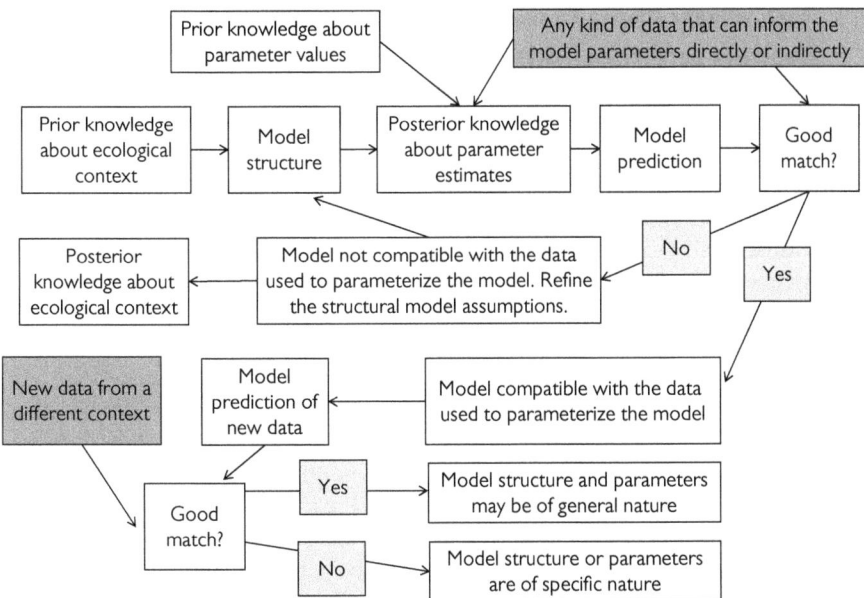

Figure 1.2 A flow diagram illustrating the interplay between models and data by combining elements from the forward and inverse approaches.

may be uncertain. We use in Figure 1.2 the terminology of Bayesian inference (see Appendix B.2); hence we refer to this information as prior knowledge about parameter values. In the Bayesian context, parameter estimation corresponds to the computation of the posterior distribution of the parameters. A key difference to the forward approach is that in the inverse approach any kind of data can be used to inform the model parameters. For example, even if the model parameters are formulated at the individual level, also population level data can be informative about the parameter values.

Parameter estimation should always be followed by model validation. To start with, it is always possible to compare model predictions to the data used to parameterize the model. If the model is not able to reproduce the data used to parameterize it, the problem is not in the parameter values but in the structural model assumptions, and thus in the assumptions made when describing the ecological context. If the model is able to reproduce the data used to parameterize it, it is encouraging, but does not necessarily mean that the mechanisms incorporated in the model actually operate in reality. To move one step forward, one may submit the model to a harder test, which is to make a prediction in a context that is different from the one used for parameter estimation. If the model is successful in predicting the new data as well, there is a stronger support that the assumptions behind the model reflect reality. If this is not the case, one may think about how to refine the model so that it simultaneously fits both the original data and the new data.

1.3 The many choices with mathematical and statistical models and methods

When describing an ecological system with the help of a mathematical model, there is a large number of options about the mathematical framework to use. Examples of modelling frameworks commonly used in ecology and evolutionary biology include simulations of individual-based models (e.g. Grimm, 1999; Zurell et al., 2010; Grimm and Railsback, 2013), the use of deterministic dynamical systems such as systems of differential equations (Mangel, 2006), and the use of stochastic models such as Markov chains and Markov processes (e.g. Black and McKane, 2012). The construction of any specific model involves a large number of choices. To illustrate, Table 1.1 lists some of the choices involved in modelling the dynamics of an ecological system as a function of time. To start with, the state variable that evolves over time can be either discrete (e.g. the number of individuals) or continuous (e.g. the amount of biomass). The state variable may be represented by a single number, such as the total number of individuals, or by a vector of numbers, such as the numbers of individuals in different habitat patches or age categories, or the numbers of individuals belonging to different species. The state variable may alternatively have a continuous structure, such as the distribution of individuals over continuous space, or the distribution of body masses. Time may advance either continuously or in discrete steps, the latter being a more natural choice if, e.g. modelling the yearly dynamics of a population in a seasonal environment. Further, the model may be deterministic or stochastic. The broad distinction between these two is that if 'running' the same model twice, a deterministic model will always give the same answer, whereas a stochastic model will yield a different realization each time unless the realizations coincide by chance.

Ideally, the choice of the mathematical modelling framework should be based on the underlying biological context and the question to be addressed. In practice, the choice is often influenced by what kind of models and tools the modeller is familiar with, and what kind of a modelling framework is most convenient for performing the mathematical analyses. Some choices during the modelling process will not make a big difference in the sense that they do not influence the qualitative results, but other choices can be crucially important. Thus, it is important to learn what kinds of choices are and are not likely to make a qualitative difference in the results. Unfortunately, there are no general rules of thumb that would tell what kind of choices matter and what not. It is something that one learns best from experience, i.e. by applying different modelling frameworks to the very same problem and by comparing the results. This is indeed what we will do in many parts of this book.

After a model has been formulated, the next step is to use mathematical methods to analyse how it behaves, e.g. to make model predictions. Each mathematical class of models is associated with its own toolbox of methods, some being analytical, some numerical. Instead of reviewing the many kinds of mathematical models and methods here, we have compiled a tutorial of them in Appendix A. We cover there in particular those fields of mathematics (some areas of matrix algebra, calculus,

Table 1.1 Broad classification of some common mathematical frameworks for dynamical systems used to address ecological and evolutionary questions.

Variable	Space	Time	Stochasticity	Model type(s)	References
Discrete	No	Discrete	No	—	
			Yes	Markov chain	2.2, 2.3, 2.5, 3.5, 4.5
		Continuous	No	—	
			Yes	Markov process	3.2, 3.5
	Discrete	Discrete	No	—	
			Yes	Multidimensional Markov chain, discrete-time IBM on grid or patch network, discrete-time SPOM	3.3, 3.4, 3.5, 5.2, 5.3
		Continuous	No	—	
			Yes	Multidimensional Markov process, continuous-time IBM on a grid or patch network, continuous-time SPOM	2.4, 3.3
	Continuous	Discrete	No	—	
			Yes	Discrete-time individual-based model in continuous space	
		Continuous	No	—	
			Yes	Spatiotemporal point process	3.2, 3.3, 3.4, 4.2, 4.3, 4.4, 4.5, 5.4
Continuous	No	Discrete	No	Difference equation	5.4
			Yes	Stochastic difference equation	3.3
		Continuous	No	Differential equation, integral equation	3.2, 3.3
			Yes	Stochastic differential equation	
	Discrete	Discrete	No	System of difference equations	
			Yes	System of stochastic difference equations	
		Continuous	No	System of differential equations	
			Yes	System of stochastic differential equations	4.2
	Continuous	Discrete	No	Integro-difference equation	
			Yes	Stochastic integro-difference equation	
		Continuous	No	Partial differential equation, differential equation with convolution	2.2, 2.3, 2.4, 2.5, 3.2
			Yes	Stochastic partial differential equation	

SPOM, Stochastic patch occupancy model. References are to subsections of this book.

random variables, and stochastic processes) that are needed for understanding the models considered in this book.

Like a mathematical ecologist facing the choice among manifold mathematical models and methods, a statistical ecologist performing data analyses faces similar choices among manifold statistical models and methods. Widely applied statistical approaches include e.g. generalized linear mixed models, generalized additive models, regression trees and ordinations, to mention but a few. We discuss in Appendix B.1 in some detail the framework of generalized linear mixed models (GLMMs), because we use this modelling framework in many examples throughout this book, and because it is very flexible and can thus be applied to almost any kind of ecological data.

With mathematical models, one can always separate the model definition (a set of equations) from the methods used to analyse the model. With many statistical models, such as GLMMs, it is also possible and desirable to write down the model as a set of equations, and consider the model parameterization as a separate step from model definition. The same model can be connected to data in many ways. In Appendix B.2, we briefly review the two most widely applied methods for statistical inference, i.e. those of maximum likelihood and Bayesian inferences. With some statistical approaches, it can be difficult to separate the model definition from model parameterization, or to use the model for predictive purposes. For example, with ordinations the main aim is to provide informative summaries of multivariate data, and with the use of test-statistics the main aim is to examine whether a particular null hypothesis can be rejected. In this book our main focus is on predictive models, and hence we refer the reader interested in ordination approaches to, e.g. the books of Legendre and Legendre (2012) and Borcard et al. (2011), and the reader interested in test-statistics and non-parametric methods to, e.g. the books of Myles and Wolfe (1999) and Corder and Foreman (2009).

1.4 What a biologist should learn about modelling

A non-trivial question for an ecologist is how much and what kind of mathematics and statistics should one learn? While it is likely that any kind of mathematical, statistical, and computational skills will be useful, the trade-off is the investment in time and effort. Learning the basics of mathematics, statistics, and programming will almost surely be worth the investment for anyone considering a research career in ecology and evolutionary biology. These include, e.g. the basics of calculus, random variables, and stochastic processes; the basics of model fitting, model selection, and model validation; and the ability to write computer programs that utilize loops and functions. These skills will be useful also for those who do not apply modelling directly in their own research, as they help to understand research done by others as well as to conduct interdisciplinary collaborations. Indeed, a majority of early-career biologists consider that they would have benefitted from a better level of mathematical and statistical training (Barraquand et al., 2014). This is partly so because research papers from almost any subfield of biology involve ideas and models formulated with the help of equations. Students and researchers with no formal training in mathematics may sometimes feel that models and equations distract the reading process, and are safer to skip over. However, after spending some effort to learn the mathematical language, one may realize that equations can greatly help to understand and assimilate the underlying biology.

In Chapters 2–5 of this book, we aim to provide insights about ecological modelling and the interplay between models and data. One particular aim is to illustrate how the very same modelling approaches apply in different fields of ecology and evolutionary biology. To do so, we start from a single individual's point of view in Chapter 2, which focuses on movement ecology. We then use the movement models as one building block to construct single-species models of population dynamics in Chapter 3, which models we further expand to the multispecies context in Chapter 4 and to the evolutionary context in Chapter 5. In all chapters, we start from individual-based models, as it is often most intuitive to make the model assumptions at the individual level. However, the drawback of individual-level models is that they can be usually analysed by simulations only. To be able to derive analytical insights, as well as to compare the behaviours of different kinds of models, we show how the individual-based simulation models can be simplified by making further assumptions, e.g. to yield models formulated directly at the population level.

2

Movement ecology

2.1 Why, where, when, and how do individual organisms move

All organisms move at some point during their lives, either under their own locomotion or through being transported by e.g. wind, water, or other organisms (Begon et al., 1996). To be able to move, organisms have evolved morphological and physiological traits (Dickinson et al., 2000). Movement is defined as a change in the spatial location of the individual in time. In a very general formulation, illustrated in Figure 2.1, the movement process can be written as

$$X_{t+\tau} = f(X_t, S, L, \tau), \qquad (2.1)$$

where the location X of an individual at time $t + \tau$ is a function of its location at time t, the set of characteristics S of the individual, the set of characteristics L of the environment through which it is moving, and the duration of the time interval τ. In its broadest sense, studying the movements of organisms thus focuses on studying how and why the interactions between X_t, S, L, and τ combine to produce $X_{t+\tau}$. For example, the variable S may indicate that the individual is hungry, and the environment L may include information on the spatial distribution of food resources. Thus, combining these two pieces of information will lead to the prediction that $X_{t+\tau}$ is likely to be located in a place where food resources are abundant. But it also depends on the previous location: $X_{t+\tau}$ is likely to be close to X_t due to constraints and energetic costs of the movement process itself.

Movement determines the location of individuals within the landscape they inhabit, and thus plays a major role in determining their fate, as it determines the environmental conditions to which an individual is exposed and the conspecifics (other individuals of the same species) and heterospecifics (individuals of other species) it encounters. There are many kinds of drivers for movement, e.g. the need for finding food resources and mates, and the need for avoiding adverse environmental conditions and predators. The movement process therefore has a fundamental influence on many ecological and evolutionary processes, and is thus key to understanding the dynamics and spatial structure of populations and communities (Turchin, 1991; Ovaskainen and Hanski, 2004b; Nathan et al., 2008; Nathan and Giuggioli, 2013). The study of organismal movement, or aspects related to it, has increased drastically over the past decades. This is partly because of the recognition of

Quantitative Ecology and Evolutionary Biology. Otso Ovaskainen, Henrik Johan de Knegt & Maria del Mar Delgado. © Otso Ovaskainen, Henrik Johan de Knegt & Maria del Mar Delgado 2016. Published 2016 by Oxford University Press. DOI 10.1093/acprof:oso/ 9780198714866.001.0001

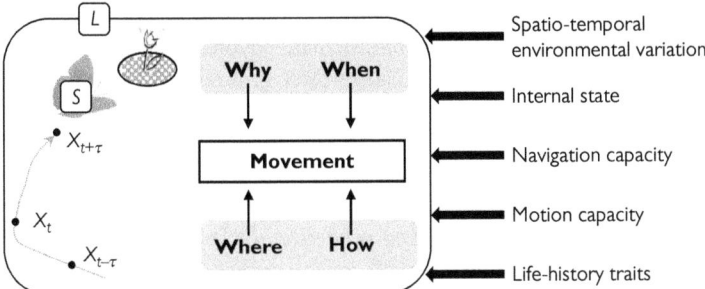

Figure 2.1 Factors influencing the movement of organisms through their environment. Consider an individual S that moves (dotted line) through an environment L, with its spatial locations being recorded at three points in time ($X_{t-\tau}$, X_t, and $X_{t+\tau}$). The individual has a set of characteristics S, including its life-history traits, internal state, motion and navigation capacities. To fulfil requirements such as maintenance, survival, and reproduction, the individual interacts with its spatiotemporally varying environment. In the figure, the internal state S of the butterfly drives it to refill its energy resources by moving towards a nectar plant that is part of the environment L. A central problem in movement ecology is to understand *why, when, where*, and *how* the individual moved.

its importance per se and for its implications for management and conservation, but also fuelled by technical advances that enable the accurate measurement of movement as well as the characteristics of the individual S and its environment L. One of the present challenges of movement ecology is to integrate detailed studies of individual movements into a broader context: how macroscopic phenomena such as the dynamics and evolution of populations and communities emerge from microscopic behaviour at the individual level (Holyoak et al., 2008).

In order to integrate the various ways of studying movement of different kinds of organisms, Nathan et al. (2008) proposed a conceptual framework that holds for organisms of all kinds, from microbes to trees and elephants. This framework describes the basic components needed to understand the mechanisms underlying movement and their consequences (Getz and Saltz, 2008; Nathan et al., 2008; Nathan and Giuggioli, 2013), i.e. the *why, where, when*, and *how* questions related to movement (Holden, 2006; Nathan et al., 2008). The organism's internal state, its intrinsic motivation to move, and their evolutionary origins relate to the question of *why* to move, its motion capacity relates to the question of *how* to move, whereas its navigation capacity relates to the questions of *when* and *where* to move. As we discuss later, all of the *why, where, when*, and *how* questions are fundamentally influenced by spatiotemporal variation in the environment.

2.1.1 Internal state: why to move

The internal state of an organism, which accounts for its physiological and (where appropriate) psychological state, is critical in answering the question *why to move?*

(Nathan et al., 2008). This question addresses both proximate and ultimate causes of movement. Proximate causes relate to immediate payoffs, such as the reward of encountering a patch of food. Ultimate causes refer generally to evolutionary payoffs. For example, emigration from a natal patch may not have any immediate payoffs for the individual, but it can lead to decreased competition with close relatives, and to decreased risk of inbreeding (Dingle, 1996; Bowler and Benton, 2005). In many situations, the proximate and ultimate causes are difficult to tell apart. For example, encountering a patch of food, escaping predation, or social interactions with conspecifics not only yield immediate payoffs for the individual, but can have long-term fitness consequences too (Nathan et al., 2008).

The intrinsic motivation driving movements is difficult to measure directly and thus it often remains largely hidden for the researcher (Getz and Saltz, 2008). But one can attempt to infer the proximate and ultimate goals of movement indirectly, e.g. by linking the positions and movement decisions of the individuals to variation in environmental covariates (Kearney et al., 2010), or to the locations of their predators or prey species (Wittemyer et al., 2008). However, this can be challenging as the costs and benefits associated with movement vary in space and time (Bowler and Benton, 2005; Bonte et al., 2011), and as the individual may pursue several goals simultaneously (Martin et al., 2013).

2.1.2 Motion capacity: how to move

The motion capacity of an individual accounts for its ability to move, either actively through its own locomotion, e.g. by walking, running, swimming, flying, or gliding; or passively through external agents, e.g. carried by wind, water, or other organisms (Nathan et al., 2008). The ability to move may be fixed throughout an individual's lifetime, or it may differ among the life-stages. For example, walking caterpillars turn into flying butterflies, and wind-dispersed seeds root in the ground and become sessile plants. The feasibility or efficiency of movement depends on the spatiotemporal heterogeneity of the environment. For example, water or wind currents can essentially determine the movement speed and direction of passively transported propagules, whereas dense vegetation, topographic relief, or deep snow cover can decrease movement speed or increase energy expenditure of actively moving organisms. Linking spatial variation in environmental conditions to observed movement tracks can yield insight into how different habitat types facilitate or impede movements. However, it is not a general rule that movements are the slowest in the least suitable habitats. It can also be the opposite: organisms are expected to slow their movements in high-quality habitat, e.g. because they are foraging there (de Knegt et al., 2007). Spatial or temporal variation in movement speed or distance should thus be interpreted carefully, as it may relate equally well to variation in the capacity to move (*how* to move?), or to variation in the motivation for movement (*why* to move?).

2.1.3 Navigation capacity: when and where to move

In the context of Eq. (2.1), the question of *where* to move is about the direction and length of the movement step from X_t to $X_{t+\tau}$, whereas the question of *when* to move is about the timing of the initiation (t) and duration (τ) of movement. Both of these questions relate fundamentally to spatiotemporal heterogeneity in the abiotic and biotic environmental conditions—in a homogeneous environment there would be no need to change location. Movements can be headed towards specific targets (e.g. food resources, nesting sites, potential mates), or away from threats (e.g. predators) or places with high competition. Navigation in space and time requires the ability to collect information about the spatiotemporal structure of the environment, e.g. the locations of resource patches or continuous gradients provided, e.g. by chemical cues. Mueller and Fagan (2008) identified three classes of cognitive processes that guide organismal movements: (i) the use of local information to adjust the speed and direction of movement, (ii) the use of large-scale cues to move towards distant targets even if they cannot be sensed directly, and (iii) the utilization of prior experience and knowledge stored in the organism's memory. All of these can lead to a pattern of habitat selection, i.e. the individual spending more time per unit area in high-quality habitats compared to poor habitats (de Knegt et al., 2007).

In temporally varying environments, the organisms need to sense not only the present availability of resources, but also to forecast their future availability (McNamara et al., 2011). The organism's ability to forecast the future availability of resources depends critically on the predictability of the environmental fluctuations (Mueller and Fagan, 2008). Often it is beneficial to move already before the conditions become adverse, e.g. before habitat quality declines due to over-depletion of resources or due to seasonal changes in the environment. This phenomenon is called pre-emption (Dingle and Drake, 2007). Pre-emption cannot rely on proximate cues such as food shortage, but it can be a response to either endogenous rhythms (i.e. internal clocks) or surrogates that forecast environmental change, such as increased population density (Dingle and Drake, 2007; Ramenofsky and Wingfield, 2007).

2.1.4 Different types of movement

Depending on its supposed goal, movements can be classified into a plethora of categories such as migration, dispersal, foraging, within-patch movement, station-keeping, wandering, ranging, nomadism, searching, home-ranging, and flocking. In particular, the term migration is applied very widely to many movement types (Dingle, 1996), including propagule dispersal of plants, vertical movements of zooplankton, seasonal excursions of birds and butterflies, and so on. Migration is generally defined as the movement that is persistent, directional, undistracted by resources that would normally halt it, with distinct departing and arrival behaviours, and energy reallocated to sustain it (Dingle, 1996; Bauer and Klaassen, 2013). Migration is an adaptation to resource fluctuations, it is often pre-emptive (Dingle and Drake, 2007), and it often involves travel in a periodically and geographically predictable way (Fryxell and Sinclair, 1988; Sugden, 2006). Dispersal in its broadest sense means movement away from the birthplace, and is often divided into the stages emigration (i.e. departure), transfer (i.e. the actual movement), and immigration (i.e. the eventual settlement) (Clobert et al., 2001; Bowler and Benton, 2005; Sugden, 2006; Bonte et al., 2011). Home-ranging generally points to movements restricted to a specific area to which the individual becomes familiar with (Moorcroft et al., 2006; Kie et al., 2010; Fagan et al., 2013), while foraging and searching refer to the search for resources without prior knowledge on their locations (Bell, 1991; Viswanathan et al., 2011).

2.1.5 Approaches to movement research

In the previous sections we have briefly summarized some key aspects of movement ecology to make the point that the movement process results from the combined effect of factors that are intrinsic to the organism and factors that relate to the external environmental conditions. As organisms live in environments that have multidimensional spatiotemporal heterogeneity in the resources needed for reproduction and survival, it is generally challenging to understand and model factors influencing the movement process, the factors of which are hidden in Eq. (2.1).

Research on movement can be grouped into four main paradigms depending on which of these factors are emphasized (Holyoak et al., 2008; Nathan et al., 2008). First, the optimality paradigm contrasts movement decisions to their immediate and future costs and benefits (Fronhofer et al., 2012) in terms of when (Charnov, 1976; McNamara and Houston, 2008) and where (Fretwell and Lucas, 1969; Fryxell et al., 2004) to move. The early studies on foraging theory (e.g. Fretwell and Lucas, 1969; Charnov, 1976) played a key role in the development of behavioural ecology. Studies within this paradigm highlight how the ultimate causes of movement are shaped by evolutionary mechanisms. As the ultimate payoffs resulting from movement are only visible at higher levels of biological organization, we will discuss them in the subsequent chapters focused on populations (Chapter 3), communities (Chapter 4), and evolutionary processes (Chapter 5). This chapter will thus focus primarily on the immediate proximate causes and payoffs of movement for individuals.

Second, the biomechanical paradigm focuses on the machinery used for locomotion, thus emphasizing the physiological adaptations and mechanics required for terrestrial, aquatic, or avian locomotion (e.g. Dickinson et al., 2000). Although the mechanisms of movement machineries and movement capabilities will receive only little direct attention in this book, they are implicit in most if not all models of movement and thus also the ones dealt with here. For example, when modelling plant dispersal through seed dispersal kernels (Section 2.2), the spatial length scale of the kernel describes how far the seed is likely to fly. This distance may depend on the height at which the seed is released, as well as its aerodynamic properties such as the terminal velocity of the seed, which in turn depends on morphological and physiological factors such as seed size (Thomson et al., 2011).

Third, the cognitive paradigm focuses on the different ways in which cognition informs movement decisions and navigation through the use of environmental cues, social information, and spatial memory (Guttal and Couzin, 2010; Tsoar et al., 2011; Painter, 2013; Fagan et al., 2013). For example, the organism's cognitive abilities can allow it to exhibit edge-mediated behaviour at patch boundaries, or to perform gradient-following behaviour of chemical cues. While we will not construct explicit models of cognition, we will model these kinds of behaviours in the subsequent sections.

The fourth paradigm emphasizes the stochastic nature of the movement process, e.g. by viewing it as a random walk process. The rationale for this paradigm is that movements are intrinsically stochastic: if we could release the same individual twice to the same environment at the same time, it would not be likely to follow an identical movement track. Additionally, when modelling movements, we have only partial knowledge on the *why, where, when,* and *how* questions, and the missing information can often be most conveniently modelled through random terms (Viswanathan et al., 1996, 2011; Bartumeus et al., 2005; Codling et al., 2008). Most simplistic random walk models assume that the movement process is Markovian, i.e. that the location at time $t + \tau$ only depends on the individuals' location at time t but not on its previous locations. More complex models also incorporate non-Markovian processes that may result, e.g. from utilization of information stored in the memory of the organism. In this chapter we will extensively discuss random walk models, and show how biologically relevant mechanisms such as habitat selection or directional bias can be built in them. Most of this chapter will thus focus on this fourth paradigm, although also the other paradigms will be addressed, either explicitly or implicitly.

2.1.6 Outline of this chapter

In the rest of this chapter, we discuss mathematical and statistical approaches to analysing movement processes and the resulting movement patterns. Section 2.2 sets the baseline by discussing stochastic movement models in homogeneous space. In this context we introduce the Lagrangian and Eulerian approaches to modelling by deriving the diffusion approximation (an Eulerian model) of random walk (a Lagrangian model). We then sequentially add biological realism to the baseline model of random walk, such as the tendency for forward persistence and home-ranging behaviour. In Section 2.3 we show how movement models can be constructed for spatially heterogeneous landscapes, and how they can be applied to assess, e.g. functional connectivity and the influence of movement corridors. As an important special case of spatial heterogeneity, we focus in Section 2.4 on highly fragmented landscapes consisting of discrete networks of patches surrounded by unsuitable matrix, a critical prerequisite for understanding metapopulation dynamics to be considered in Chapter 3. In the final section of this chapter, we illustrate how statistical tools can be used to link mathematical models to movement data. Much of the modelling work presented in this chapter is motivated by the long-term study of the Glanville fritillary butterfly conducted by Ilkka Hanski in the Åland Islands in SW Finland (Figure 2.2; Box 2.1).

16 • Movement ecology

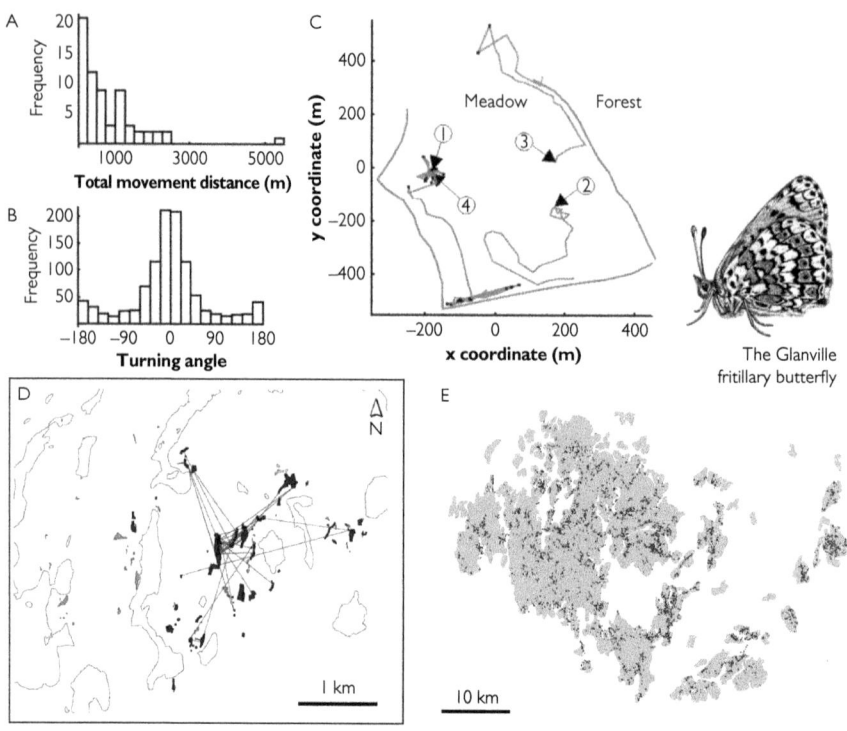

Figure 2.2 Movement ecology of the Glanville fritillary butterfly (*Melitaea cinxia*). (A–C) illustrate small-scale movement data acquired by tracking the flight paths of 66 females with harmonic radar in an experimental study focusing on evolution of dispersal (Ovaskainen et al., 2008a). Shown are the distribution of total movement distances (A), the distribution of turning angles (B), and flight tracks of four individuals (C). In (C), the numbered arrows show the starting locations of the individuals, and the outer line shows the border between the meadow and the surrounding forest. (D) illustrates capture-mark-recapture data collected by Hanski et al. (1994) for 518 female and 1,204 male Glanville fritillaries. The colours refer to habitat patches visited during the capture-mark-recapture study (black), patches not visited (dark grey), the matrix between the patches (light grey), and lakes (white). The straight lines show the observed movements among the habitat patches. (D) is modified from Harrison et al. (2011), who used these data to parameterize a movement model. The patch network shown in (D) is a small subset of the entire network of approximately 4,000 patches in the Åland Islands in SW Finland (E). The illustration of the butterfly was drawn by Zdravko Kolev. (A–C) reproduced with permission from US National Academy of Sciences. (D) reproduced with permission from Ecological Society of America.

Box 2.1 An empirical example of movement ecological research

The Glanville fritillary metapopulation has been extensively studied by Ilkka Hanski (see e.g. Ehrlich and Hanski, 2004; Hanski, 2011), and it has motivated many aspects of the modelling work presented in this book. In the Åland Islands in SW Finland, the Glanville fritillary butterfly inhabits a network of habitat patches surrounded by unsuitable matrix (Figure 2.2E). In this network, the species follows classical colonization–extinction dynamics, and thus to be able to survive in the long term, it needs to compensate local extinctions by recolonizations. While we will return to metapopulation dynamics in more detail in Chapter 3, the point here is that the recolonizations can only be made by individuals that successfully move between the habitat patches, making the species an interesting study object from the point of view of movement ecology.

At the small scale, the flight tracks of the individuals (Figure 2.2C) show a clear forward persistence in movement direction, with an occasional tendency to turn back to the direction from which the individual came from (Figure 2.2B). The movements depend much on the structure of the environment, the butterflies being e.g. reluctant to cross the edge between a meadow and the neighbouring forest (Figure 2.2C). There is great variation among individuals in movement activity (Figure 2.2A), providing raw material for natural selection to act upon, an issue to which we will return in Chapter 5. The frequency and distance-dependency of movements between habitat patches can be quantified by network-scale capture-mark-recapture studies, such as that illustrated in Figure 2.2D.

2.2 Movement models in homogeneous environments

When modelling individual movements, an ecologist is faced with an overwhelming variety of models, ranging from mathematically tractable diffusion models to complex individual-based simulation models (Schick et al., 2008). Movement models vary greatly regarding their underlying assumptions and levels of complexity, but most of them share the common feature of being stochastic. Why do we need stochastic models to analyse individual movements? The most obvious reason is that movements are usually intrinsically stochastic. For example, if attempting to predict the location of a butterfly one day after its original sighting, it would make sense to estimate, e.g. the probability that the butterfly is at most 1 km away from its original location, rather than trying to predict at which exact location the butterfly will be. Further, as we have discussed in the previous section, movements are influenced by a myriad of factors. Because it is usually impossible to consider all these factors and their interactions explicitly, a practical approach is to describe the un-modelled factors implicitly through stochastic terms (Smouse et al., 2010).

The simplest stochastic model for moving organisms is the uncorrelated random walk, where the direction and length of each step taken are considered completely independent of the previous steps (Okubo and Levin, 2001; Codling et al., 2008). The well-known mathematical abstraction of random walk is the Brownian motion, in which the steps taken are infinitely short and movement speed is infinite.

While an empirical ecologist probably would not believe in the idea of solely random movements, we posit that random walk can be very useful for understanding movement ecology. First, for any modelling exercise, it is important to understand the behaviour of the simplest models, as that allows one to assess which features of more complex models are predicted already by the simplest models. In terms of data analysis, one can gain insights into the ways in which organisms move non-randomly by examining the deviation between data and the prediction of a random walk model, which in this context can be considered as a null model (Bergman et al., 2000). Second, in some circumstances (for some organisms, for some spatial and temporal scales) random walk can be a surprisingly good description of real movements. For example, when studying the lifetime movements of female cabbage butterflies (*Pieris rapae*), Root and Kareiva (1984) found a good correspondence between observed and predicted displacements during the oviposition phase, but not during nectar-feeding phase. Thus, deviations from random walk model predictions may reveal specific movement behaviours employed by certain life-history stages, certain individuals, or certain species.

Random walk is also a good model for introducing two classic approaches to ecological modelling, or actually, to modelling in almost any field of science, namely the Lagrangian and Eulerian points of view (Smouse et al., 2010). A Lagrangian model describes the rules that an individual follows, whereas the Eulerian approach describes the statistical properties of ensembles of such realizations. Thus, the Lagrangian approach allows one to generate random realizations of a stochastic process, whereas the Eulerian approach is suited for performing mathematical analyses of the model behaviour. In case of movements, a description of distributions of turning angles, waiting times, and step lengths is an example of the Lagrangian approach. Such a model is often the most intuitive starting point for a field biologist, as describing the model coincides with thinking of what the organism is doing. The Eulerian version of many random walk models will be a diffusion model, as discussed in more detail later. While in the Lagrangian random walk model the state variable is the current location of the individual, in the Eulerian diffusion model the state variable is the probability density of the individual's location.

2.2.1 The Lagrangian approach

Let us first consider in some detail a very simple random walk model. Among the many technical choices one could make (Table 1.1), we build our model in discrete time and in 2-dimensional continuous space. In the Lagrangian description, the state of the movement process at time t is described by the location of the individual, denoted here by the (x, y) coordinates. The model is simply a description of where the individual will be in the next time step $t + \tau$. We model movement at regular intervals of $\tau = 1$ day, and assume that during each time step the individual takes a step of a fixed length of $L = 1$ km. The direction of this step is random, i.e. anywhere between 0° and 360°. A Lagrangian description of a model gives a straightforward starting point for writing a code for simulating the process.

A simulation generated by such a code is illustrated in Figure 2.3A. In this simulation, the individual has moved 100 steps from its original location $(x, y) = (0, 0)$ to the location $(x, y) = (-8.1, -6.8)$, making a route of length 100 km and a net displacement distance $\sqrt{(-8.1 - 0)^2 + (-6.8 - 0)^2} = 10.6$ km. But this is clearly not a general result. If we run the simulation again, we will obtain different values, with the individual ending up in a different location. As is generally the case with stochastic processes, it is not very interesting to examine in detail a single realization of the process, but to consider what happens in a collection of realizations, and thus to ask what kind of statistics characterize the behaviour of the movement model. This is the aim of an Eulerian approach.

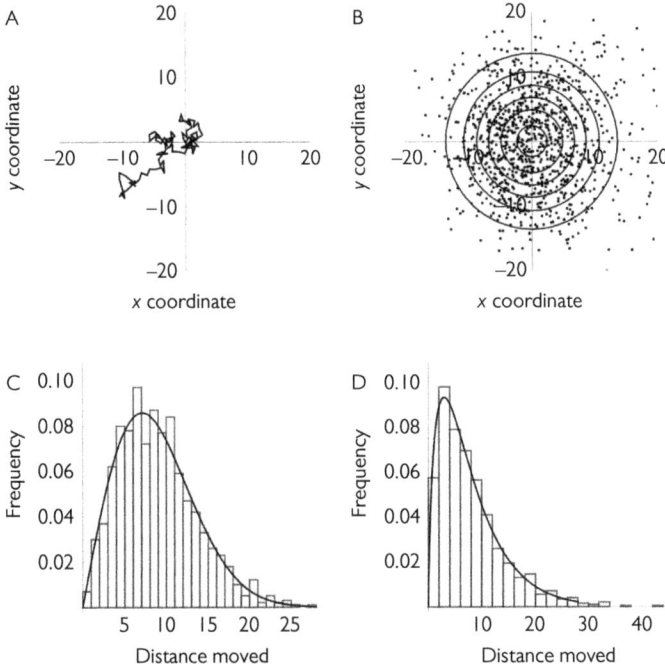

Figure 2.3 Illustration of random walk model and its diffusion approximation. (A) A simulated trajectory for an individual starting from the origin $(x, y) = (0, 0)$ at time $t = 0$, and performing a random walk until $t = 100$ with constant step length $L = 1$, constant step duration $\tau = 1$, and uniformly distributed turning angles. (B) The points show the distribution of final locations of 1,000 replicate simulations of the random walk illustrated in (A), and the lines show the contours of the probability density predicted by the diffusion approximation. (C) The distribution of distances to the origin after 100 steps, based on simulations of the random walk model (the bars) and the diffusion approximation (the line). (D) The distribution of distances to the origin after stopping in a model where the per step stopping probability is $r = 1/100$, based on simulations of the random walk model (the bars) and the diffusion approximation (the line).

Before writing down the model itself in an Eulerian description, let us run the Lagrangian simulation model multiple times. We repeated the simulation shown in Figure 2.3A for 1,000 times, and recorded for each simulation the final location at time $t = 100$, shown in Figure 2.3B. As an example of a movement statistic, we may ask how far on average the random walker will be away from its starting point in 100 time steps. From the simulations, the mean displacement is 8.8 km. Or we may be interested in the frequency of long-distance dispersal events, say how often the random walker is displaced further than a given distance from its origin. For example, 94 out of the 1,000 simulation replicates (9.4%) reached at least the distance of 15 km. Or we may be interested in the entire distribution of displacements, shown in Figure 2.3C. From this figure it appears the individual remains almost always within a distance of 25 km from the origin, though theoretically they could travel up to 100 km (e.g. if all steps are towards north).

The statistics generated in the previous paragraph illustrate how more general insights can be achieved by considering an ensemble of realizations rather than a single realization, i.e. by taking an Eulerian point of view. The results stemming from the Eulerian point of view can often be phrased in terms of distributions, probabilities, expectations, and variances. For example, we used simulations to generate the distribution of movement distances, to learn that the expected displacement is 8.8 km, and that the probability of the displacement being at least 15 km is 0.094.

2.2.2 Translating the Lagrangian model into an Eulerian model

In many situations it is possible to write down the model directly in an Eulerian formulation (Turchin, 1998). The random walk model discussed in the previous section is a well-known example of such a situation. Assuming that the number of steps that the random walker takes is sufficiently large, the behaviour of random walk (a Lagrangian model) is approximated by diffusion (an Eulerian model). In the diffusion model, the state variable is the probability that the individual is at location (x, y) at time t, denoted by $v(x, y, t)$. As we construct our model in continuous space, v is actually not a probability but technically a probability density, an issue to which we will return later. The diffusion model (Turchin, 1998; Moorcroft and Lewis, 2006; Ovaskainen and Crone, 2009) can be written as the partial differential equation (PDE)

$$\frac{\partial v(x, y, t)}{\partial t} = D \left(\frac{\partial^2 v(x, y, t)}{\partial x^2} + \frac{\partial^2 v(x, y, t)}{\partial y^2} \right), \qquad (2.2)$$

where the parameter D is called the diffusion coefficient, and for the random walk model described earlier it obtains the value $D = L^2/(4\tau) = 1/4$. The left-hand side of the equation is the partial derivative of the probability density $v(x, y, t)$ with respect to time, and thus the equation describes how the probability density evolves with time. As we measured spatial units by kilometres and temporal units by days, the unit of the diffusion coefficient is square kilometres per day. We may think that D measures how fast the area where the individual may have moved to increases with time.

Before wondering where Eq. (2.2) came from and what it precisely means, let us start by explaining why it is useful. This particular model is simple enough to be solved analytically, the solution being

$$v(x, y, t) = \frac{\exp(-(x^2 + y^2)/(4Dt))}{4\pi Dt}, \tag{2.3}$$

where we have assumed that the random walker starts from the origin. This solution yields many kinds of insights, including the results that we generated earlier with simulations. For example, we may immediately create a plot of the probability density of where the individual is expected to be after 100 time steps, as shown by the contour lines in Figure 2.3B. For another example, we may use a simple transformation of Eq. (2.3) (see Appendix A) to plot the distribution of movement distances, shown by the line in Figure 2.3C. Further, by integrating Eq. (2.3) over the relevant part of the space (Appendix A), we may compute that the expected displacement is 8.9 km, and that the probability of the displacement being at least 15 km is 0.11.

An analytical approach is more powerful than a simulation approach for several reasons. To start with, it allows one to examine how the results depend on the parameter values. For example, the general formula for the mean displacement after t time steps is $\sqrt{Dt\pi}$. Setting $D = 1/4$ and $t = 100$ yields the specific result of the expected mean displacement being 8.9 km. But now we know the mean displacement for any diffusion parameter (and thus for any combination of the parameters L and τ, as we have the relationship $D = L^2/(4\tau)$) and any duration t of the random walk, without the need to do additional simulations.

We have demonstrated some benefits of taking an Eulerian approach and thus translating a random walk model into a diffusion model. But taking an Eulerian approach usually requires more mathematical knowledge than what is required for simulating a Lagrangian approach. For readers with biological rather than mathematical training, Eq. (2.2) may look like an inaccessible monster that is safer to skip over. Even though it may take some effort before the monster is tamed, we assure it is worth the effort! Being able to read equations is useful not only for those ecologists who plan to apply models in their own research, but for all ecologists because many ecological concepts and ideas are most efficiently expressed in terms of equations. We thus encourage those readers who are not familiar with partial differential equations to have a look at the Appendix A.2, where we provide some background needed to read and interpret Eq. (2.2). Additionally, for those brave enough to take further steps, we suggest to go through the Appendix 'Diffusion for ecologists' of the book by Turchin (1998), where excellent answers can be found to questions such as how Eq. (2.2) can be derived from the Lagrangian description of the random walk model, and how the solution of Eq. (2.3) can be obtained.

2.2.3 Dispersal kernels

Consider a species in which the adult stage is sedentary and the propagules disperse, such as plants or those animals which disperse as juveniles and establish a home range as adults. For such organisms, a dispersal kernel can be defined as the probability distribution of where the dispersing individual will settle, relative to its natal location. Dispersal kernels are often assumed to be radially symmetric, meaning that settling to any direction from the natal site is equally likely.

Assuming that the individual follows the diffusion model, and moves for a fixed time t, the resulting dispersal kernel is given directly by Eq. (2.3), and thus illustrated in Figure 2.3BC. But let us consider a variant of the model in which the individual does not stop after a fixed time but each day makes a decision of stopping with probability r, or continuing with probability $1 - r$. To simulate the Lagrangian version of the new model, we simply randomize at each time step whether the individual settles, and record the final location where the individual decides to settle. In the Eulerian description, inclusion of the stopping rate modifies the diffusion equation as

$$\frac{\partial v(x,y,t)}{\partial t} = D\left(\frac{\partial^2 v(x,y,t)}{\partial x^2} + \frac{\partial^2 v(x,y,t)}{\partial y^2}\right) - rv(x,y,t), \qquad (2.4)$$

where the new term $-rv(x,y,t)$ represents the rate of stopping. This term decreases the probability density proportionally to its current value. To illustrate why this is the case, assume that the individual is still moving at time t with probability $p = 0.1$. If the daily stopping probability is, e.g. $r = 0.01$, the probability that the individual stops the next day is $rp = 0.001$, and that it continues moving is $(1 - r)p = 0.099$. Note that this reasoning holds also if we replace 'stopping' with 'dying', and thus Eq. (2.4) can equally well be used to analyse the movements of an individual with mortality rate r, as we will illustrate later when connecting movement models to data.

The solution to the diffusion equation with stopping rate r is

$$v(x,y,t) = \exp(-rt)\frac{\exp\left(-\left(x^2 + y^2\right)/(4Dt)\right)}{4\pi Dt}. \qquad (2.5)$$

Here the term $\exp(-rt)$ is the probability that the individual is still moving, whereas the remaining part (which is the same as Eq. (2.3), the solution to the diffusion equation without stopping rate) describes the probability density for its location, conditional that the individual is still moving.

Let us define the dispersal kernel $u(x,y)$ as the probability density of where the individual will eventually settle in the random walk model with stopping rate r. At time t, the probability by which the individual is still alive and located at (x,y) is $v(x,y,t)$. If this is the case, it will settle at rate r, and thus the product $v(x,y,t)r$ gives the contribution to the dispersal kernel at time t. The individual will eventually settle at some point t in time, and thus we obtain the dispersal kernel by integrating (summing over) all possible stopping times t,

$$u(x,y) = \int_{t=0}^{\infty} v(x,y,t)r dt. \qquad (2.6)$$

This integral can be computed analytically, yielding

$$u(x,y) = \frac{rK_0\left(\sqrt{r(x^2 + y^2)/D}\right)}{2D\pi}, \qquad (2.7)$$

where K_0 is a one particular mathematical function called the modified Bessel function of the second kind. The corresponding 1-dimensional distribution of dispersal distances is illustrated in Figure 2.3D, which also shows that it matches well with simulations of the corresponding random walk model. At first sight, this is rather similar to the distribution of dispersal distances resulting from fixed movement period (Figure 2.3C). However, the distribution in Figure 2.3D is more leptokurtic ('fat-tailed') than that in Figure 2.3C. This means that the distribution in Figure 2.3D has more weight on very short and very long distances, and thus less weight on intermediate distances. This is expected to be the case, as the probabilistic stopping rule allows the random walker to settle already before the fixed settling time, resulting on a shorter movement distance, or to continue for longer time than with the fixed settling time, resulting on a longer movement distance.

2.2.4 Adding directional persistence: correlated random walk models

Movement ecologists increasingly recognize the need for more 'realistic' models. This need partly comes from empirical works on movement, which highlight the mismatch between predictions derived from simple random walk models and observed movement paths (Morales, 2002). As a first step of bringing more realism into the random walk model, we may assume that the step lengths are not fixed quantities, but sampled from some distribution (Figure 2.4A). Further, turning angles often follow a distribution that peaks around 0 (Figure 2.4B), so that the individual has the tendency of continuing moving in the same direction. Incorporating these assumptions into the random walk model makes the simulated tracks (Figure 2.4C) look different from those of uncorrelated random walk (Figure 2.3A), as they show a higher level of persistence in direction, thus resembling, e.g. the real movement tracks of the Glanville fritillary butterfly (Figure 2.2).

In the Eulerian description, the diffusion equation (Eq. (2.2)) holds also for this more general correlated random walk model. However, while in the case of the simple random walk the diffusion coefficient is given by $D = L^2/(4\tau)$, in the correlated random walk model it is given by (Patlak, 1953a, b)

$$D = \frac{E[L^2] + E[L]^2[2\kappa/(1-\kappa)]}{4E[\tau]}. \tag{2.8}$$

In this equation, $\kappa = E[\cos(\theta)]$ is the mean (expected) cosine of the turning angle. For the uncorrelated random walk it holds that $\kappa = 0$, whereas for increasingly linear movement κ approaches the value $\kappa = 1$. $E[\tau]$ is the mean duration of the time step, $E[L]$ is the mean length of the movement step, and $E[L^2]$ is the mean of the squared length of the movement step. Note that if assuming uncorrelated random walk ($\kappa = 0$) with constant step length ($E[L] = L, E[L^2] = L^2$) and constant time interval ($E[\tau] = \tau$), Eq. (2.8) gives $D = L^2/(4\tau)$, which is the special case we started from.

The ability to convert any distributions of turning angles, step lengths, and step durations into a diffusion coefficient (Eq. (2.8)) is very useful for many applications.

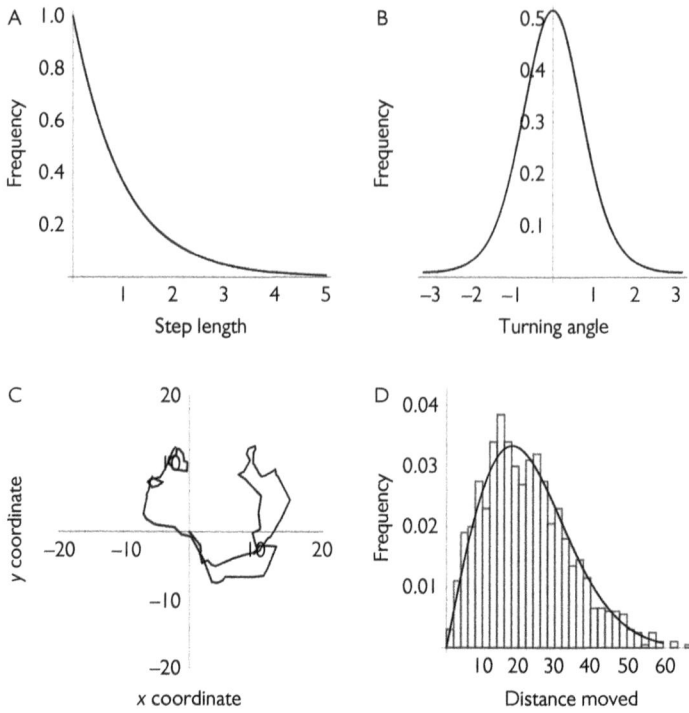

Figure 2.4 Illustration of a correlated random walk model and its diffusion approximation. (A) The step lengths are assumed to be gamma distributed with parameters (1,1), yielding $E[L] = 1$, $E[L^2] = 2$. (B) The distribution of turning angles is assumed to be von Mises distributed with parameters (0,2), yielding $\kappa = 0.70$. The step duration is assumed to be constant $\tau = 1$, so Eq. (2.8) yields the estimate $D = 1.65$ for the diffusion coefficient. (C) A simulated trajectory for an individual starting from the origin $(x, y) = (0, 0)$ at time $t = 0$, and performing a random walk until $t = 100$. (D) The distribution of distances to the origin after 100 steps, based on simulations of the correlated random walk model (the bars) and the diffusion approximation (the line).

First, Eq. (2.8) shows how the attributes of the movement behaviour influence the diffusion coefficient and thus the effective rate of movement. For example, the term $2\kappa/(1-\kappa)$ measures how the diffusion coefficient increases when moving from uncorrelated movements ($\kappa = 0$, resulting in $2\kappa/(1-\kappa) = 0$) to highly correlated movements ($\kappa \to 1$, resulting in $2\kappa/(1-\kappa) \to \infty$). Second, for both theoretical statistical distributions (such as those used in the simulation of Figure 2.4) and empirically observed distributions, it is enough to compute $E[L]$, $E[L^2]$, $E[\tau]$, and κ to obtain an estimate for D. As discussed earlier and illustrated in Figure 2.4D, being able to compute D means that one can predict the distribution of movement distances, either for a fixed time period or until the random walker stops based on the assumption of a constant movement stopping rate r. Third, but related to the second point, the diffusion coefficient D gives a unified currency that can be used to compare the empirical distributions of step length and turning angles measured for different organisms. Finally, many other kinds of movement models besides correlated random walks can also be approximated by diffusion (Turchin, 1998; Gurarie and Ovaskainen, 2011). Consequently, if the aim is to understand long-term movement statistics of many movement models, it is sufficient to understand the behaviour of the diffusion model.

It is important to realize that the diffusion model is an approximation of the underlying random walk model. It captures accurately the long-term behaviour of the model, but fails to characterize short-term behaviour, such as persistence in direction. If the short-term movement characteristics are of interest, additional statistics need to be considered, e.g. the characteristic time-scale of movement that measures how fast the persistence in direction (and movement speed, if that is assumed to vary) decays in time (Gurarie and Ovaskainen, 2011).

2.2.5 Adding directional bias: home-range models

Many organisms conduct exploratory movements at short time-scales, while at longer time-scales they may return to previously visited patches or conduct dispersal events that result in the finding of new patches (Börger et al., 2008; Gautestad and Mysterud, 2010). Movement behaviours that include directional bias towards familiar areas are called home-range or area-restricted space use behaviours. Such behaviours can result from interactions with other individuals and with environmental variability, and they can be central to understanding how individuals are distributed in space and time (Moorcroft and Barnett, 2008; Benhamou, 2011).

Movement models that contain a consistent bias to a preferred direction or towards a given target are termed biased random walks, or biased and correlated random walks if forward persistence is also included (Benhamou, 2006; Codling et al., 2010). The diffusion model can be extended into a diffusion–advection model to account for a bias as follows (Turchin, 1998; Ovaskainen, 2008):

$$\frac{\partial v(x,y,t)}{\partial t} = D \left(\frac{\partial^2 v(x,y,t)}{\partial x^2} + \frac{\partial^2 v(x,y,t)}{\partial y^2} \right) + \frac{\partial}{\partial x} \left(b_x(x,y,t) v(x,y,t) \right)$$
$$+ \frac{\partial}{\partial y} \left(b_y(x,y,t) v(x,y,t) \right). \tag{2.9}$$

Here the new terms that involve the first spatial derivatives are called the advection terms. The parameters $b_x(x,y,t)$ and $b_y(x,y,t)$ measure the amount and direction of movement bias in location (x,y) at time t. For example, if setting $b_y(3,2,5) = 2$, we assume that at time $t = 5$, an individual at location $(3,2)$ is biased towards South (to the negative y direction) with speed 2.

To illustrate home-range behaviour, we modify the basic random walk model so that after each random step, the individual moves towards the centre of the home range, which we set to the origin. We assume that the tendency to return to the

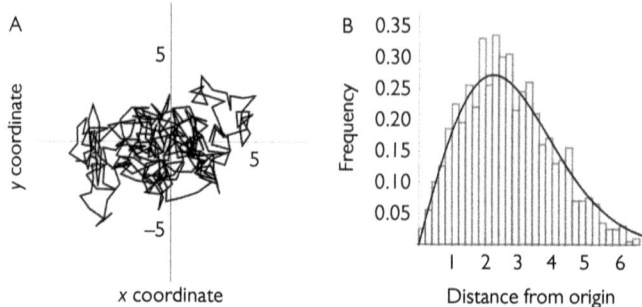

Figure 2.5 Illustration of a home-range random walk model and its diffusion approximation. (A) A simulated trajectory generated by otherwise the same model as in Figure 2.3A but the individual is assumed to take after each random step a deterministic step towards the origin, with length 0.05 of its distance to the origin. (B) The stationary distribution of the individual's distance from the origin, based on simulations of the random walk model (the bars) and the diffusion approximation (the line).

origin increases with increasing distance, implemented so that the length of the movement step towards origin is 5% of the current distance from origin. As illustrated in Figure 2.5A, the random walk part of the model makes the individual drift away from the origin, which is balanced by the tendency towards return to the origin. Running the model for a long time generates the stationary space use distribution illustrated in Figure 2.5B. As the origin is the point of attraction and thus the most likely single location, the reader may wonder why the individual spends more time, e.g. at distance 2 from the origin than at the origin. The explanation here is simply the geometry of 2-dimensional space: the area of locations within the distance range (0.0, 0.1) from the origin is smaller than the area of locations within the distance range (2.0, 2.1) from the origin. Thus, while the origin is the most likely location, the most likely distance from the origin is not 0.

To incorporate the tendency to return to the origin into the diffusion–advection model of Eq. (2.9), we set $b_x(x, y, t) = bx$ and $b_y(x, y, t) = by$, where the constant b is set to $b = 0.05$ to correspond to the assumption that the distance that the individual moves to the origin is 5% of the current distance. The reader may verify that inserting the formula

$$v(x, y, t) = \frac{\exp(-(x^2 + y^2)/(D/b))}{\pi D/b} \tag{2.10}$$

into Eq. (2.9) yields $\partial v(x, y, t)/\partial t = 0$, showing that Eq. (2.10) is the stationary solution to the home-range model. The line in Figure 2.5B shows that this analytical prediction matches well with space use obtained by simulating the random walk model. Equation (2.10) shows that the stationary solution has a Gaussian shape, with variance (which reflects the size of the home range) proportional to D/b. Thus, as expected, the size of the home range increases with increasing random component of movement (D), and it decreases with increasing tendency to return to the origin (b).

2.3 Movement models in heterogeneous environments

Movements generally vary across space and time due to the individuals responding to spatiotemporally varying environmental cues (Schick et al., 2008). For example, in highly fragmented landscapes the individuals usually spend more time in habitat fragments (i.e. areas where resources for survival and reproduction are available) than in the matrix (i.e. unsuitable areas between the habitat fragments; Forman and Godron, 1986; de Knegt et al., 2007). As another example, the movement activity of many animals correlates strongly with the time of the day and with the weather conditions (Polansky et al., 2010; Penteriani and Delgado, 2011).

In this section, we will build spatial heterogeneity into movement models. We start by simulating a random walk process in which we assume that the individuals respond to the local environment through simple behavioural movement rules. As in Section 2.2, we then move from the computational random walk simulations to mathematical diffusion models, which we use to analyse the consequences of environmental heterogeneity on movement probabilities and the times that the individuals are expected to spend in different parts of the landscape.

2.3.1 Random walk simulations in heterogeneous space

Consider a hypothetical butterfly population that inhibits the heterogeneous landscape shown in Figure 2.6. For the sake of simplicity, this landscape consists of just two habitat types that differ in their suitability for the species: patches of suitable habitat (e.g. meadows) located within an unsuitable matrix (e.g. forest). A general tendency observed for many organisms is to move fast and in a straightforward manner in unsuitable environments, but slow and in a tortuous manner in areas of high suitability (Frair et al., 2005; Conradt and Roper, 2006; Klaassen et al., 2006; de Knegt et al., 2007; Fryxell et al., 2008). To simulate such behaviour, we have assumed in Figure 2.6 that the butterflies take on average larger steps and that their turning angles are on average smaller if they are located in the matrix than if they are located in a patch. Another general tendency observed in many organisms is that of habitat selection, meaning that the individuals actively select to move to some habitat types and thus avoid other ones (Lele et al., 2013). For example, the Glanville fritillaries followed by harmonic radar in Figure 2.2C clearly prefer the meadows more than the surrounding forests. In Figure 2.6, we have considered two model variants. In the first model variant (shown by the upper panels) we have assumed that the individuals cross the boundaries between the patches and the matrix without any specific behavioural response. In the second model variant (shown by the lower panels), we have assumed that the individuals exhibit edge-mediated behaviour (Schultz and Crone, 2001; Crone and Schultz, 2008; Schultz et al., 2012), i.e. that they avoid emigrating out of the habitat patches, and that they have a tendency to move towards the habitat patches when they are in the matrix.

To generate the simulated movement trajectories, we used the simple but efficient forward particle-sampling approach (Tremblay et al., 2009; Potts et al., 2014).

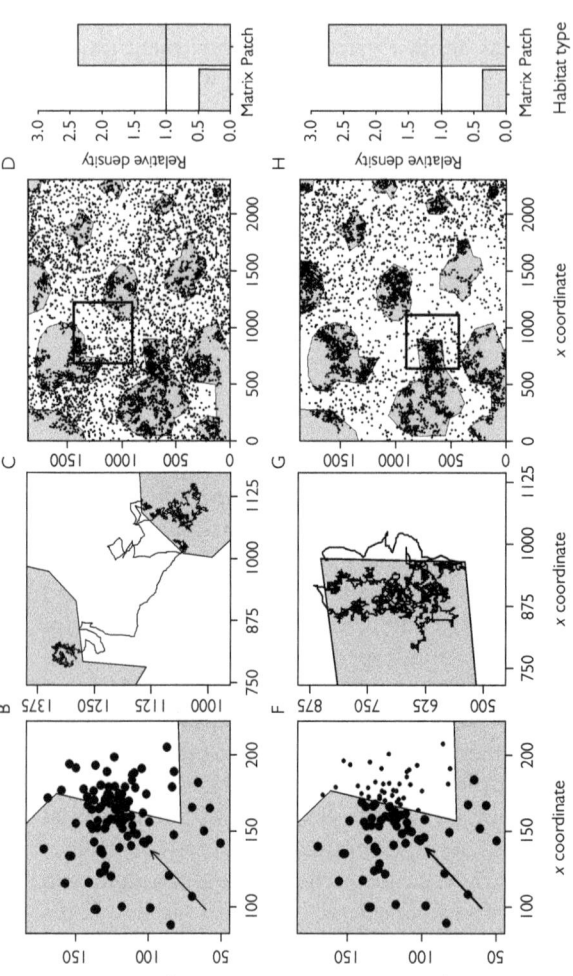

Figure 2.6 Random walk simulations in heterogeneous space. The maps show a landscape consisting of an unsuitable matrix (white) and patches of suitable habitat (grey). The random walk model simulated in the figure is a discrete time model (with minute as the time unit), in which the step lengths are sampled from the Weibull distribution with shape parameter β and scale parameter α that depend on the habitat type, indexed by the subscripts m for matrix and p for patch: $\beta_m = 1.5$, $\beta_p = 2$, $\alpha_m = 5$, $\alpha_p = 2$. The turning angles are sampled from a zero-centred wrapped Cauchy distribution, with habitat-dependent concentration parameters $\omega_m = 0.9$ and $\omega_p = 0.5$. (A) and (E) illustrate the forward particle-sampling approach, where candidate positions are generated by randomly drawing step lengths and turn angles relative to the direction of the previous step, depicted by the arrow. In the top-row panels the simulated individuals do not exhibit edge-mediated behaviour, in which case any of the candidate locations is selected randomly (A). In the bottom-row panels, edge-mediated behaviour is simulated by weighting the candidate locations fourfold (as illustrated by the larger dots in E) if they fall within a patch compared to those that fall within the matrix. (B) and (F) show simulated trajectories of individuals moving for a 1-hour period. (C) and (G) illustrate the space use for 25 individuals moving for 1 week, with locations sampled every hour. The patterns of habitat selection are quantified in (D) and (H), where the bars show the relative densities (number per unit area) of individuals in the patches and in the matrix (see Table 2.1 on page 53 on how these are calculated).

Without edge-mediated behaviour, we simply simulated the next location using habitat-specific step length and turning angle distributions. With edge-mediated behaviour, we first simulated a set of candidate locations, out of which we selected the next actual location by weighting candidate locations within patches four times more compared to candidate locations within the matrix (Figures 2.6A, E). Figures 2.6B and F illustrate 1-hour-long movement tracks generated without and with edge-mediated behaviour, respectively. These simulations illustrate that movements are faster and more directed in the matrix than in the patches, and that with edge-mediated behaviour the individuals are reluctant to leave the patch.

To assess the long-term consequences of the assumed movement behaviours, we placed 25 individuals into random positions within the landscape, and followed their movements for the time period of 1 week. The hourly sampled locations of these individuals are shown in Figures 2.6C and G for the cases where the individuals do not (Figure 2.6C) or do (Figure 2.6G) exhibit edge-mediated behaviour. Habitat selection analyses (Lele et al., 2013) quantify these kinds of patterns by asking how much time the individuals spend within a given habitat type compared to the availability of that habitat type in the landscape. Even without edge-mediated behaviour, the simulated butterflies spend more time per unit area within the patches than in the matrix (Figure 2.6D). Thus, a pattern of habitat selection emerges already from the fact that the individuals move faster within the matrix than within the patches. As edge-mediated behaviour leads to active avoidance of moving to the matrix, it further increases the relative time spent within the patches (Figure 2.6H).

In Section 2.5, we will illustrate statistical approaches to movement ecology by applying them to the data shown in Figure 2.6. But before continuing with this example there, we next discuss how heterogeneous-space random walk models can be implemented and analysed within the framework of diffusion–advection-reaction models.

2.3.2 Diffusion models with continuous spatial variation in movement parameters

To simplify the notation, we denote in this section the 2-dimensional location of the individual by the vector x with the elements $x = (x_1, x_2)$. The general form of the diffusion–advection–reaction model can then be written as

$$\frac{\partial v(x, t)}{\partial t} = \text{diffusion} + \text{advection} + \text{reaction} \qquad (2.11)$$

where

$$\begin{cases} \text{diffusion} = \sum_{i,j=1}^{2} \frac{\partial^2 \left[D_{ij}(x, t) v(x, t) \right]}{\partial x_i \partial x_j}, \\ \text{advection} = \sum_{i=1}^{2} \frac{\partial \left[b_i(x, t) v(x, t) \right]}{\partial x_i}, \\ \text{reaction} = -c(x, t) v(x, t). \end{cases}$$

The diffusion term involves the second spatial derivatives, and it models the random component of movement. In the general case of anisotropic diffusion, the diffusion coefficient is not just a simple number D, but a symmetric 2×2 matrix $\mathbf{D}$ with elements D_{ij}. In this case the elements D_{11} and D_{22} describe the amounts of diffusion in the x_1 and x_2 directions, respectively, whereas the element $D_{12} = D_{21}$ describes a possible correlation between these two. We will return to anisotropic diffusion in the context of multivariate evolution in Section 5.3, but assume here isotropic diffusion, in which case $D_{11} = D_{22} = D$ and $D_{12} = D_{21} = 0$. As explained in Section 2.2, the advection term involves the first spatial derivatives, so that the parameters b_1 and b_2 measure the amount of non-random tendency to move in the x_1 and x_2 directions, respectively. As we model here individual movements rather than population dynamics, the reaction term includes only mortality and not reproduction, for which reason there is the negative sign in front of the mortality rate parameter c. The key in Eq. (2.11) is that all model parameters (D_{ij}, b_i, c) can depend on space (x) and time (t). As we will illustrate later, this modelling framework is very flexible by allowing one to incorporate many kinds of biological mechanisms. But to simplify the treatment, we will focus in this section on spatial rather than temporal heterogeneity, and thus we do not consider parameter values that vary in time.

While most ecologically relevant movement models are formulated in 2- or 3-dimensional space, we start by illustrating some of the key concepts with a 1-dimensional hypothetical case study. Thus, consider a butterfly that lives in an island, consisting of two hills and the intervening valley shown in Figure 2.7A. In 1-dimensional space and with temporally constant parameters, Eq. 2.11 reduces to

$$\frac{\partial v(x,t)}{\partial t} = \frac{\partial^2 \left[D(x) v(x,t) \right]}{\partial^2 x} + \frac{\partial \left[b(x) v(x,t) \right]}{\partial x} - c(x) v(x,t). \qquad (2.12)$$

To avoid the butterfly leaving the island, we assume that it turns back if hitting either of the boundaries ($x = 0$ or $x = 1$). In the framework of the diffusion–advection–reaction model of Eq. (2.12), this assumption is incorporated as a reflecting boundary condition (Turchin, 1998; for mathematical details, see e.g. Ovaskainen, 2008). We note that another option would have been to assume absorbing boundary conditions, in which case the individual would die if it hits the boundary.

We assume that, from the butterfly's point of view, habitat quality increases with altitude $a(x)$, and thus its optimal habitats are found at the hill tops. We further assume that the diffusion rate is lower in high-quality habitats than in low-quality habitats. As discussed in Section 2.2, this may be because in high-quality habitats the individual takes shorter steps, it takes them less frequently, or it turns back and forth instead of moving in a directed manner. We assume the dependence to be inversely proportional, so that $D(x,t) = z_D/a(x)$, where in the numerical example illustrated in Figure 2.7D we have set scaling constant to $z_D = 1$. For the time being, we assume that there is no advection ($b(x) = 0$) nor mortality ($c(x) = 0$).

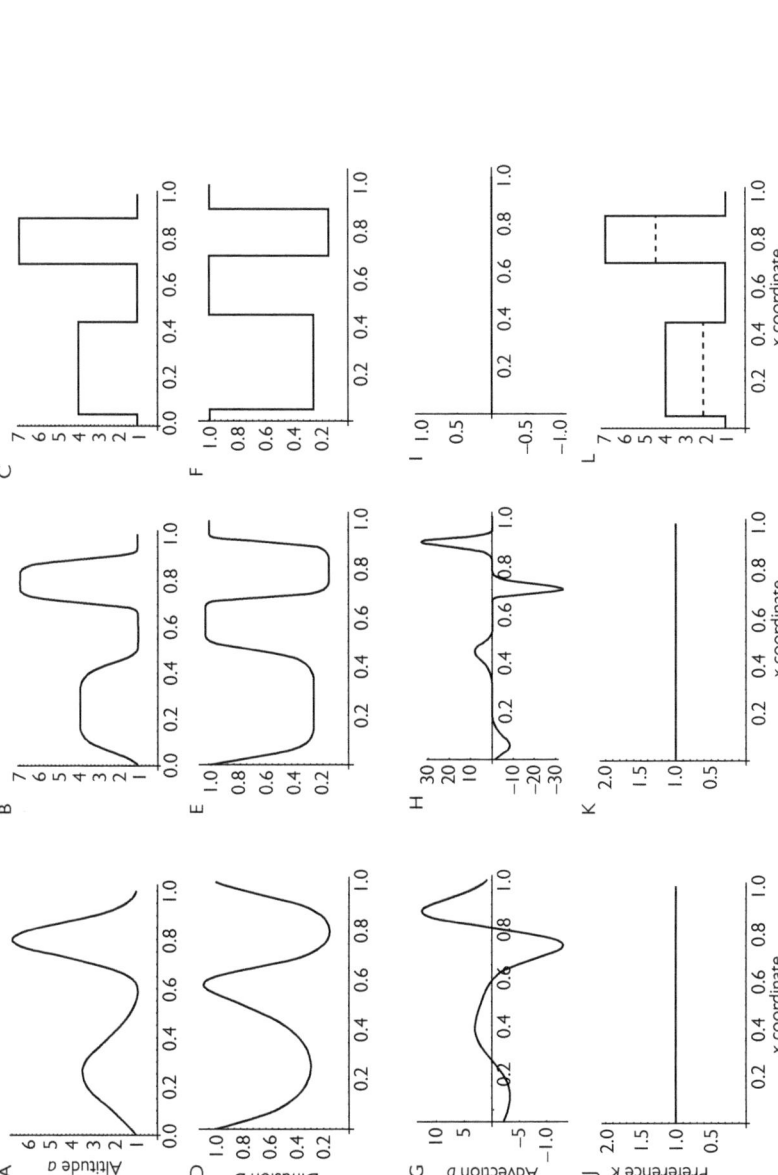

Figure 2.7 A 1-dimensional example illustrating movement models in heterogeneous space. We assume that the individual prefers high altitude habitats, and consider the three altitude profiles shown in (A–C). In (D–F) we assume that the preference for high altitude is manifested through a diffusion rate that is inversely proportional to altitude, $D(x) \propto 1/a(x)$. In (G–I) we assume that the preference for high altitude is manifested through an advection term that is proportional to the negative of the slope of altitude, $b(x) \propto -a'(x)$, so that the individual biases its movements uphill. The right-hand panels present a limiting case in which the landscape consists of discrete habitat types. In this case, we model explicitly the individual's preference to different habitat types, the scenario corresponding to variation in diffusion rate shown by continuous lines and the scenario corresponding to the variation in advection rate shown by dashed lines.

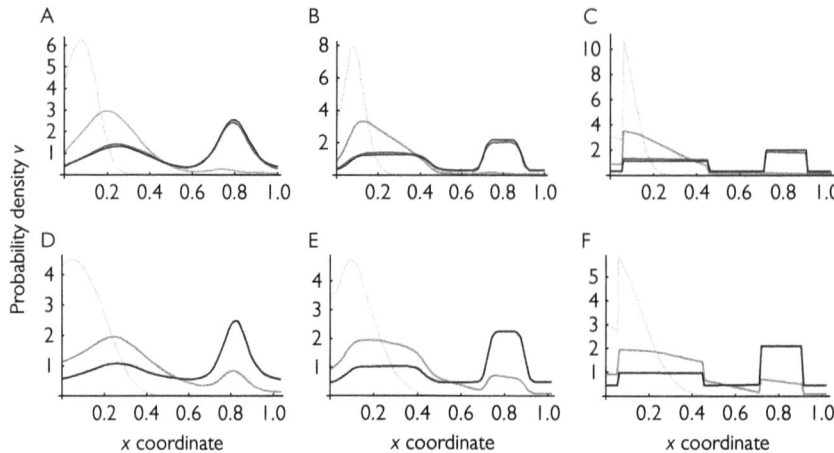

Figure 2.8 The probability density of the individual's location after $t = 0.01, 0.1, 1$, and 10 time units since it started moving from the left-hand edge of the islands shown in Figure 2.7. Increasing time is shown as increasingly dark and narrow lines. The columns correspond to the three altitude profiles of the islands, shown as columns of Figure 2.7. In (A–C) diffusion is assumed to follow the profiles shown in Figures 2.7D–F, whereas advection is set to 0. In (D–F) advection is assumed to follow the profiles shown in Figures 2.7G–I, whereas diffusion is set to the constant $D(x) = 1$. In (C) and (F), habitat preference is assumed to follow the continuous and dashed lines in Figure 2.7L, respectively. In all panels, mortality is set to 0, and reflecting boundary conditions are assumed at the edges of the islands. Numerical solutions were computed with the finite element method (Ovaskainen, 2008).

Let us release the butterfly on the left-hand edge of the island, and ask where its location will be in the future times. Figure 2.8A answers this question by showing how the solution to Eq. (2.12), i.e. the probability density $v(x, t)$ for the individual's location, evolves over time. As is intuitive, initially the individual is most likely to be found near its release site, but as time increases, uncertainty about its location increases. Eventually, the probability density converges to a limiting shape $v^*(x)$, as seen by the two thinnest lines being very close to each other in Figure 2.8A. Such convergence indicates that the process has reached its stationary state. The limiting shape $v^*(x)$ can be obtained directly by solving Eq. (2.12) for $\partial v(x, t)/\partial t = 0$ (Turchin, 1998). In Figure 2.8A, the limiting shape resembles the altitude profile of the island. This is not a coincidence, as in this model the stationary state is inversely proportional to the diffusion rate, $v(x) \propto 1/D(x)$. As we assumed that diffusion is inversely proportional to altitude, it follows that the probability density is directly proportional to altitude, $v^*(x) \propto a(x)$.

Let us then assume that the preference for high altitude does not influence the diffusion rate, so that the diffusion rate is constant over the whole island. Instead, we assume that the butterfly shows hill-topping behaviour (Alcock, 1987), i.e. a tendency to move uphill. Such a tendency can be incorporated into the advection term $b(x)$. To do so, we note that in an uphill the derivative of the altitude $a'(x) = da(x)/dx$ is positive (if moving from left to right), whereas in a downhill the slope is negative. As a positive value of $b(x)$ makes the individual move to the left, we can make the individual move uphill by assuming that the bias is proportional to the negative of the slope, $b(x) = -z_b a'(x)$, where in the numerical example (Figure 2.7G) we have set the constant $z_b = 1/4$. Note that the advection term is 0 at the bottoms of the valleys and at the hill tops, as in those locations the tendencies to move either left or right cancel each other. Figure 2.8D shows how the probability density for the individual's location evolves in this model. We note that the solution is qualitatively similar to the case where we assumed diffusion depends on altitude (Figure 2.8A): the individual ends up spending disproportionally much time on high altitude.

Also the advection model is simple enough for the stationary probability density $v^*(x)$ to be solved. In the general case it is (Ovaskainen and Cornell, 2003)

$$v^*(x) \propto \exp\left(-\int^x \frac{b(x')}{D(x')} dx'\right), \tag{2.13}$$

where the lower limit of integration influences only the proportionality constant and is thus arbitrary. With the assumption of $b(x) = -z_b a'(x)$, and a constant diffusion rate D, we obtain $v^*(x) \propto \exp(z_b a(x)/D)$. Note that the stationary solution depends not only on the altitude profile but also on the diffusion rate. This is intuitive, as the random component of movement will smooth out the pattern generated by the deterministic tendency to climb uphill.

Our assumption of the diffusion rate being inversely proportional to altitude was rather arbitrary. We could have more generally assumed that the diffusion rate is any decreasing function of altitude. Similarly, we could have assumed that the bias term is not directly proportional to the negative of the slope but any decreasing function of the slope of the altitude. One such equally arbitrary assumption is that the bias equals the diffusion rate times the negative derivative of the logarithm of altitude, $b(x) = -D\frac{d}{dx}\log(a(x))$. Inserting this assumption into Eq. (2.13) results in $v^*(x) \propto a(x)$, and so with this assumption the stationary probability density becomes directly proportional to altitude. This is exactly the same result which we obtained by assuming that there is no advection, but that the diffusion rate is inversely proportional to altitude. This is one example of the general observation that multiple mechanisms can lead to an identical pattern. Consequently, if we observe that the population density of butterflies is higher in high altitudes than in low altitudes, we cannot say if this is because the individuals move faster in low altitude, because they bias their movements uphill, or because some other mechanisms. Additional information is needed to resolve these competing hypotheses.

2.3.3 Diffusion models with discrete spatial variation in movement parameters

In the previous example, we assumed continuous variation in the environmental feature (altitude) that influenced movement behaviour. As we did in the example of Figure 2.6, it is often natural to classify the landscape into a number of habitat types, and assume that movement behaviour is homogeneous within each habitat type, but that the movement parameters are habitat-specific. As a bridge between models with discrete and continuous variation in habitat quality, we modify the altitude profile of the butterfly island (Figure 2.7A) so that spatial variation is still continuous, but the two hills become essentially two discrete patches separated by unsuitable matrix (Figure 2.7B). Out of the two patches, the left-hand one is larger but of lower quality, whereas the right-hand one is smaller but of higher quality. Thus, effectively there are three habitat types: matrix, low-quality patch, and high-quality patch. Assuming again that the diffusion rate is inversely proportional to altitude, the diffusion rate becomes essentially constant within each of the three habitat types, as illustrated in Figure 2.7E. This makes the stationary probability density also essentially constant within the habitat types, as illustrated in Figure 2.8B. The probability density is highest in the high-quality patch, as the butterfly moves there the slowest, and consequently ends up spending most time per unit area (or in the 1-dimensional case, per unit length). Similarly, if assuming that the bias term is proportional to the negative of the altitude slope, the bias is essentially zero within the three habitat types, and non-zero only at the narrow boundary zone between the habitat types (Figure 2.7H). Also this assumption leads to the probability density being essentially constant within the habitat types (Figure 2.8E).

Let us then move to the case where the landscape consists of completely discrete habitat types without any transition zone between them (Figure 2.7C). If assuming that diffusion rate depends on habitat, we then have three distinct diffusion coefficients (Figure 2.7F): $D^{(M)}$ in the matrix, $D^{(L)}$ in the low-quality patch, and $D^{(H)}$ in the high-quality patch. Next, we need to describe what happens at the edges between the habitat types. As the width of the edge is zero, there is no space where to follow edge-mediated behaviour, and hence we set $b^{(M)} = b^{(L)} = b^{(H)} = 0$. Instead we model the influence of habitat selection with the help of a matching condition which describes how the probability density changes across the edge between two habitat types. As suggested by the transition from smooth to discrete variation in Figure 2.8, the probability density becomes discontinuous across the boundary between two habitat types (Ovaskainen and Cornell, 2003). To translate this assumption into a matching condition that is applicable for diffusion models, we denote the relative probability densities (to be called habitat preferences) in the three habitat types by $k^{(M)}$, $k^{(L)}$, and $k^{(H)}$. For example, if crossing the boundary from matrix to high-quality habitat, the relative value of the probability density jumps from $k^{(M)}$ to $k^{(H)}$, so that in the high-quality habitat it is $k^{(H)}/k^{(M)}$ times higher than in the matrix. As only the relative values of the k parameters matter, we set $k^{(M)} = 1$ as the reference level, so that $k^{(L)}$ and $k^{(H)}$ describe how much higher the probability density is within the two patches than in the matrix (Figure 2.7F). In addition to the condition describing the discontinuity of probability density, also a second matching condition is needed (Ovaskainen and Cornell, 2003). In mathematical terms, the second matching condition we have assumed here describes that the flux (essentially, the derivative of the probability density) is continuous across habitat edges. In biological terms, this means that the edges do not act as sinks or sources, i.e. that the individuals do not die or reproduce when crossing the edges.

Figure 2.8C shows the model prediction in the case in which the diffusion rate depends on altitude, $D(x) = z_D/a(x)$, there is no advection, and the habitat preferences are inversely proportional to diffusion parameters (continuous line in Figure 2.7L), $k^{(L)} = D^{(M)}/D^{(L)}$, $k^{(H)} = D^{(M)}/D^{(H)}$. In this model the probability density becomes discontinuous across the edge even if the edge is invisible for the individual, because the individual spends more time on that side of the edge where it moves slower. This case corresponds to the 2-dimensional random walk model considered in the upper panels of Figure 2.6. In contrast, Figure 2.8F shows the model prediction in the case in which diffusion rate is assumed to be constant $D=1$, there is no advection, and the habitat preferences are those generated by a biased behaviour at the boundaries (dashed line in Figure 2.7L): $k^{(L)} = \exp(z_b a^{(L)}/D)/\exp(z_b a^{(M)}/D)$, $k^{(H)} = \exp(z_b a^{(H)}/D)/\exp(z_b a^{(M)}/D)$. In this model, the probability density is discontinuous across the edges because the individual shows edge-mediated behaviour.

As discussed earlier, the observation that the stationary density of individuals is higher in the preferred habitat is not enough to tell whether this is due to the individuals moving slower in the preferred habitat or because of the individuals showing edge-mediated behaviour at the edges between the habitat types. However, in the former case, the density difference is inversely proportional to the diffusion rate, whereas in the latter case the density difference is arbitrary, as it depends on the strength of the edge-mediated behaviour. Thus, if we measure both diffusion rates and habitat preferences (as will be illustrated in Section 2.5 for capture-mark-recapture data), we can test whether the density difference is inversely proportional to the diffusion rate. If not, this provides evidence for edge-mediated behaviour.

At a more mechanistic level, edge-mediated behaviour can be due to multiple factors, such as the tendency to move back if crossing the boundary from the preferred to un-preferred habitat (Schultz and Crone, 2001; Conradt and Roper, 2006; Crone and Schultz, 2008). But at a larger scale, e.g. when analysing landscape level capture-mark-recapture data, the mechanistic details are not necessarily relevant because almost any kind of edge-mediated behaviour is expected to lead to the same end result, i.e. discontinuity of the probability density across the edge.

2.3.4 Using movement models to define and predict functional connectivity

Connectivity is a key concept in movement ecology, population ecology, and conservation biology (Calabrese and Fagan, 2004; Kool et al., 2012 and references therein). Connectivity has been defined and can be measured in many different ways (Bélisle, 2005), but in general it measures how the structure of the landscape facilitates (e.g. through corridors) or impedes (e.g. due to the presence of barriers) the movements of individuals, and consequently how it influences the demographic and genetic structure of populations (Moilanen and Nieminen, 2002). Structural connectivity refers to physical structures of the landscape, whereas functional connectivity asks how these physical structures influence the movements and dynamics of a particular species. For example, consider that an overpass is constructed to enable wildlife to cross a highway. Almost by definition, such an overpass provides structural connectivity. Whether the overpass provides also functional connectivity depends on whether the organisms use it. Indeed, highway overpasses are known to facilitate the movements of some species, but not for others (Corlatti et al., 2009). Thus, whether a structural corridor is a functional corridor depends on the species in question.

There are many ways to measure functional connectivity (Moilanen and Nieminen, 2002). For example, one may ask how landscape structure influences the probability that the individual will move from one location to another location, or the amount of time the individual will spend in different parts of the landscape. We will next develop and apply such measures of functional connectivity in the context of the diffusion model. In examples considered earlier, we illustrated how the probability density $v(x, t)$ for an individual's location evolves over time in a heterogeneous landscape. In those examples, we omitted for simplicity the initial location from the mathematical expressions. But as it will be needed in what follows, we now denote by $v(x, t; y)$ the probability density for the individual being at location x at time t, assuming it was initially (at time $t = 0$) at location y. The time-dependent solution $v(x, t; y)$ contains the full information about the movement process. However, it is often interesting to summarize parts of the information more compactly. We will do so by defining four variables which relate to movement probabilities or time-use, and can thus be considered as measures of functional connectivity: the occupancy time density $u(x; y)$, the occupancy time $T_X(y)$, the hitting probability $p^C(y)$, and the quasi-stationary solution $q(x)$.

The occupancy time density $u(x; y)$ is defined as the amount of time that the individual is expected to spend during its lifetime around location x, assuming that it starts from the location y. The occupancy time density is thus defined as the integral of the fundamental solution over all times:

$$u(x; y) = \int_0^\infty v(x, t; y) dt. \tag{2.14}$$

We used the term occupancy time density instead of just occupancy time because $u(x; y)$ defines the time that the individual spends around location x per unit area. If using metre as the spatial unit and second as the temporal unit, u measures the time use in seconds per square metre, assuming the model is formulated in 2-dimensional space.

2.3 Movement models in heterogeneous environments • 37

To obtain the occupancy time $T_X(y)$, we integrate occupancy time density over a region X of interest,

$$T_X(y) = \int_X u(x; y)\, dx. \qquad (2.15)$$

Occupancy time $T_X(y)$ tells how much time the individual is expected to spend in the region X during its lifetime, assuming that it is initially located at y. If using seconds as the temporal unit, the unit of $T_X(y)$ will be seconds, independent of whether we consider movements in 1-, 2-, or 3-dimensional space.

The hitting probability $p^C(y)$ is defined as the probability that an individual that is initially located at y will, during its lifetime, visit some region denoted here by C. As $p^C(y)$ is a probability, it is a unitless number bounded to be between 0 and 1.

Finally, the quasi-stationary distribution $q(x)$ is the stationary probability density for the individual's location, conditional on the individual not having died yet. Technically, $q(x)$ can be defined as

$$q(x) = \lim_{t \to \infty} \frac{v(x, t; y)}{\int_\Omega v(x, t; y) dx}. \qquad (2.16)$$

This equation involves the limit of time going to infinity. Thus, we ask where the individual is likely to be eventually, i.e. after such a long time that knowledge of its initial location y has become irrelevant. The denominator of Eq. (2.16) is the integral of the probability density over all space Ω, and it gives the probability that the individual is still alive. Assuming that there is no mortality, then $\int_\Omega v(x, t; y) dx = 1$. In this case, the quasi-stationary solution reduces to the stationary solution, which we denoted in section 2.3.2 by $v^*(x)$.

If there is no mortality, none of the quantities defined in the previous paragraphs bring any real insights. This is because an immortal individual will move around for an infinitely long time, and consequently it will also visit every possible region and spend an infinite amount of time within any finite region (this is mathematically true in 1- and 2-dimensional spaces; Giuggioli et al., 2012). Therefore, to make the analysis of the measures of functional connectivity defined earlier meaningful, we will assume that the mortality rate is positive, $c(x, t) > 0$.

Let us then return to the 1-dimensional island model. To keep the treatment short, we will consider here only the model where diffusion depends on altitude and there is no advection, and the island consists of three discrete habitat types. We recall that for this model variant, in the absence of mortality, the time-evolution of probability density is illustrated in Figure 2.8C. We now modify the model by setting the mortality rate parameter to $c(x) = 5$, in which case the lifetime of the individual is exponentially distributed with parameter 5, and thus its expected lifetime is $1/5 = 0.2$ time units. As shown by Figure 2.9A, including mortality does not change the shape of the probability density $v(x, t; y)$, but it makes it to decay exponentially to 0. This is the case because we assumed that the mortality rate is constant in space and time.

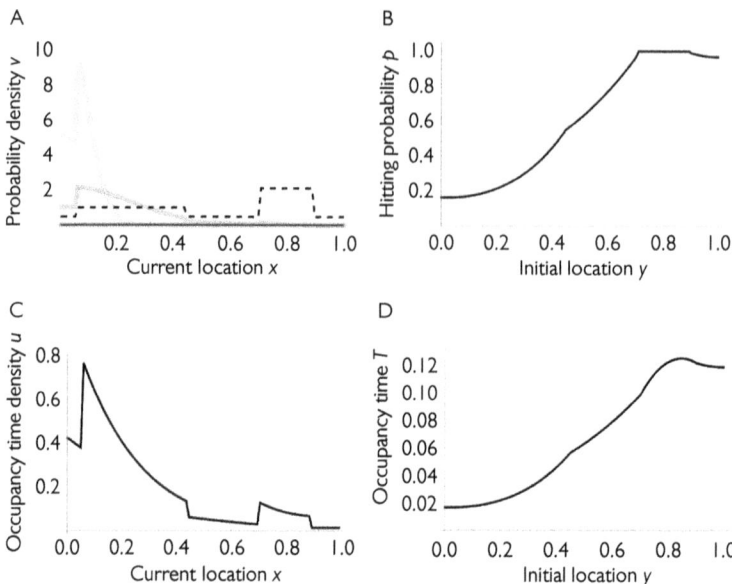

Figure 2.9 Illustration of different measures of functional connectivity in the island model with discrete variation in landscape structure (Figure 2.7C). In all panels, we assume that diffusion varies in space (Figure 2.7F), and that mortality rate attains the constant value of $c(x) = 5$. (A) shows the time-evolution of the probability density, assuming that the individual starts from the left-hand side of the island (line types as in Figure 2.8), and the quasi-stationary solution (dashed line). (B) shows, as a function of the initial location, the probability by which the individual will visit within its lifetime the region $0.7 < x < 0.9$, i.e. the high-quality patch. (C) shows the occupancy time density, assuming that the individual starts from the left-hand side of the island. (D) shows, as a function of the initial location, the occupancy time that the individual is expected to spend within its lifetime in the region $0.7 < x < 0.9$.

The behaviour of the hitting probability $p^C(y)$ is illustrated in Figure 2.9B as a function of the initial location y. We have set the target region to $C = (0.7, 0.9)$, and thus this figure illustrates the probability that the individual will reach the high-quality patch within its lifetime. If the individual starts from the region C, it will hit it certainly, and thus $p^C(y) = 1$ in this region. The closer the target patch the individual starts, the higher is the likelihood that it will reach it. In the absence of spatial variation in the diffusion rate, the hitting probability would decay exponentially with distance. As we have assumed that the diffusion rate is different in different parts of the domain, the rate of that exponential decay varies in space. For example, the hitting probability decreases faster within the other patch ($0.05 < y < 0.45$) than in the matrix, because the individual moves slowly within the other patch.

The behaviour of the occupancy time density $u(x; y)$ is illustrated in Figure 2.9C. The occupancy time density is a function of both the initial location y and the present location x, out of which we have fixed the initial location to $y = 0$, and thus assume that the individual is initially at the left-hand side of the island. For this reason the occupancy time density decreases as a function of the x-coordinate: the individual is likely to spend more time close to its initial location than far away from it. As the occupancy time density $u(x; y)$ is the time integral of the probability density $v(x, t; y)$, the shapes of u and v resemble each other. In particular, in the case of discrete variation in habitat quality, $u(x; y)$ is discontinuous across habitat boundaries, corresponding to the fact that the individual spends more time in high-quality habitats than in low-quality habitats.

Figure 2.9D illustrates the behaviour of the occupancy time $T_X(y)$, shown as a function of the initial location. The occupancy time is also a function of the region X within which we are interested in the time use by the individual. In this example, we have set $X = (0.7, 0.9)$, and thus we ask how much the individual is expected to spend in the high-quality patch within its lifetime, given its initial location. The occupancy time $T_X(y)$ peaks if the individual starts from the middle of this patch, i.e. at $y = 0.8$. If it would stay within this patch its entire lifetime, the expected time it would spend there would equal to its expected lifetime, which, with $c = 5$, is 0.2 time units. But as the individual may be drifted out from the patch, it spends there on average less than that time (Figure 2.9D). If starting from outside the region X, the occupancy time is a product of two factors: the probability of moving to the region X, and the time the individual will spend there after it has reached it. For this reason, outside the region X the profile of the occupancy time resembles the profile of the hitting probability.

The major advantage of the diffusion–advection–reaction modelling framework is that it enables the use of mathematical and numerical tools that would not be applicable for the underlying random walk models. To generate the illustrations of Figure 2.9, instead of running a large number of replicate simulations that would require a lot of time and computational power, the quantities p^C, u, T_X, and q can be solved directly from differential equations (Ovaskainen, 2008). As an example, the hitting probability p^C can be solved from the equation

$$D(y)\frac{d^2 p^C(y)}{dy^2} - c p^C(y) = 0, \qquad (2.17)$$

supplemented with the boundary condition that $p^C(y) = 1$ for $y \in C$. Solving Eq. (2.17) numerically yields $p^C(y)$ simultaneously for all locations in a fraction of a second. In contrast, generating Figure 2.9B by a simulation approach would have required us to conduct a large number of replicate simulations separately for all initial locations, and to track in which cases the simulated individual reached the region C before it died.

A mathematical explanation of where Eq. (2.17) comes from is given in Ovaskainen (2008). To provide here an intuitive explanation, consider a discretized version of the domain $\Omega = [0, 1]$, with $y_i = i/n$, where $i = 1, \ldots, n$, and n denotes the number of grid cells to which the domain is split. Let p_i^C denote the probability that the individual will hit the region C if it starts from the grid cell i. Assuming that the grid cell i belongs to the region C, we have $p_i^C = 1$, corresponding to the boundary condition described in the previous paragraph. Assuming that the grid cell i does not belong to the region C, we denote by $m_{i \to i-1}$ the probability that the individual presently in i will move to grid cell $i - 1$ before it dies and before it moves to any other grid cell. Similarly, we denote by $m_{i \to i+1}$ the movement probability to grid cell $i + 1$.

The probability by which the individual will hit the target region from the location i depends on the probabilities by which it moves from location i to the neighbouring locations, and by the probabilities by which it will hit the target region from those neighbouring locations. This reasoning yields the equation

$$p_i^C = m_{i \to i-1} p_{i-1}^C + m_{i \to i+1} p_{i+1}^C. \qquad (2.18)$$

The differential equation Eq. (2.17) can be derived from Eq. (2.18) by taking the limit where the duration of the time step and the size of the grid cell go to 0.

2.3.5 The influence of a movement corridor

We next apply the previously defined functional connectivity measures to examine how a movement corridor might influence movements. Our example is motivated by the empirical study of Ovaskainen et al. (2008c), where a movement corridor was constructed by cutting an opening area through forest, with the aim of facilitating the movements between two local populations of the Clouded Apollo butterfly. Capture-mark-recapture studies conducted before and after the construction of the corridor showed that the butterflies were willing to enter the corridor region. Despite this, and somewhat counterintuitively, the frequency of movements between the local populations did not increase. To see why this may have been the case, we consider the simplified example shown in Figure 2.10, in which the spatial scale and the model parameters correspond roughly with the Clouded Apollo capture-mark-recapture study. In Figure 2.10A, the landscape consists of two habitat patches surrounded by unsuitable matrix. In the case of the Clouded Apollo butterfly, the habitat patches represent breeding habitats which contain larval host plants, whereas the matrix consists of forests and cultivated fields, with no resources relevant for the species. The butterflies move slower within the patches than in the matrix, and they show edge-mediated behaviour at boundaries between the patches and the matrix, with preference for the patches over the matrix.

To quantify functional connectivity, we consider the right-hand patch of Figure 2.10A as the target region C, and computed the hitting probability $p^C(y)$ by which the individual will reach the target patch if starting from the initial location y. To compute $p^C(y)$ simultaneously for all initial locations y, we solved a 2-dimensional version of Eq. (2.17) numerically with the finite-element method, as detailed in Ovaskainen (2008). As shown by the contour lines in Figure 2.10A, the probability of the individual hitting the target patch (the right-hand patch) decreases with increasing distance from it. However, not only distance matters, but also the direction: the probability of reaching the target patch is smaller if the individual starts near the other patch (the left-hand patch) than if it starts at the same distance from

2.3 Movement models in heterogeneous environments • 41

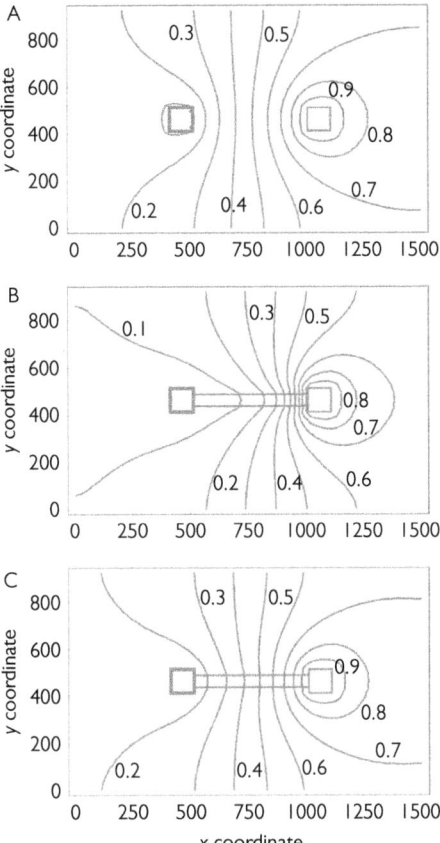

Figure 2.10 The effect of a movement corridor on movement probabilities. The two squares represent two patches of breeding habitat. The contour lines show the probability of an individual reaching the right-hand patch before it dies, as a function of the initial location. In (A), the breeding habitat is surrounded by unsuitable matrix. In (B), the two patches are connected by a corridor that consists of breeding habitat. In (C), the corridor is assumed to consist of movement habitat. The mortality parameter is assumed to be $c = 0.1$ in all habitat types. The diffusion coefficients and the habitat preference parameters (D, k) are $(10^3, 1)$ for the breeding habitat, $(10^4, 10^{-1})$ for the movement habitat, and $(10^5, 10^{-3})$ for the matrix. The external boundaries of the landscape have been set to be reflecting.

the target patch but not near the other patch. This is because the other patch acts as a 'trap': if the individual enters this patch, it is reluctant to leave, and thus the probability of reaching the target patch decreases.

In Figure 2.10B, the two patches are connected by a corridor, which is assumed to consist of the breeding habitat. While one may intuitively assume that adding a corridor consisting of the preferred habitat would increase the probability that the butterflies would move between the patches, this is actually not the case. While in Figure 2.10A the probability of an individual moving from the left-hand patch to the right-hand patch is 8%, in Figure 2.10B it is only 3%. While the individuals move readily to the corridor area, they move so slowly there that they are unlikely to reach the other end. Thus, in this case the corridor does not provide functional connectivity.

In Figure 2.10C, the two patches are again connected by a corridor, but now the corridor is assumed to consist of movement habitat. In case of the Clouded Apollo butterfly, the movement habitat consists of open habitats that include nectar plants but not larval host plants. We assume that the individuals cross the boundaries between the breeding habitat and the movement habitat without any edge-behaviour, and that they move faster in the movement habitat than in the breeding habitat. The probability of moving from left to the right patch is now 10%. Thus, with these parameters corridor increases movement probability and provides functional connectivity.

Figure 2.11 illustrates how the effectiveness of a movement corridor depends on landscape geometry, namely the sizes of the habitat patches and the distance between them. As expected, the probability of moving from one patch to another patch decreases with increasing distance between the patches, whether the patches are not connected by a corridor (Figure 2.11A) or connected by a corridor consisting of movement habitat (Figure 2.11B). The effect of patch area is less trivial, because while larger patches are easier to find, the individual is also less likely to leave a large patch.

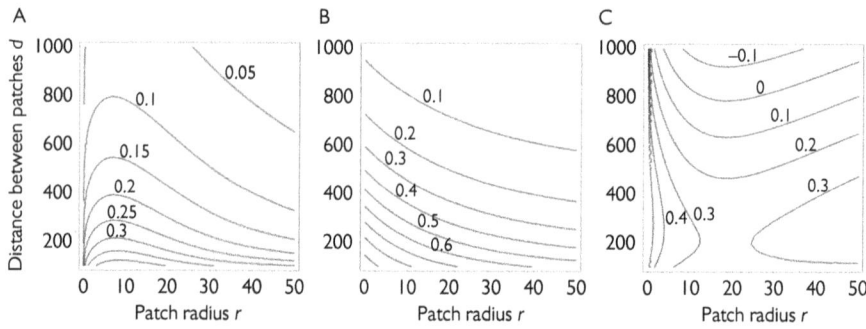

Figure 2.11 Dependency of movement probabilities on landscape geometry. (A) and (B) show how the probability that an individual moves from one patch to another depends on patch size and the distance between the patches. (A) shows the movement probability p_U, which assumes that the patches are surrounded by unsuitable matrix, thus corresponding to the landscape shown in Figure 2.10A. (B) shows the movement probability p_C, which assumes that the patches are connected by a corridor consisting of movement habitat, thus corresponding to the landscape shown in Figure 2.10C. (C) shows the ratio $\log_{10} p_C/p_U$. The movement probabilities have been computed analytically using the approximations derived by Ovaskainen et al. (2008c).

As a result, without the movement corridor the probability of moving from one patch to the other one is maximized for an intermediate size patch (Figure 2.11A). With the directing effect of the movement corridor, the probability of finding a patch is almost independent of patch size, but emigration rate still decreases with increasing patch area. As a result, the probability of moving between the patches decreases with increasing patch area (Figure 2.11B).

Figure 2.11C compares the movement probabilities with and without the corridor, thus asking in which kind of situations the corridor increases movement probability and thus provides functional connectivity. The corridor increases movements in the case that the patches are near to each other, as in this case the lowered movement speed in the corridor area (compared to that of the matrix) is not a major problem. The corridor also increases movements if the patches are very small, as in such case the directing effect of the corridor is critical for the individuals to find the target patch. In contrast, an intermediate-sized target patch that is far away from the source patch can be more effectively found without a corridor than with a corridor. Thus, in this case, the structural corridor works as a functional barrier!

Let us then return to the empirical study of Ovaskainen et al. (2008c). The results from the simplified example help to explain why the construction of the corridor did not increase movements between the local populations even if the individuals entered the corridor area. This is because the individuals preferred the corridor habitat so much that they moved there slowly, making the probability of reaching the other end of the relatively long corridor low.

2.4 Movements in a highly fragmented landscape

As we illustrated in the previous sections, diffusion–advection–reaction models can be applied to landscapes with arbitrary variation in habitat structure, whether it is of continuous or discrete nature. An important special case is a fragmented landscape consisting of discrete patches surrounded by matrix, such as that inhabited by the Glanville fritillary metapopulation in the Åland Islands (Figure 2.2). Figure. 2.12 shows the habitat patch network within one island where the movements of the Glanville fritillary butterfly have been studied (Kuussaari et al., 1996; Ovaskainen et al., 2008b; Zheng et al., 2009b). Individual movements among patches connect local populations, enabling recolonizations of previously empty patches. Thus, understanding how the frequency of movements between patches depends on the structure of the patch network (e.g. inter-patch distances, patch areas, and patch qualities) is essential for assessing the distribution, dynamics, and the persistence of sets of local populations, i.e. metapopulations (Hanski, 1999; Hanski and Ovaskainen, 2000).

In highly fragmented landscapes, it is natural to think about movements as transitions between the patches. For example, when we will model metapopulation dynamics in Section 3.3, it will not be necessary to know the detailed movement

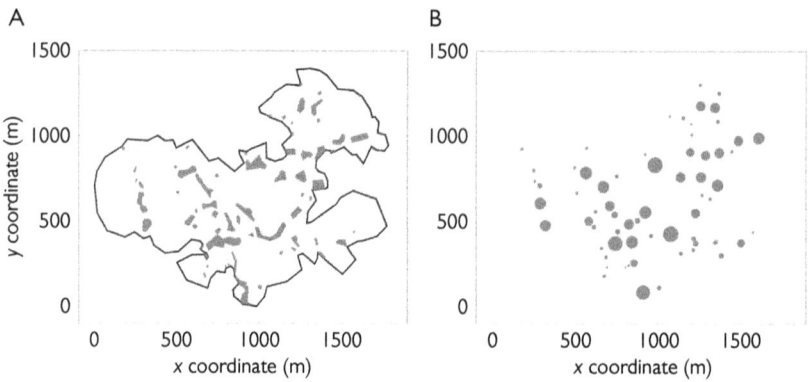

Figure 2.12 The highly fragmented habitat patch network of the Husö Island located within the Åland Islands in SW Finland. The black line in (A) shows the boundary of the island, and the grey polygons show dry meadows that are suitable habitat for the Glanville fritillary butterfly (Kuussaari et al., 1996). In (B), the patches are approximated with circles of the same area, the centre of each circle located in the centre of mass of each polygon.

tracks of the individuals, just to estimate the patch-to-patch movement probabilities and the times that the individuals spend within the patches. As we learnt in Section 2.3, patch-to-patch movement probability is an example of hitting probability, whereas the time spent in a patch is an example of occupancy time. Thus, we could apply the machinery developed so far to compute these quantities for the landscape shown in Figure 2.12, calculate the hitting probabilities and occupancy times numerically, and plug them into a metapopulation model. However, as we will discuss later, given the special nature of highly fragmented landscape structures, it is possible to derive these variables also mathematically, and thus avoid the need to resort to numerical solutions.

Before modelling movements in the patch networks shown in Figure 2.12, we will first consider the case of a single habitat patch surrounded by matrix. This example will provide basic insights into how occupancy times and immigration and emigration probabilities depend on patch size, patch quality, and distance to the patch. In addition, the single-patch results are the needed bricks to build the more general results for networks of patches.

2.4.1 The case of a single habitat patch

Consider a patch of radius r surrounded by unsuitable matrix that extends over an infinitely large area, and an individual moving within this landscape according to the diffusion–advection–reaction model. If the individual is in the patch, its diffusion rate is denoted by D_P, whereas for the matrix it is denoted by D_M. The mortality rates for the patch and the matrix are denoted by c_P and c_M, respectively. We normalize the habitat preference for the matrix to $k_M = 1$, so that k_P measures the individual's preference to the patch, relative to that of the matrix. Thus, the habitat preference parameter k_P measures how many times higher the probability density for the individual's location is within the patch than in the matrix. In the absence of edge-mediated behaviour, the density difference $k_P = D_M/D_P$ emerges due to a difference in movement rate, but k_P can attain also any other value if the individual shows active habitat selection, and thus we consider it as a free parameter.

We define the immigration probability $P_I(r, s)$ as the probability by which the individual, being initially at distance s from the boundary of the patch, will ever visit the patch during its lifetime. As derived by Ovaskainen and Cornell (2003),

$$P_I(r, s) = \frac{K_0(\alpha_M(r + s))}{K_0(\alpha_M r)}. \tag{2.19}$$

Here K_0 is a modified Bessel function of the second kind, which we saw to emerge as a 'kernel' of 2-dimensional random walk in Section 2.2. The parameter α_M is the square root transformed ratio of the mortality and diffusion parameters, $\alpha_M = \sqrt{c_M/D_M}$. We call α_M the movement scale parameter, as its inverse determines the length scale of dispersal in the matrix. To see this, note that $1/\alpha_M$ has the unit of metre (or whichever is used as the spatial unit), and it increases with increasing diffusion rate (allowing the individual to travel faster) and decreasing mortality rate (allowing the individual to travel longer).

The immigration probability decreases with increasing distance from the patch and with decreasing size of the patch (Figure 2.13A). This is intuitive, as small patches faraway are especially difficult to find. Note that the immigration probability does not depend on movement parameters within the patch (D_P or c_P) as we only count whether the individual hits the patch boundary, but we do not follow its movements after that. Further, it does not depend on the habitat preference parameter k_P, as this parameter relates to the probability by which the individual will enter the patch after hitting the boundary.

Let us then consider the emigration probability $P_E(r, s)$, defined as the probability that an individual initially in a patch with radius r will ever move at least to the distance s from the patch. Also $P_E(r, s)$ can be solved analytically but, as its expression is somewhat complex, we refer to Ovaskainen (2004) rather than copying it here. The emigration probability decreases with increasing size of the patch r and with increasing habitat preference k_P (Figure 2.13B). This is because in a large patch the individual encounters the boundary less often than in a small patch, and a high preference makes the individual reluctant to leave the patch. Clearly, the emigration probability decreases with increasing distance s: if the individual will ever move to the distance of 2 km, it will surely have moved also to the distance of 1 km. The emigration probability also depends on the movement scale parameters $\alpha_M = \sqrt{c_M/D_M}$ in the matrix and $\alpha_P = \sqrt{c_P/D_P}$ in the patch, as fast movements in the matrix make the individual more likely to move far away, and fast movements within the patch make the individual more likely to hit the boundary of the patch and thus emigrate to the matrix.

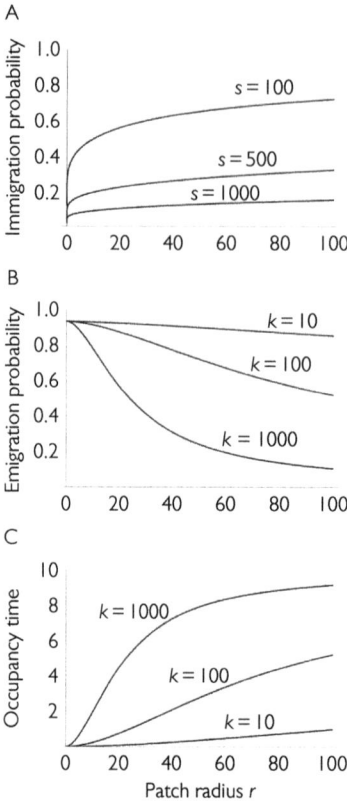

Figure 2.13 Immigration probability, emigration probability, and occupancy time for a circular patch surrounded by matrix. The immigration probability $P_I(r,s)$ is defined as the probability that an individual at distance s from a patch of radius r will ever reach the patch. The emigration probability $P_E(r,s)$ is defined as the probability that an individual initially in a patch of radius r will ever reach a distance s from the patch. The occupancy time $T(r)$ is defined as the time an individual initially in a patch of radius r is expected to spend within the patch in its lifetime. Parameters $D_M = 10^5$, $D_P = 10^3$, $c_M = c_P = 0.1$. The parameters $k = k_P/k_M$ and s are varied in the panels. In (B), $s = 500$.

Finally, we consider the occupancy time $T(r)$, defined as the time that an individual initially in a patch of radius r is expected to spend in the patch during its lifetime. Here we sum up the durations of multiple visits if the individual makes a tour in the matrix and then returns back to the patch. $T(r)$ can also be solved analytically, the mathematical expression being given in Ovaskainen (2004). If the individual would spend all of its time in the patch, $T(r)$ would equal $1/c_P$, the expected lifetime of an individual with mortality rate c_P. But as the individual may emigrate from the patch, $T(r)$ is generally smaller than $1/c_P$, and it is influenced by the same factors that determine the emigration rate. Thus, the individual spends less time in a small patch than in a large patch, and it spends more time in the patch if it is reluctant to cross the boundary (Figure 2.13C).

2.4.2 The case of a patch network

Let us then move from a single patch to a network consisting of a number n of patches. We assume that the individual moves within the patch network according to the diffusion model, with habitat-specific parameters. Thus, the model parameters consist of the diffusion D_M and mortality c_M parameters in the matrix, and the diffusion D_i, mortality c_i, and habitat preference parameters k_i for each patch $i = 1, \ldots, n$. We normalize the preference for the matrix to $k_M = 1$, so that k_i measures the preference for patch i, relative to the matrix.

Following Zheng et al. (2009b), we define R_{ij} as the probability that an individual currently in patch i will ever visit patch j within its lifetime. As another measure of movement probability, we define P_{ij} as the probability that an individual currently in patch i will visit patch j before it visits other patches and before it dies. Further, we define $P_i = \sum_{j \neq i} P_{ij}$ as the probability that the individual makes a successful movement event in the sense of visiting any other patch than its present patch i, and $\Phi_i = 1 - P_i$ as the probability that it will die before doing so. As a measure of occupancy time, we define F_i as the time that an individual is expected to spend in its present patch i before it dies or visits any other patches. We may also condition F_i on the patch that the individual will visit next, so that $F_{i|j}$ is the mean time that the individual will spend in patch i, conditional on visiting next patch j, and $F_{i|0}$ is the mean time that the individual will spend in patch i, conditional on it dying next.

Ovaskainen and Cornell (2003) and Zheng et al. (2009b) present analytical formulae for these variables characterizing movements in a patch network. As the formulae are too complex to provide straightforward insights, we do not replicate them here. Briefly, they consist of a combination of 'movement kernels' that stem from the single-patch results, such as Eq. (2.19), and of matrix algebra. The matrix algebra is needed because movements among the patches are not independent. For example, an individual that would otherwise move from patch i to patch j may encounter on its way another patch k and thus visit that first instead of patch j. As a flavour of that matrix algebra, the probabilities R_{ij} and P_{ij} are connected by

$$R_{ij} = P_{ij} + \sum_{k \neq i,j} P_{ik} R_{kj}. \qquad (2.20)$$

This equation says that, an individual that is currently in patch i, will eventually move to patch j either by moving there straight away (the term P_{ij}), or by moving next to any another patch k, and then eventually moving from patch k to the target patch j. If one knows either kinds of probabilities (R_{ij} or P_{ij}) for all pairs of patches, the other kinds of probabilities can be solved from Eq. (2.20).

Three assumptions are needed to be made for deriving analytical solutions for movement probabilities in highly fragmented landscapes (Ovaskainen and Cornell 2003; Zheng et al., 2009b). First, to be able to utilize the single-patch results presented earlier, the patches are assumed to be circular, and thus the landscape of Figure 2.12A is approximated by that of Figure 2.12B. Second, the patches are assumed to be small compared to the inter-patch distances, i.e. the landscape is assumed to be highly fragmented. This assumption is needed because, otherwise, it becomes important on which side of the patch the individual initially is, and that is ignored in the derivations. Third, the matrix between the patches needs to be homogeneous, and infinitely large. All of these assumptions are only partially valid for the landscape of Figure 2.12A, as the patches are of irregular shapes, as some of the patch-to-patch distances are of the same order as patch radii, and as the movements among the patches cannot take arbitrary routes but are restricted to be within the island. However, as shown by Zheng et al. (2009b), the analytical formulae are still good approximations.

Given that we did not even bother writing down the analytical formulae for the movement probabilities and occupancy times, one may wonder what are they needed for. Their value lies in the fact that they enable one to compute movement characteristics numerically. To illustrate, we will next apply them to simulate movements in a patch network, and to explore how movement rates depend on landscape structure and on movement parameters. For the sake of illustration, we assume that there are two kinds of habitat patches, high-quality patches and low-quality patches, shown in Figure 2.14 by dark grey and light grey colours, respectively. To parameterize the model with values realistic for butterflies, we set the diffusion coefficient to

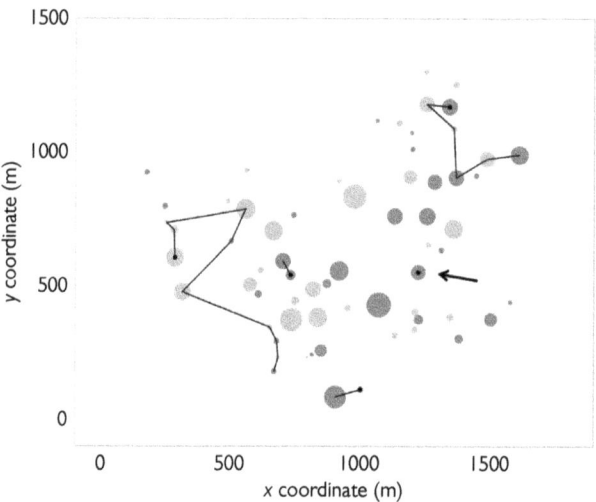

Figure 2.14 Simulated movements in a patch network. The network is identical to that shown in Figure 2.12 except that we have randomly assigned each patch either as a high-quality patch (shown by dark grey) or low-quality patch (light grey). The lines show movement simulations of five individuals, generated as described in the main text. The dots show the initial locations. Note that one out of the five individuals never did move outside its natal patch. Parameters $D_i = 10^3$, $D_M = 10^5$, $c_i = c_M = 0.1$, $k_i = 1000$ in high-quality patches and $k_i = 100$ in low-quality patches. The patch marked with the arrow is further analysed in Figure 2.15.

$D_i = 1,000$ (m²/day) for all habitat patches, and $D_M = 100,000$ (m²/day) for the matrix. We set the mortality rate to $c_i = c_M = 0.1$ (1/day) for both the patches and the matrix, so that on average the individuals live for 10 days. We assume that habitat preference k, relative to that of the matrix, is $k_i = 1000$ in high-quality patches and $k_i = 100$ in low-quality patches.

Figure 2.14 shows movement trajectories simulated for a few individuals. As has been observed with capture-mark-recapture studies on the Glanville fritillary butterfly (Figure 2.2), also in our simulations some individuals stay their entire lifetime within their natal patch, whereas others visit a couple of neighbouring patches. To conduct the movement simulations, we used the probabilities P_{ij} to determine whether an individual currently in patch i dies (with probability Φ_i) or moves next to another patch j (with probability P_{ij}). When connecting the movement model to metapopulation dynamics in Section 3.3, we will also need to know how much time the individuals will spend within the patches, as that will determine their

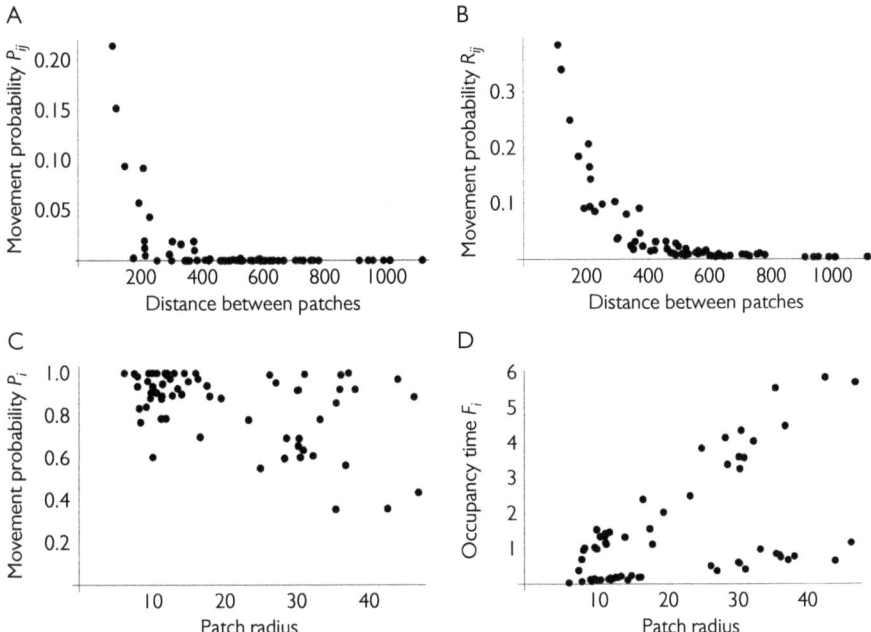

Figure 2.15 Movement probabilities and occupancy times in the patch network of Figure 2.14. (A) and (B) show, respectively, the next movement probability P_{ij} and the eventual movement probability R_{ij}. In these panels, the target patch j has been fixed to that shown by the arrow in Figure 2.14, and black and grey dots refer, respectively, to source patches i with high and low quality. (C) shows, as a function of patch size, the probability P_i by which an individual initially in patch i will move to any other patch before it dies. (D) shows the time F_i that the individual is expected to spend in its present patch i before dying or moving to any other patches.

reproductive output. To simulate these, we use the conditional occupancy times $F_{i|j}$ or $F_{i|0}$, depending on whether the individual moved to another patch j or died. As the diffusion model is a Markov process, we assume that the occupancy times are exponentially distributed, with the expectation set to the values given in the previous paragraph.

Figure 2.15 illustrates how movement characteristics depend on the distances between the patches, as well as on the sizes and qualities of the patches. The movement probability P_{ij} is almost 0 for many pairs of patches (Figure 2.15A), illustrating that the individual typically moves next to one of the neighbouring patches, even if it can eventually (R_{ij}) move also to many other patches (Figure 2.15B). In addition to distance between the patches, both the eventual hitting probability R_{ij} and the next hitting probability P_{ij} depend on patch sizes and qualities, as well as on the locations of other patches in between the source patch i and the target patch j. In particular, the probability P_i of visiting any other patch than the present patch i decreases with patch size and quality (Figure 2.15C). This is simply because the time that an individual is expected to spend in a patch (F_i) is higher for a large and high-quality patch than for a small and low-quality patch (Figure 2.15D), and thus the probability of emigration decreases with patch size and quality.

2.5 Statistical approaches to analysing movement data

Thus far we have discussed mathematical approaches for modelling movement. We hope to have illustrated that mathematical models can be useful even in the absence of any empirical data, as they clarify the links between the underlying processes and the resulting patterns, and they can give rise to hypotheses that can be tested empirically (Chapter 1). In this section, we will illustrate how movement data can be analysed. We will do so both with solely statistical approaches, as well as with a more mechanistic model-based approach. Solely statistical approaches allow one to summarize major trends in the data as well as to test hypotheses about relationships between variables, whereas more mechanistic approaches can be useful for inferring unobserved parameters or processes that have given rise to the data.

Technological advances have drastically increased the ability to collect movement data. Nowadays, most data types involve 'tracking', i.e. determining where an organism is in space at consecutive time points. Some tracking techniques allow one to determine the location of the organism at predefined regular time intervals, such as satellite-based GPS and Argos systems (Hebblewhite and Haydon, 2010). Other tracking techniques provide location information only irregularly, e.g. when a transmitter is close enough to a receiver, as is the case with very high frequency (VHF) radio-waves (Capaldi et al., 2000) and radio frequency identification (RFID) techniques (Krause et al., 2013). Analyses of the latter kind of data are more complicated not only because the data come at irregular intervals, but also because they are influenced by the observation process. For example, VHF tracking may disturb individuals, potentially causing a change in their normal movement behaviour (White and Garrott, 1990). As another example, some areas within a study area may be more accessible to the researchers than others, leading to biased results if the observation process is not accounted for. For example, a disproportionate fraction of observations may be found near roads, which may be the case either because the species prefers roadside habitats, or because the researchers were driving along roads when attempting to observe the individuals (Millspaugh and Marzluff, 2001).

Another major movement data type is capture-mark-recapture (Péron et al., 2010), which is based on resightings of marked individuals, such as ringed birds (Calvo and Furness, 1992), butterflies with a number written on their wing (Morton, 1982), repeated identification of the same individuals from camera-trap (O'Connell et al., 2011), or genetic data (Lukacs and Burnham, 2005). With these data types, accounting for the observation process becomes especially important, as the recaptures are typically attempted only at certain times and locations.

As movement data are collected with many kinds of methods, each of which comes with its own sources of error and bias, there are also many kinds of statistical methods to analyse movement data (Gurarie et al., 2015). We illustrate the analysis of movement data with two contrasting examples. We start with high-resolution data which are not influenced by the observation process, which data we label here as 'GPS data'. After that we consider low-resolution data that are much influenced by the observation process, which we label here as 'CMR data' to abbreviate capture-mark-recapture data. We do not analyse real data, but return to the simulated movement trajectories that were illustrated in Figure 2.6. We sampled from these trajectories both GPS data and CMR data (Figure 2.16), and now our aim is to utilize these data to learn about the underlying movement process. The benefit of using simulated instead of real data is that we know the data-generating process, and thus we can validate whether the inferences that we derive from the data are accurate.

For the GPS data, we assume that we know everything: that we have acquired GPS data for all of the 25 individuals at the same 1-minute frequency that we used to simulate the process, and that the GPS yields the locations with negligible spatial error. In the case of CMR data, we assume that the underlying population consists of 250 individuals and thus we have simulated an additional 225 individuals on top of those shown in Figure 2.6. Mimicking a typical CMR study of butterflies (such as that shown in Figure 2.2D), in these simulations we accounted for mortality, assuming a mortality rate of 0.1/day so that the average lifetime of the individuals was 10 days. As is commonly the case with CMR data, we assumed that we have sparse observations of the movements. Thus, during the 20-day study period that we simulated for the CMR data, each of the 25 capture sites (Figures 2.16C, D) was visited twice per day. During the visits, we assumed that an individual that was located within the capture site at the time of the visit was observed with 80% probability.

2.5.1 Exploratory data analysis of GPS data

Exploratory analysis of movement data involves the calculation of various kinds of movement statistics as well as partitioning variation in these statistics to different

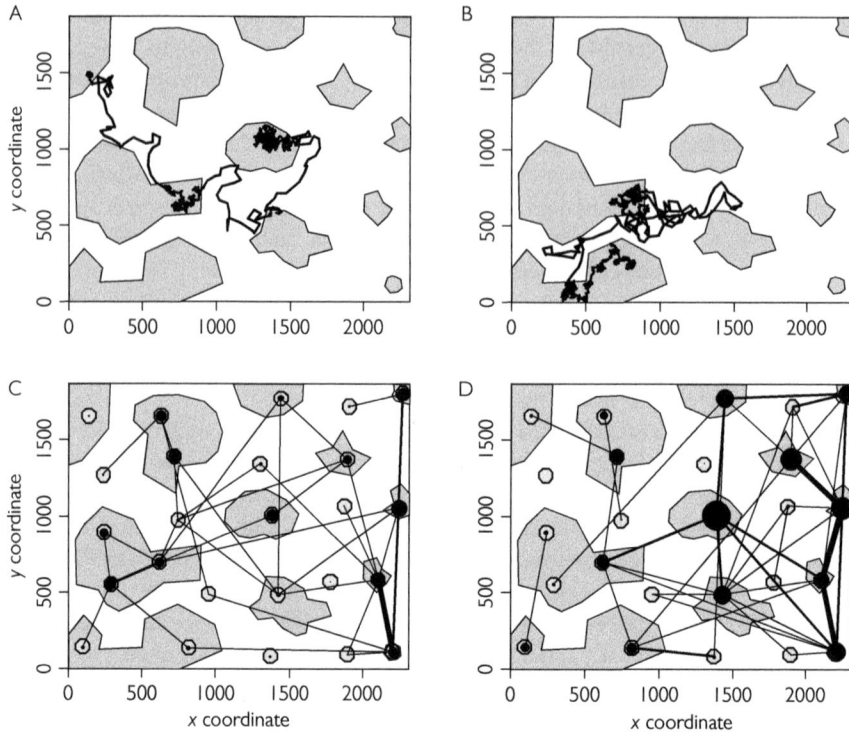

Figure 2.16 Simulated GPS data and capture-mark-recapture (CMR) data used to illustrate statistical analyses. (A, B) show GPS data sampled at 1-min interval for one individual, whereas (C, D) illustrate CMR data. (A, C) correspond to the scenario without edge-mediated behaviour (the upper row of simulations in Figure 2.6), whereas (B, D) correspond to the scenario with edge-mediated behaviour (the lower row of simulations in Figure 2.6). The landscape consists of unsuitable matrix (white) and patches of suitable habitat (dark grey). The light grey octagons in the lower panels depict capture sites. The size of the black dot within each capture site is proportional to the number of sightings made at that site, whereas the thickness of the line between two capture sites is proportional to the number of observed movements. In the scenario without edge-mediated behaviour, a total of $n = 116$ individuals (out of the simulated population of 250 individuals) were observed at least once. For these, the mean number of recaptures per individual was $r = 1.83$, out of which for the fraction $\rho = 0.28$ the recapture was in a different capture site than the previous capture. In the scenario with edge-mediated behaviour, the corresponding numbers were $n = 123$, $r = 2.14$, and $\rho = 0.28$.

sources, such as among individuals, among population, or among spatiotemporally varying environmental conditions. We will illustrate here some commonly used statistics: habitat selection ratios, distributions of turning angles and step lengths, diffusion coefficients and mean squared displacements, autocorrelation functions, and first passage times.

Table 2.1 The calculation of habitat selection ratios.

Case	Habitat	Observations	Availability	Selection ratio	k
W/O (GPS)	Patch	239,984	1,407,637	2.39	1
	Matrix	132,528	3,809,327	0.49	0.20
W/O (CMR)	Patch	145	640	1.44	1
	Matrix	12	360	0.21	0.15
W (GPS)	Patch	245,531	1,407,637	2.72	1
	Matrix	89,098	3,809,327	0.36	0.13
W (CMR)	Patch	201	640	1.45	1
	Matrix	15	360	0.19	0.13

The cases where data have been generated without and with edge-mediated behaviour are denoted by W/O and W, respectively. In the case of GPS data, *observations* is the number of locations observed in the focal habitat type, and *availability* is the total area of the focal habitat type. In case of capture-mark-recapture (CMR) data, *observations* is the number of times individuals were observed in a capture site located in the focal habitat, whereas *availability* is the number of times captures were attempted in the focal habitat. *Selection ratio* is computed as the proportion of observations (out of total observations) in the focal habitat type divided by the proportion of availability (out of total availability) of the focal habitat type. The parameter k is computed by dividing the selection ratio for the focal habitat by the selection ratio of the patch, and it thus gives the relative density of individuals in the focal habitat compared to that in the patch.

Habitat selection ratios measure the degree to which the usage of habitats differs from their availability (see Lele et al., 2013 for an overview on habitat selection analyses). Assuming that all sites in the landscape are equally available to the organisms while moving, the selection ratio for each habitat type can be computed from the location data following the logic of Table 2.1. A selection ratio with value greater than 1 indicates that the habitat type is favoured, as is the case here for the patches, whereas a value smaller than 1 indicates that the habitat type is avoided, as is the case here for the matrix. Without edge-mediated behaviour, the underlying process does not involve any active component of habitat selection, and thus the apparent favouring of the patches simply follows from the individuals moving faster in the matrix than in the patches (see Section 2.3). With edge-mediated behaviour, the pattern of habitat selection is amplified. With infinite amount of data, the habitat selection ratios computed from GPS and from CMR data would coincide, and thus the deviation between them in Table 2.1 is because the CMR have fewer data points and contain more sampling error.

One challenge in quantifying habitat selection ratios from movement data is the question of how to define availability. Assume that we made our study in a large landscape of which Figure 2.16 shows only a part, and that the large landscape includes a third habitat type. Assuming that the individuals would have stayed within the part of the landscape shown in Figure 2.16, the habitat selection analysis of Table 2.1 would have suggested that the individuals strongly avoid that third habitat type. But maybe they would have even preferred that habitat type, just never encountered the part of the landscape where it was available. One version of habitat selection analysis that accounts for such considerations is called step selection analysis. In step selection analysis the habitats that the individual chooses are compared to the availability of habitat types in the locations to which the individual could have moved to from its present location (Thurfjell et al., 2014).

As illustrated in Figure 2.17, *distributions of step lengths and turning angles* can be straightforwardly computed from GPS data (Getz and Saltz, 2008). As the simulated movement process behind the GPS data was a correlated random walk process, we basically recover the distributions that were used when simulating the data. In addition to plotting histograms of the empirically observed distributions (as we did with the Glanville fritillary data in Figure 2.2B), statistical distributions can be fitted to these data using, e.g. maximum likelihood estimation (MLE) or Bayesian estimation methods (see Appendix B.2). Distributions, the support of which covers the positive real axis, e.g. the Weibull, lognormal, exponential, and Levy distributions, are suitable for the step lengths L. Turning angles θ can be fitted to any circular distribution, such as von Mises, wrapped normal, and wrapped Cauchy distributions. In Figure 2.17 we

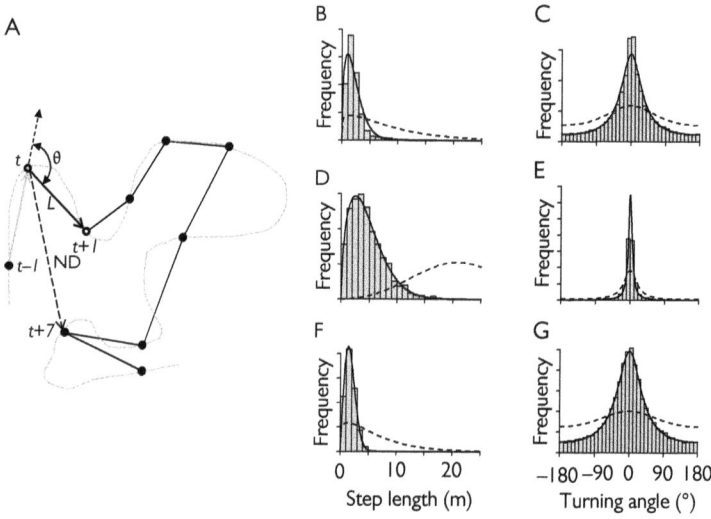

Figure 2.17 Inferring step lengths, turning angles, and net displacements from movement data. (A) shows 10 observations (black dots) of an underlying movement trajectory (grey line) sampled at a regular sampling interval. For the movement step $t \rightarrow t + 1$, the step length is denoted by L and the turning angle by θ. The net displacement (ND) from location t to location $t + 7$ is shown by the dashed arrow. The distributions of step lengths (B, D, F) and turning angles (C, E, G) are shown for the movement trajectory depicted in Figure 2.16B. The bars show the empirical distributions at 1-min sampling interval, solid lines show the distributions (Weibull for the step lengths and wrapped Cauchy for turning angles) fitted to the data which are summarized by the histogram bars, and dashed lines show the same distributions fitted to data thinned to a 5-min sampling interval. (B, C) include all data, (D, E) the subset of data collected in the matrix, and (F, G) the subset of data collected in the patches.

have used MLE to fit the observed data to the same distributions (Weibull distribution for the step lengths and the wrapped Cauchy distribution for the turning angles) that were used to generate the data. Thus, it is not surprising that the assumed distributions fit well to the data collected from each habitat type. When the data are pooled without distinguishing from which habitat type it was collected, the empirical distributions are a compromise between the habitat-specific ones (Figure 2.17).

Figures 2.17 and 2.18 illustrate the effect of the sampling interval on the distributions of observed step lengths and turning angles. With a longer sampling interval, the sampled step lengths obviously increase, whereas the turning angles become less concentrated around 0. With long enough sampling interval, the direction of the individual in the previous step is not anymore expected to correlate with the direction taken in the present step, in which case the turning angle distribution becomes uniform. The confidence intervals of the estimated parameters are wider at longer sampling intervals, simply because there are less replicates. The sampling frequency also influences the estimated total length of the movement track. This is because the assumption that the individuals move along straight lines between the observed locations yields obviously an underestimate of the total distance travelled, the level of underestimation increasing with decreasing sampling frequency (Turchin, 1998; Codling and Hill, 2005; Rowcliffe et al., 2012).

While the full distributions of turning angles and step lengths can be of interest by themselves, the most important information that they involve is captured by a couple of key parameters. As we discussed in Section 2.2, these include the mean step length $E[L]$ and the mean of the squared step length $E[L^2]$. These can be combined into the step length variability index

$$\lambda = \frac{E[L^2]}{E[L]^2}, \qquad (2.21)$$

which attains its smallest value $\lambda = 1$ if the step lengths are constant (i.e. $E[L]^2 = E[L^2]$), and increases with increasing variability of the step lengths (Gurarie and Ovaskainen, 2011). Consistent with this interpretation, the step length variability index is greater when pooling the data among the habitat types than if analysing them separately (Table 2.2).

The most essential information from the turning angle distribution can be captured by the mean cosine of the turning angles, $\kappa = E[\cos(\theta)]$, or alternatively by other similar statistics which quantify the concentration of the angular distribution about its mean (Fisher, 1993; Shimatani et al., 2012). If step directions are not correlated, $\kappa = 0$, whereas for increasingly linear movements κ approaches the value $\kappa = 1$. Table 2.2 shows that the empirically observed mean cosines of turning angles are consistent with the theoretical expectations based on the distributions used in the simulations.

In Section 2.2 we translated a random walk model into a diffusion model, showing how the distributions of step lengths and turning angles can be converted into the diffusion coefficient D. Conveniently, Eq. (2.8) shows that the diffusion coefficient D can be computed from the summary statistics discussed earlier, as we have done

Table 2.2 Movement statistics that summarize the distributions of step lengths and turning angles.

Case	Habitat	E[L](m)	E[L²](m²)	λ	κ	$\log_{10}D$(m²/day)
W/O (GPS)	Patch	1.77 (1.77)	3.99 (4.0)	1.27 (1.27)	0.50 (0.5)	3.56 (3.57)
	Matrix	4.49 (4.51)	29.53 (29.77)	1.46 (1.46)	0.90 (0.9)	5.14 (5.15)
	All	2.74	13.08	1.74	0.64	4.15
W (GPS)	Patch	1.77 (1.77)	3.98 (4.0)	1.27 (1.27)	0.50 (0.5)	3.56 (3.57)
	Matrix	4.50 (4.51)	29.61 (29.77)	1.46 (1.46)	0.90 (0.9)	5.13 (5.15)
	All	2.50	10.81	1.74	0.60	4.03

The values are shown as empirical (theoretical), where the empirical estimates are based on the observed distributions shown in Figure 2.17, whereas the theoretical expectations are derived from the distributions used to simulate the data. As we assumed habitat-specific parameters, the theoretical expectations are not available for the data pooled over the habitat types.

in Table 2.2 and Figure 2.18D. As illustrated by Figure 2.18D, such obtained estimate of the diffusion coefficient is essentially independent of the sampling interval, making it a robust quantity to compare movement data acquired at different resolutions. In contrast, the dependency of step length and turning angle distributions (Figures 2.18A,B,C) on sampling frequency makes their use problematic, as e.g. their direct comparison between studies is meaningful only if the sampling intervals have been the same.

A commonly computed movement statistic which is tightly linked to the diffusion parameter is the *mean squared displacement* (MSD) (Nouvellet et al., 2009). A straightforward way to estimate the MSD from data is to measure the net displacement (ND) after n steps (Figure 2.17), square this distance, and average the result over the pairs of locations that are n steps apart from each other. Thus obtained MSDs are illustrated in Figure 2.19. MSD can also be predicted from the key movement statistics discussed earlier in this section. For correlated random walk, the MSD is given by (Kareiva and Shigesada, 1983)

$$\text{MSD}(n) = E[R_n^2] = nE[L^2] + 2E[L]^2 \frac{\kappa}{1-\kappa}\left(n - \frac{1-\kappa^n}{1-\kappa}\right). \quad (2.22)$$

In Section 2.2, we noted that in the diffusion model the MSD after time t behaves as

$$\text{MSD}(t) = 4Dt. \quad (2.23)$$

Assuming that the steps are taken at constant intervals of $\tau = 1$ so that the time t equals the number of steps taken, we can use Eq. (2.8) to convert MSD(t) = $4Dt$ into

$$\text{MSD}(n) = nE[L^2] + 2E[L]^2 \frac{\kappa}{1-\kappa} n \quad (2.24)$$

Equation (2.24) is similar to Eq. (2.22) but not identical. The reason here is that Eq. (2.22) is an exact result for correlated random walk, whereas Eq. (2.24) holds for the diffusion approximation of correlated random walk. These two coincide when the number of steps taken is large, i.e. when $n \to \infty$, and thus eventually MSD increases

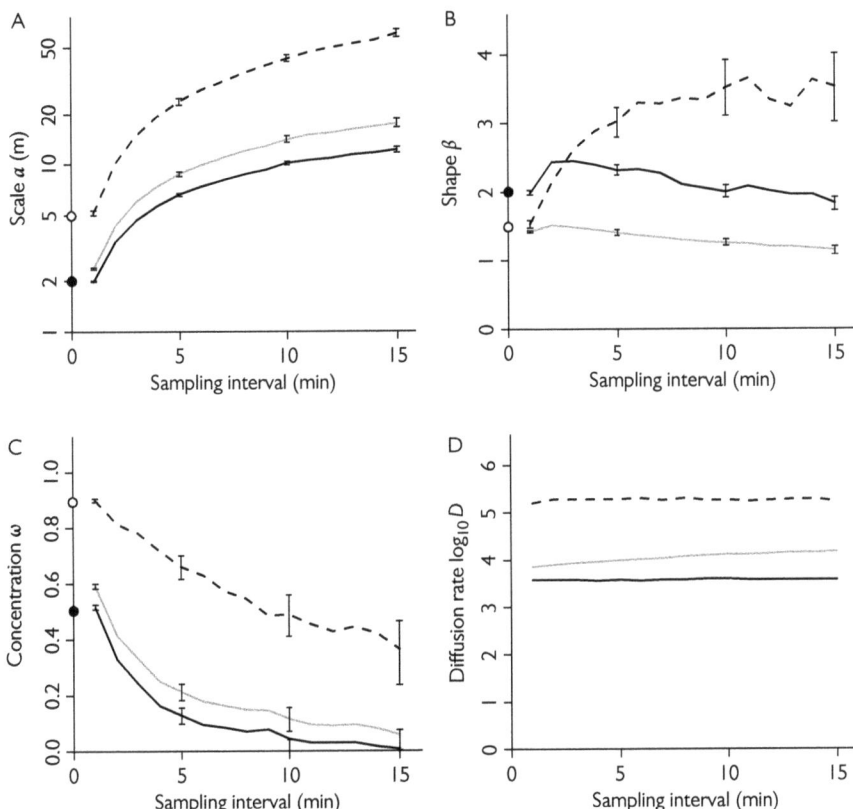

Figure 2.18 The dependency of movement parameters on sampling interval. As in Figure 2.17, we fitted the Weibull distribution to step length data, yielding estimates of the scale parameter α (A) and the shape parameter β (B). We fitted a zero-centred wrapped Cauchy distribution to the turning angle data, yielding estimates of the concentration parameter ω (C). (D) shows the diffusion coefficient D derived from these parameters by Eq. (2.8). The line types indicate whether movement data are pooled over the two habitat types (grey lines), sampled only from the patches (black solid lines), or sampled only from the matrix (black dashed lines). The filled (patches) and empty (matrix) dots show the underlying true parameters for movements simulated at 1-min intervals. The error bars show the 95% confidence intervals for selected sampling frequencies for those parameters that were directly inferred from the data.

linearly with the number of steps taken. However, Eq. (2.22) shows that initially the MSD increases non-linearly, as illustrated in Figure 2.19. Both the diffusion coefficients (Table 2.2) and the MSDs (Figure 2.19) show that the individuals move faster through the matrix than within the suitable patches. Equations (2.22) and (2.24) show that this is the case if in the matrix the step lengths are longer or if the individuals move there in a more directed manner—with our simulated movements we assumed both of these mechanisms.

58 • *Movement ecology*

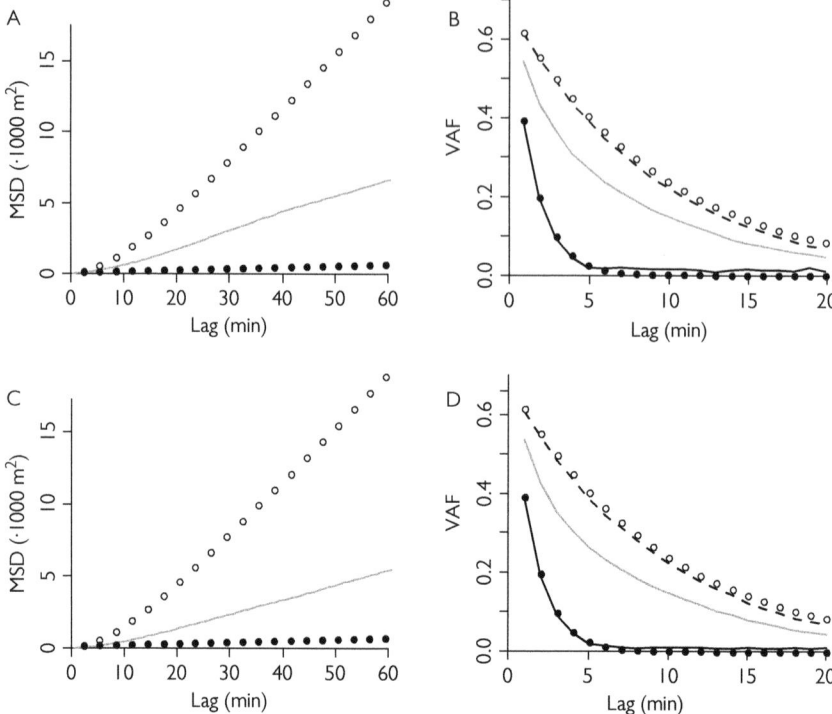

Figure 2.19 Dependency of mean squared displacement (MSD) and velocity autocorrelation function (VAF) on temporal lag between sampling points. As in Figure 2.18, the line (point) types indicate whether the data are pooled over the habitat types (grey lines), sampled only from the patches (black solid lines and closed points), or sampled only from the matrix (black dashed lines and open points). The lines show the empirically estimated MSDs (A, C) and VAFs (B, D) based on movement trajectories of the 25 simulated individuals shown in Figure 2.6. The dots show the theoretical expectations that we derived from the underlying movement model for habitat-specific MSDs and VAFs. (A, B) correspond to the scenario without edge-mediated behaviour, whereas (C, D) correspond to the scenario with edge-mediated behaviour. The empirical VAFs were calculated using Eq. (2.25), whereas the theoretical expectations for the MSDs were calculated using Eq. (2.22) and for the VAFs using Eq. (2.27).

While the diffusion coefficient D and the MSD are informative on movements over large spatial and temporal scales, analyses of autocorrelation are informative about movement behaviour over short temporal scales (Boyce et al., 2010). Autocorrelation can be computed for speed, direction, or velocity (the vector that includes both speed and direction). The *velocity autocorrelation function* (VAF) quantifies how fast the tendency of forward persistence vanishes as a function of the time lag Δt. VAF is defined by (Alt, 1990, Takagi et al., 2008)

$$\text{VAF}(\Delta t) = \frac{\langle v(t + \Delta t) \cdot v(t) \rangle}{\langle |v(t)|^2 \rangle} \quad (2.25)$$

2.5 Statistical approaches to analysing movement data

where $v(t)$ is the velocity vector at time t, the dot $\cdot$ denotes the dot product between two vectors, $|v(t)|$ is the length of the velocity vector, i.e. movement speed, and the brackets $\langle \cdot \rangle$ stand for average taken over initial times t. For many models the VAF decreases either exactly or approximately exponentially as

$$\text{VAF}(\Delta t) = \exp\left(-\frac{\Delta t}{\tau}\right), \quad (2.26)$$

where τ is the characteristic time-scale of the movement process (Gurarie and Ovaskainen, 2011). For example, for the discrete time correlated random walk that we used to generate the GPS data, for large time lags Δt it holds that

$$\text{VAF}(\Delta t) = \frac{1}{\lambda}\kappa^{\frac{\Delta t}{\Delta T}}, \quad (2.27)$$

where ΔT is the duration of one movement step, λ is the step length variability index, and κ is the mean cosine of the turning angle distribution. Figure 2.19 shows that the VAF decays slower in matrix where the movements are more straightforward than in the patches where the individuals are turning more back and forth. The figure also illustrates that the VAF estimated from parts of movement tracks restricted to either habitat type follows the theoretical expectation of Eq. (2.2). For the VAF computed from the entire data, the empirically computed VAF is a compromise between the habitat-specific VAFs.

With the help of the characteristic temporal scale of movement τ (Eq. (2.26)), the diffusion coefficient D can be transformed into the characteristic spatial scale of movement σ defined by $\sigma = 2\sqrt{D\tau}$ (Gurarie and Ovaskainen, 2011). These two parameters (σ and τ) are sufficient for predicting key aspects of movement both at short scales (through the autocorrelation function) and at long scales (through diffusion coefficient).

Movement trajectories recorded at a high temporal frequency can be explored using sliding window analyses to identify times at which the movement behaviour changes. For example, one can compute the amount of time an individual spends within a circle centred at the focal location (Figure 2.20), the quantity of which is called the *first-passage time* (FPT) (Johnson et al., 1992; Fauchald and Tveraa, 2003). Analysis of FPT can be used to segment movement trajectories into different modes and to quantify spatial and temporal scales at which movement behaviour switches, e.g. from explorative movements to area-restricted search. The FPT analysis of Figure 2.20B shows that an individual spends less time around its present location when it is in the matrix than when it is in a patch. This is consistent with the analyses conducted earlier, as it means that the individual moves faster or in a more straightforward manner when it is within the matrix than when it is within a patch. FPT is scale dependent as its measurement requires one to specify the radius of the circular window. The rate at which the mean FPT increases as a function of the radius of the focal window, illustrated in Figure 2.20C, is an alternative measure of the diffusive rate of movement. Plotting the variance in FPT as a function of the radius of the focal window can reveal the spatial scale at which the animal concentrates its search effort (Fauchald and Tveraa, 2003). In the case of the simulated movement trajectories,

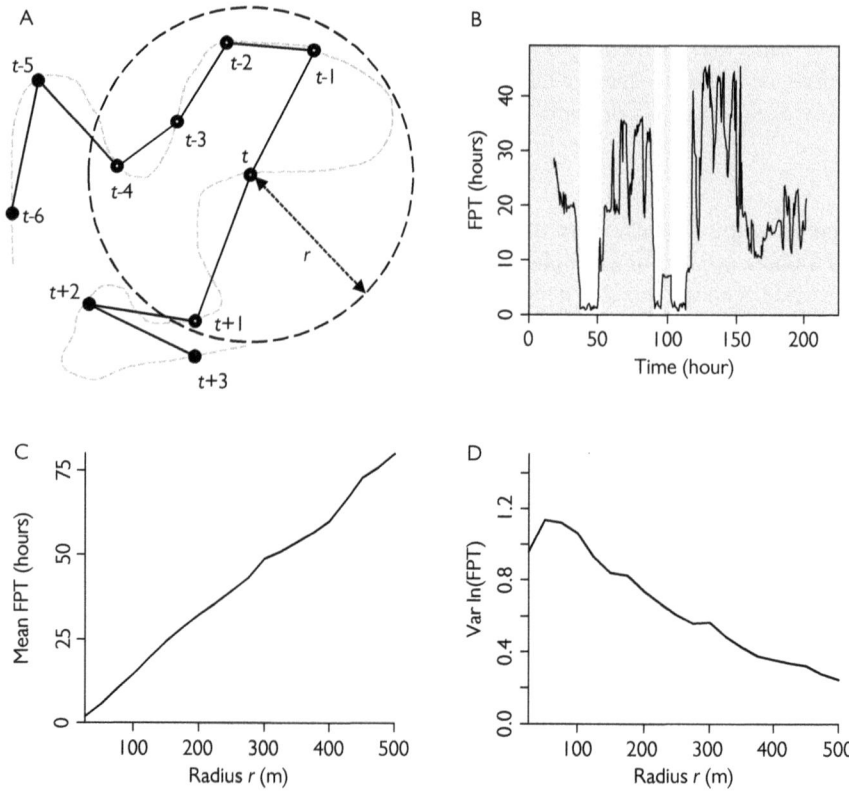

Figure 2.20 First passage time (FPT) analyses. (A) illustrates the computation of FPT for location t: a circle with radius r is centred on location t, and the time that the individual resides within the circle is measured; here the time from $t - 4$ to $t + 1$ equals 5 hours. The FPT is illustrated in (B) for the movement trajectory shown in Figure 2.16B, using $r = 100$ m. Movements through the matrix are indicated by white, while movements within a patch are indicated by grey. (C) shows the mean FPT of all simulated trajectories as a function of the radius r, whereas (D) shows the variance of ln(FPT) as a function of the radius r.

the variance of FPT peaks around window size of 100 m (Figure 2.20D), which reflects the patch size in the part of the simulated landscape where the movement data come from.

2.5.2 Fitting a diffusion model to capture-mark-recapture data

In case of high-resolution GPS data considered in Sections 2.5 and 2.5.1, we had access to full information about the movement tracks, and the question was how to summarize that information. In the case of the CMR data, we only have sparse observations of the movement process. Further, the data are influenced not only by the structure of the landscape and the movement behaviour of the species, but also

2.5 Statistical approaches to analysing movement data • 61

by the observation process itself: where and when individuals were attempted to be captured, and how successful the capture process was. For this reason, inferring the movement process from CMR data is not straightforward. For example, the distribution of observed movement distances may tell equally much about the observation process as it tells about the movement process.

In Table 2.1 we have followed a simple procedure to compute habitat selection ratios based on the CMR data. Unfortunately, the estimation of movement rates is not equally simple. We will next attempt to estimate the movement rates by fitting a movement model to the CMR data. We will take a state-space approach, which explicitly combines a process model with an observation model (Patterson et al., 2008; Jonsen et al., 2013). In the context of movement analysis, the process model describes what we assume about the movement process. In the present case, we will assume the heterogeneous-space diffusion model described in Section 2.3. The parameters of this model are the habitat selection parameter k_M for the matrix (relative to that of the patches, which we normalize to 1, $k_P = 1$), the diffusion rates in the matrix (D_M) and in the patches (D_P), and the mortality rate c. We could have assumed that also the mortality rate is habitat-specific, but we have not done so as the CMR data would have very limited resolution to estimate habitat-specific mortality rates. The observation model describes where and when the recaptures are attempted, and it includes the capture probability parameter p. The capture probability is defined as the probability of observing an individual during a capture attempt, conditional on the individual being in the capture site.

We followed the method of Ovaskainen et al. (2008b) to estimate the model parameters using Bayesian inference. As discussed in Appendix B.2, a key ingredient for Bayesian inference is the likelihood of the data. Once one figures out how to compute the likelihood of the data, estimating the parameters with Bayesian inference (or with maximum likelihood) can be considered just a technical step that may be challenging to perform in practice, but that involves no biological assumptions or insights. So how to compute the likelihood of the CMR data for a given parameter vector (k_M, D_M, D_P, c, p)? Unfortunately, this is not very simple to do with spatially explicit capture-mark-recapture data, and we refer for the full details to Ovaskainen et al. (2008b). To illustrate the basic principles, we have assumed in Figure 2.21A that an individual has been marked on day 0 in the capture site shown by the arrow, and utilized the tools from Section 2.3 to compute the probability density for the individual's location at day 5. Let us note in passing that, as expected, the probability density decreases with increasing distance from the location where the individual was marked and released, and it is greater in the patches than in the matrix. Now assume that on day 5 the individual was observed in one of the capture sites. The probability by which the individual should have been there can be derived by integrating the probability density over the capture site. Multiplying this probability by the observation probability gives the probability by which we should have observed the individual in the site, i.e. the likelihood of the data point. Additionally, one needs to account for all those cases where an individual has not been seen in a capture attempt, as such negative observations are also informative about the movement parameters

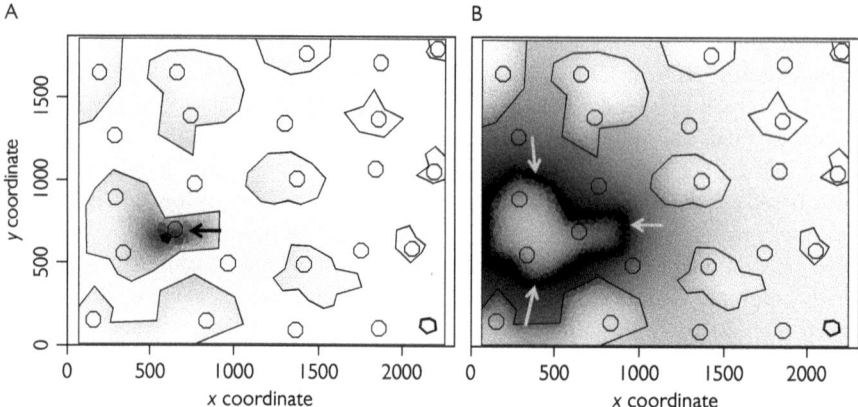

Figure 2.21 Predictions of movement probabilities based on the fitted diffusion model. (A) shows the probability density v at time $t = 5$ days, assuming that the individual was released at time $t = 0$ to the site marked by the arrow. The lightest colour corresponds to $v = 1.8 \times 10^{-8}/m^2$ and the darkest colour to $v = 1.8 \times 10^{-6}/m^2$. (B) shows the hitting probability p by which the individual will visit within its lifetime the boundary of the patch shown by three arrows, as a function of its initial location. The lightest colour corresponds to the value $p = 0.20$ and the darkest colour to $p = 1.0$ (obtained if the individual is initially at the boundary of the target patch). Parameter values were set to the posterior median estimates derived from the CMR data with edge-mediated behaviour ($k_M = 0.26, D_P = 2900, D_P = 180000, c = 0.11$). The triangles seen in the panels were used to solve the diffusion model numerically with the finite-element method (see Ovaskainen, 2008).

(Ovaskainen et al., 2008b). Combining the probabilities of all observations (positive and negative), we obtain the likelihood of the entire data set.

Figure 2.22 summarizes the parameter estimates obtained with Bayesian inference by showing the marginal posterior distributions for the model parameters. Reassuringly, the true underlying parameter values are located within the core areas of the posterior distributions. In the simulations, we assumed a mortality rate of $c = 0.1$ and a capture probability of $p = 0.8$, and the estimates of Figure 2.22 include these values. We did not assume directly any diffusion coefficient, as we simulated the movements with a correlated random walk with particular distributions of turning angles and step lengths. But as discussed earlier, such distributions can be converted into a corresponding diffusion coefficient (Table 2.2). Figure 2.22 shows that we were able to estimate those diffusion coefficients from the CMR data, separately for the matrix and for the patches. It is not straightforward to compute a theoretical prediction for the relative density of the individuals in the matrix (k_M), but Figure 2.22 shows that the values that we estimated from the CMR data using the diffusion model match with the selection ratios that we computed from the high-resolution GPS data.

The key point is that by accounting explicitly for the observation process we have been able to obtain essentially unbiased estimates of the diffusion model from the CMR data. With the parameterized model, it is possible to make predictions

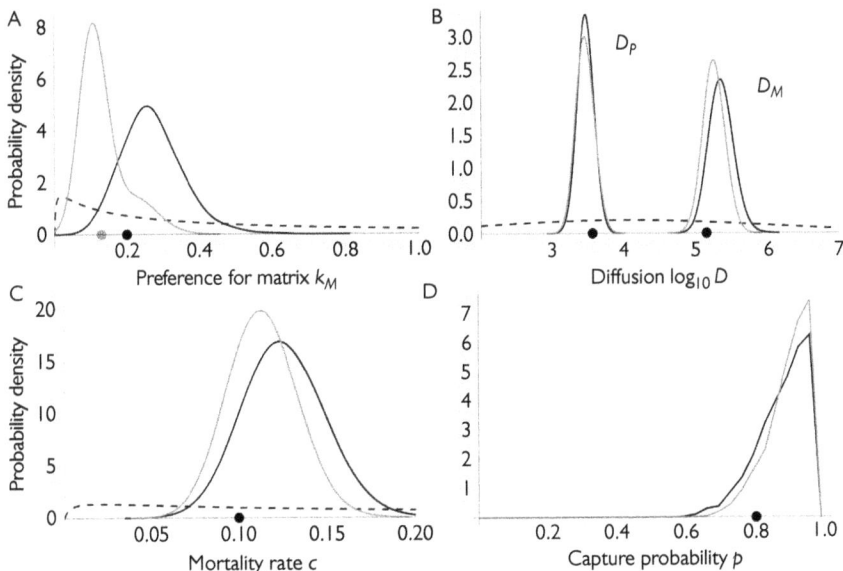

Figure 2.22 Parameter estimates of the diffusion model based on the CMR data. Marginal posterior distributions are shown for the habitat selection parameter (A), diffusion parameters (B), mortality rate (C), and capture probability (D). The solid lines show the posterior densities, whereas the dashed lines show the prior densities. The black lines show the parameter estimates for the data generated without edge-mediated behaviour, whereas the grey lines show the parameter estimates for the data generated with edge-mediated behaviour. In (B) the estimates for the diffusion coefficient D are shown both for patches (D_P) and matrix (D_M). The dots represent the true values used when generating the data (C, D) or the emergent statistics derived from the movement trajectories (A, B; see Tables 2.1 and 2.2).

about movement behaviour. As an example, Figure 2.21B shows how the probability of hitting a specific patch depends on the initial location. As discussed in Section 2.3, constructing movement predictions such as those shown in Figure 2.21 is computationally much more effective with the diffusion model than with the underlying correlated random walk model. Such predictions are useful when analysing population dynamics, as movement probabilities are a key ingredient for predicting colonization rates (see Section 3.3). They can also be useful in the context of conservation planning, as they allow one to examine how the functional connectivity of the landscape depends on habitat configuration.

2.6 Perspectives

In this chapter, we have presented a selection of mathematical and statistical modelling approaches in the context of movement ecology. Clearly, we have only scratched the surface by focusing on the simplest possible class of models, namely random walks and their diffusion approximations. These models provide a backbone for quantitative movement ecology (Turchin, 1998), and their extensions have enabled the development of many important model classes, e.g. home-range models (Moorcroft et al., 1999; Moorcroft and Lewis, 2006) and dispersal redistribution kernels (Morales and Carlo, 2006; Chipperfield et al., 2011). However, random walk and diffusion models are only a rough caricature of the complex movement behaviours that many organisms perform. We next discuss the many limitations and hidden assumptions behind random walk and diffusion models and provide some pointers to other kinds of models that overcome those limitations.

2.6.1 Limitations and extensions of random walk and diffusion models

While random walk and diffusion models can be good approximations of real movement behaviours at a particular spatial and temporal scale, they generally fail to describe behaviours over multiple scales. This is because the rates of movement can be very different when the individual is e.g. resting, foraging, or dispersing. To characterize such heterogeneity in movement rates, models of behavioural switching identify behavioural change points along the animal's movement path, and thus they can help to understand how a broad array of internal and external factors influences movements (Gurarie et al., 2009). There is a large number of methods for identifying change points, which vary in their fundamental approach, complexity, and interpretation (Gurarie et al., 2009). As one example, between the change points the animal may be assumed to follow a correlated random walk with one set of parameters, the parameters changing at the switching points.

Another approach to study animal movement at large spatiotemporal scales is the use of Lévy walks (Viswanathan et al., 1996). Lévy walks are a special case of random walk models. Their step lengths include so much variation (the probability distribution of step lengths has a power-law tail) that their simulations resemble short-term foraging behaviour interrupted by rare large-distance movements. Lévy walks have been used extensively, e.g. for the investigation of search tactics (Viswanathan et al., 2000; Bartumeus et al., 2005).

Due to the high amount of variation, Lévy walks cannot be approximated by diffusion, but they are super-diffusive. Fitting a power-law function to the MSD provides one way to distinguish between diffusive, super-diffusive, and sub-diffusive movements. In the power-law fit MSD $\sim t^\alpha$, the exponent α will have the value of $\alpha = 1$ for diffusive movements, $1 < \alpha < 2$ for super-diffusive movements such as the Lévy walks, whereas $0 \leq \alpha < 1$ for sub-diffusive movements such as home-ranging behaviour (Codling et al., 2008). Random walks with finite mean and variance of the step length distribution result in diffusive movements and a linearly increasing MSD (Getz and Saltz, 2008). However, as discussed earlier, not all random walks belong to this class of models.

We defined correlated random walk models with the help of step length and turning angle distributions. An important, but somewhat hidden, assumption behind most discrete time-correlated random walk models is that the step lengths and turning angles are assumed to be independent of each other, and independent of the lengths and turning angles of the previous steps. Further, these distributions are not only properties of the species, but they depend also on the sampling interval (see Figure 2.18). Movement models that are defined in continuous time can often be considered more natural as they overcome many of these problems (Gurarie and Ovaskainen, 2011; McClintock et al., 2014).

Another limitation with basic random walk models is that they are technically Markov processes (see Appendix A.4), meaning that the next movement step depends only on the current location, not on where the individual was before. However, animal movements can be much influenced by the past experiences stored in the individual's memory (Fagan et al., 2013). This can make a major difference especially on heterogeneous landscapes, allowing the individuals, e.g. to revisit patches more effectively than encountering them by random, with obvious consequences to the spatial dynamics of populations. For example, consider the case of the highly fragmented landscape of Section 2.4. As we utilized a diffusion model with habitat-specific parameters and edge-mediated behaviour, our results are relevant for species which adjust their movements based on information about their local environment, such as many insects. This can be considered as an intermediate case between two extremes. At one extreme would be species that have full knowledge of the landscape, and for which movements are not costly. The movements of such species could be understood in the framework of the optimality paradigm discussed in Section 2.1, as they would simply select to move to the location that maximizes fitness. This would lead to the so-called ideal free distribution of individuals over the landscape (Fretwell and Lucas, 1969). At the other extreme is the case where the individual does not have any knowledge of its surroundings, and has no active means to control its movements. This is the case for passive propagule dispersal of many plants and fungi, the movements of which are often modelled by dispersal kernels.

In Section 2.3 we discussed in much detail how to incorporate spatial heterogeneity into movement models. We assumed that the key characteristics of movement, such as movement speed or the tendency to move to a given direction, depend on the underlying habitat type. In particular, we showed how the pattern of habitat selection can emerge either from the individuals moving slower in their preferred habitats or from the process of edge-mediated behaviour at patch boundaries. However, we note that movements are also influenced by many other factors than the underlying habitat type, such as the distributions of conspecifics and heterospecifics, spatiotemporal variation in thermal conditions, and so on. In principle, such factors can be and have been incorporated into random walk and diffusion models by making assumptions on how the movement parameters depend on them (McClintock and King, 2012).

Due to our focus on habitat heterogeneity, we discussed also functional connectivity from the point of view of how habitat structure influences movement probabilities and occupancy times. Another context where hitting probabilities are relevant is in the analysis of encounter rates (Viswanathan et al., 1999, 2011; Gurarie and Ovaskainen, 2013). The sensory and cognitive abilities of the searcher are of major importance in determining encounter rates, and consequently many species have evolved sophisticated methods for sensing resources, e.g. through vision and olfaction (Bell, 1991). Movements can be related to the rate at which a forager finds resources (e.g. food plants or prey) or to the rate at which males encounter females, both of which are fundamental ingredients for assessing population dynamical consequences of movement. Searching for scarce targets is generally most efficient using fast movements with high directional correlation, whereas the search for abundant and clustered targets requires tortuous movements for high efficiency (Viswanathan et al., 1999; de Knegt et al., 2007; Bartumeus et al., 2008; Gurarie and Ovaskainen, 2013). In some studies, the time to the first encounter can be of interest, while in others the mean encounter rate may be more relevant. Encounter rates can be computed for static targets (e.g. herbivores searching for food plants) or mobile targets (predators searching for a mobile prey).

Besides the diffusion-based methods for analysing functional connectivity, applications of graph-theoretic connectivity measures have rapidly increased in ecology and conservation (Rayfield et al., 2011; Moilanen, 2011). In graph-theoretic analyses the patches are considered as nodes which are connected by lines or 'edges' (Urban and Keitt, 2001; Cushman et al., 2013). The most popular of such connectivity measures are the least-cost path (LCP; e.g. Adriaensen et al., 2003) and measures based on circuit theory (e.g. McRae et al., 2008). Like connectivity measures based on diffusion analyses, connectivity measures based on circuit theory account for the possibility of having multiple routes that connect a source patch with a target patch. LCP analyses, on the other hand, consider only the route of least resistance, and thus implicitly assume that the moving organism has large-scale knowledge about its environment and takes the route that minimizes the cost of movement.

2.6.2 The many approaches of analysing movement data

A key point that we made when analysing movement data statistically is that the technique used for acquiring data needs to be accounted for. Some techniques enable the measurement of locations at predefined regular temporal intervals (e.g. GPS data), whereas other methods collect movement data at temporally irregular intervals (e.g. CMR data). In the latter case, the observations may contain a bias due to the opportunistic nature of the sampling. Thus, to obtain unbiased inference from the CMR data, we needed to combine the movement model with an explicit observation model. Further, some types of data are accurate about the spatial and temporal location (e.g. GPS data and CMR data), while other data types have a large positional error (e.g. geolocation data), the amount of which may further depend on the characteristics of the environment, e.g. topography or density of forest canopy (Frair et al., 2010). Positional error in the data can obviously bias movement estimates, unless accounted for by an error model (Patterson et al., 2008; Jonsen et al., 2013; Thuiller et al., 2013).

The spatiotemporal resolution of the data has a major influence on what kind of analyses are feasible. In the examples considered in this chapter, the GPS data allowed us to analyse both the large-scale (e.g. diffusion rate) and the small-scale (e.g. distributions of step lengths and turning angles, and the velocity autocorrelation function) properties of movement. In contrast, while the CMR enabled us to estimate habitat-specific diffusion rates, it was not possible to tell whether a given diffusion rate was obtained, e.g. through slow and directed or through fast and tortuous movement.

The methods we presented just scratch the surface of the many approaches developed for the analysis of movement data (Gurarie et al., 2015). To start with, one can summarize movement trajectories in various ways not considered here. For example, a number of measures have been developed to characterize search efficiency in terms of the tortuosity of the movement path (Benhamou, 2004). These measures include the ratio between the total distance moved and the area covered within a buffer distance from the movement path (de Knegt et al., 2007; Almeida et al., 2010), a straightness index given by the net displacement distance divided by the total length of the movement track (Almeida et al., 2010; Postlethwaite et al., 2013), and the fractal dimension of the movement track (Benhamou, 2004).

Habitat selection studies examine on what kind of locations the individual spent time rather than studying the geometric shape of movement trajectories. Such studies analyse the usage of habitat characteristics with respect to their availability, and can thus identify drivers of animal movement and distribution at large scales (e.g. a species' geographic range or an individual's home range; Senft et al., 1987; Manly et al., 1993; Lele et al., 2013). Step selection functions (Fortin et al., 2005; Thurfjell et al., 2014) examine habitat selection at smaller scales by contrasting the habitat type at the location where the animal actually moved to habitat types at locations where the animal could have moved to. Another method that contrasts known animal locations with potential ones is offered by point-process analyses (e.g. McDonald, 2013; Renner et al., 2015), which view movement data as a set of point locations in continuous space, and consider the intensity (number of unit area) of points as the response variable to be modelled (Renner et al., 2015).

3

Population ecology

3.1 Scaling up from the individual level to population dynamics

This chapter is about population ecology, and thus turns our attention to changes in population size over space and time, as well as on the demographic structure of populations, e.g. distributions of ages and sexes. A population can be defined as a collection of individuals of the same species that inhabit a particular region at the same time, and which function together as an ecological entity (Begon et al., 1996). Populations are characterized by features such as their size, density, and spatial distribution. Population size refers to the number of individuals in an area, whereas population density measures the number of individuals per unit area. Spatial distribution describes the locations of the individuals, and thus how individuals are distributed with respect to each other and the environment. One central concept related to spatial distribution is population dispersion, which refers to whether the spacing of individuals is more or less aggregated than expected by random.

In theory, all population-level phenomena can be derived from individual-level phenomena. For example, in a closed population, the size of the population varies over time because individuals reproduce and die. In an open population, the size of the population varies additionally because individuals can move in and out of the area (Figure 3.1). Similarly, demographic structure changes as individuals grow, age, die, or give birth to new individuals. The starting point for assessing population dynamics is to understand how individual-level processes translate to population-level processes, as well as to understand how individual-level processes are modified by the state of the population (Koehl, 1989).

A very general model for population dynamics, describing how population size $N(t)$ changes over time t, can be written as

$$N(t + \tau) = N(t) + B(t) - D(t) + I(t) - E(t). \tag{3.1}$$

Here the terms $B(t), D(t), I(t)$, and $E(t)$ denote, respectively, the number of individuals that are born, that die, that immigrate into the population, and that emigrate out of the population, between the times t and $t + \tau$.

Quantitative Ecology and Evolutionary Biology. Otso Ovaskainen, Henrik Johan de Knegt & Maria del Mar Delgado. © Otso Ovaskainen, Henrik Johan de Knegt & Maria del Mar Delgado 2016. Published 2016 by Oxford University Press. DOI 10.1093/acprof:oso/ 9780198714866.001.0001

3.1 Scaling up from the individual level to population dynamics • 69

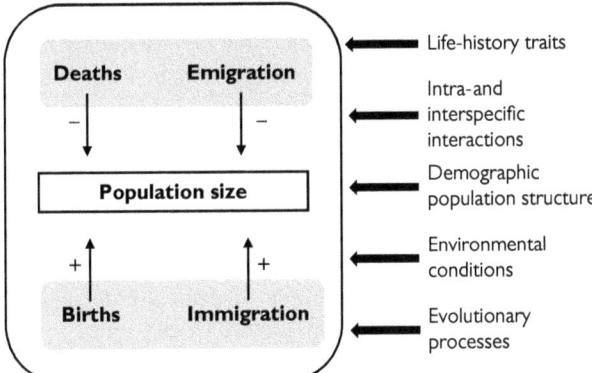

Figure 3.1 Factors influencing temporal variation in population size in a local population. The size of a local population increases due to birth and immigration, whereas it decreases due to deaths and emigration. These rates are controlled by various processes that may be of deterministic or stochastic nature, and of density-independent or density-dependent nature.

While Eq. (3.1) is a fundamental starting point for constructing models of population dynamics, it is just a bookkeeping equation. Population ecology starts by asking how the terms $B(t)$, $D(t)$, $I(t)$, and $E(t)$ are controlled by the interplay between the individuals and their environment (Figure 3.1). Factors influencing variation in population size that are due to variation in the external environment (e.g. variation in climatic conditions or habitat structure) are called exogenous, whereas factors that depend on the intrinsic dynamics of the species (e.g. dispersal or competition) are called endogenous. Population dynamics, including the extinction of local populations and colonizations of empty habitat patches, result generally from a combination of exogenous and endogenous forces (Bolker, 2003).

In a hypothetical world where resources would be unlimited, the population sizes of animals, plants, and other organisms could be in a continuous growth through time. However, real populations will eventually exhaust their resources, and thus their maintenance depends on resource renewal. Linked to this idea lies the concept of carrying capacity, which is the maximal population size that a certain area can support. The kinds of resources and environmental conditions that an organism requires for its survival and reproduction are described by the concept of ecological niche, the discussion of which we will, however, postpone until Chapter 4.

3.1.1 Factors influencing population growth through birth and death rates

Demographic processes refer to births and deaths, and they can be partitioned into density-independent and -dependent processes. Density-independent processes are those that influence births and deaths irrespectively of population density. These include e.g. the survival and reproduction rates of individuals in the absence of resource limitation. Different organisms vary greatly in these rates. For example, while a female elephant may produce offspring once in 5 years, a butterfly female may lay several hundreds of eggs during 1 week, whereas a fungal fruit body can produce millions of spores within a minute. On the other hand, the elephant may live for several decades, while the adult stage of the butterfly life cycle may last only for a week. These differences in life-history characteristics have obvious consequences to population dynamics, e.g. how fast population size can recover after a perturbation. Another important example of density-independent processes is environmental stochasticity. Environmental stochasticity refers to variation in environmental conditions, such as stochastic variation in climatic conditions or habitat availability. Variation in environmental conditions influences all individuals simultaneously, and thus it will cause the same proportional decrease or increase in the population, no matter its absolute size.

As its name implies, the influence of density-dependent processes depends on current population density. Many kinds of density-dependent processes suppress the growth rates of large populations. High local density can either increase mortality or decrease fecundity, through interference competition, through competition for resources, or through the influence of specialist predators or infectious diseases. But it is not always that population growth rate decreases with increasing population density; the relationship can be also the opposite. For example, if the individuals are very sparsely distributed, the rate of encountering mates can be infrequent, thus lowering the reproductive output (this is an example of the so-called Allee effect; Courchamp et al., 1999). As another example, the survival of individuals can be greater when they are in a group, as is the case for many birds and fishes, which escape from their predators more effectively when living in a flock (Krause and Ruxton, 2002). Processes that lead to a decreasing population growth rate with increasing population density are said to be directly density-dependent, whereas processes that lead to a decreasing population growth rate with decreasing population density are said to be inversely density-dependent.

3.1.2 How movements influence population dynamics

As the resources needed for reproduction and survival are patchily distributed in space and time, the location of an individual is an important determinant of its fate. All organisms are located where they are because they have moved there during some part of their life cycle. A passively dispersing propagule may accidentally end up in an unsuitable environment, and thus only those lucky propagules which happen to end up in a suitable environment may have a chance to establish and reproduce. In contrast, habitat selection performed by many actively moving organisms helps them to stay within favourable environments more often than would be expected by random. As we discussed in Chapter 2, some species only utilize local information: among the environmental conditions available within the perceptual range, the individual moves to the best suited one. Some other species are able to acquire large-scale information about environmental conditions, in which case the individuals may, e.g. decide to cross large areas of unfavourable conditions in order to reach favourable conditions present somewhere else. An extreme example of the latter type includes migratory birds crossing oceans and deserts to reach their wintering grounds.

In highly fragmented landscapes, resources are especially patchily distributed. The spatial structure of a species inhabiting a highly fragmented landscape can be described as a metapopulation, i.e. a network of local populations linked via dispersal (Hanski, 1999). In its essence, metapopulation theory describes the dynamics of a species as a balance between extinctions of local populations and (re)colonizations of empty patches (Levins, 1969, 1970; Hanski, 1999). Clearly, movements generate colonizations. In addition, immigration of new individuals into extant populations may help buffer them from extinction (Brown and Kodric-Brown, 1977). This rescue-effect is related to source-sink theory (Pulliam, 1988), which suggests that some local populations may persist even under conditions outside of their fundamental niche, because they receive immigrants with a sufficiently high rate. However, too much movement can also increase extinction risk. If the emigration rate is high, a local population may go extinct simply due to the last individual leaving the patch. More subtly, a high rate of movement among local populations can synchronize their dynamics, which in turn can increase the probability of simultaneous local extinctions and thus the global extinction of the entire metapopulation (Ranta et al., 1995; Liebhold et al., 2004). Habitat fragmentation increases the isolation of local populations and thus decreases colonization rates. When the rate of local extinction exceeds that of recolonization, the metapopulation reaches its extinction threshold and is doomed to global extinction (Hanski and Ovaskainen, 2000).

Like factors influencing births and deaths, factors influencing movements can also be classified into density-independent and -dependent ones. As an example of the latter type, numerous studies have reported that movement rates from sites with high population density to sites with low population density are higher than would be expected by random, either due to density-dependent emigration (Denno and Peterson, 1995; Kim et al., 2009; Fonseca and Hart, 2014) or due to density-dependent immigration (Sibly and Hone, 2002; Serrano et al., 2004; Massot et al., 2014). The influence of density on movements can also be the inverse of this expectation (Roland et al., 2000). For example, immigration rate can increase with increasing density of the local population, e.g. because high population density may indicate high habitat quality, or increase the probability for finding a mate.

3.1.3 How population structure influences population dynamics

Populations do not consist of identical individuals, but their structure can be characterized by many kinds of attributes. These include the sexes, ages, and sizes of the organisms, as well as their genotypes. Equation (3.1) ignores population structure in the sense that it describes the change in population size only in terms of numbers of individuals, irrespectively of their type. Knowledge about population structure can be essential in predicting future population dynamics. For example, if the demographic structure of the local population is dominated by individuals that have not yet reached the reproductive age, in the near future the population is likely to grow less than what would be expected solely from the number of individuals. Breeding performance and mortality vary with age in most organisms. Mortality is often high in early life, e.g. the majority of seeds or spores being doomed to die, or juveniles experiencing higher mortality rates than experienced adults (Clutton-Brock, 1988; Newton, 1989). While age plays a primary role in the development and life cycle of many organisms, for many others the size distribution of the individuals can be biologically more important. Size can influence, e.g. the individual's rate of resource acquisition, competitive ability, survivorship, sexual maturity, and reproductive output (Kenward et al., 1999).

Among species with sexual reproduction, the proportion of males and females is clearly one primary characteristic influencing the populations growth rate (Galliard et al., 2005). In many species the per capita reproductive success per female depends differently on the densities of males and females (Vahl et al., 2013). In a small population stochastic variation in the numbers of males and females may potentially drive it to the extinction (Ferrer et al., 2009). In some species the females are able to adjust the sexes of their offspring as a function of the sex ratio of the present population, thus buffering the influence of such effects (Trivers and Willard, 1973).

3.1.4 The outline of this chapter

The focus of this chapter is on single-species population ecology. We thus ignore interspecific interactions and evolutionary processes—this is because we will consider them later. In Chapter 4 we bring a community-level point of view and thus interactions among heterospecifics. In Chapter 5 we incorporate evolutionary processes, which may also feed back to the population dynamics by modifying the terms $B(t)$, $D(t)$, $I(t)$, and $E(t)$.

In this chapter we will develop two population dynamical models that differ in their assumptions about the life history of the modelled species. While the models are of general nature rather than tailored to any specific organism, for illustrative purposes we will refer to these as the butterfly metapopulation model and the plant population model. Concerning the plant population model, we will provide a motivating empirical example in the context of community ecology in Chapter 4. The butterfly metapopulation model has evolved in collaboration with Ilkka Hanski (e.g. Hanski and Ovaskainen, 2000; Ovaskainen and Hanski, 2001, 2004b; Zheng et al., 2009a; Harrison et al., 2011), and it is much inspired by Hanski's long-term empirical study of the Glanville fritillary butterfly in the Åland Islands. We have illustrated the movement ecology of the Glanville fritillary in Box 2.1 and Figure 2.2, and now return to the same species from the point of view of population ecology in Box 3.1 and Figure 3.2.

Box 3.1 An empirical example of population ecological research

As illustrated in Figure 3.2, the Glanville fritillary forms a classical metapopulation in the Åland Islands (see e.g. Hanski, 2011). The population dynamics of the species can be analysed at multiple spatial scales. At the scale of an individual habitat patch, populations vary greatly in size over time (A), as is typical for many insect species. In an especially adverse year, population size may hit zero, meaning that the population goes extinct. This happens especially frequently in small patches (Figures 3.2A, C), because there the population size is small to start with. Extinctions are compensated with colonizations, which restart the dynamics in a local patch. Colonizations are especially frequent in large patches that are close to other patches (Figures 3.2A, D), because such patches are especially easy to find by individuals that are dispersing through the matrix. At the level of patch networks, colonization–extinction dynamics cause stochastic variation in the fraction of occupied patches over time (Figure 3.2B). Networks that consist of many large and well-connected patches have generally a higher occupancy level than networks that consist of few small and isolated patches (Figures 3.2B, E). The amount and connectivity of habitat within a patch network can be measured by the metapopulation capacity (Hanski and Ovaskainen, 2000). Networks for which the metapopulation capacity is above the extinction threshold of the species are able to support a viable metapopulation, whereas metapopulations inhabiting networks with a low metapopulation capacity are prone to extinction (Figure 3.2E).

The structure of this chapter parallels that of the movement ecology chapter. Thus, we start by ignoring the effects of exogenous variation and assume that the population inhabits a homogenous environment (Section 3.2). In Section 3.3 we move to heterogeneous environments, thus allowing the components of Eq. (3.1) to vary in either time or space. As an important special case of spatial variation, we consider highly fragmented landscapes, i.e. habitat patch networks embedded in a matrix of unsuitable habitat. In Section 3.4 we use heterogeneous-space population models to examine the impacts of habitat loss and fragmentation on the dynamics and persistence of metapopulations. Finally, in Section 3.5 we discuss some statistical approaches for analysing population data. To link statistical and mathematical approaches, we apply the statistical approaches to data generated by the mathematical models of Sections 3.2–3.4.

3.2 Population models in homogeneous environments

An empirically minded ecologist might wonder why we have devoted a full section for population dynamics in homogeneous space despite the fact that most real environments are so obviously heterogeneous. There are two reasons. First, homogeneous-space models, and even non-spatial models, are sufficient to illustrate many key concepts of population dynamics, and starting directly with heterogeneous-space models would bring unnecessary confounding factors that would make it more difficult to understand these basic concepts. Second, how heterogeneous space influences population dynamics can be best understood by comparing the behaviours of

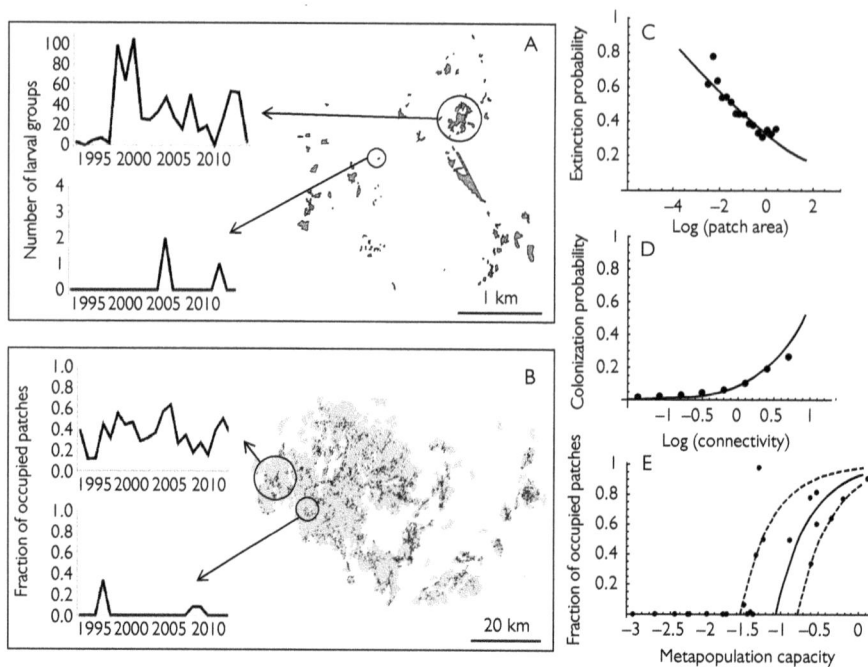

Figure 3.2 Metapopulation ecology of the Glanville fritillary butterfly (*Melitaea cinxia*). (A) illustrates population dynamics (variation in the number of larval groups) for two patches that are located within the same network, and (B) illustrates metapopulation dynamics (variation in the fraction of occupied patches) for two habitat patch networks located within the Åland Island study system. (C) shows the annual extinction probability of extant populations as a function of patch area, and (D) shows the annual colonization probability of empty patches as a function of connectivity (proximity to occupied patches). In (C) and (D), the points are averages over habitat patches belonging to the same size (C) or connectivity (D) class, and the lines show maximum likelihood fits to the data (see Ovaskainen and Hanski 2004a for details). (E) shows the relationship between network-scale patch occupancy and the metapopulation capacity of the patch network. In this panel, each dot shows data from one patch network, and the lines show the theoretical prediction of a metapopulation model fitted to these data (Hanski and Ovaskainen, 2000). (C) and (D) reproduced with permission from Oxford University Press. (E) reproduced with permission from Nature Publishing Group.

heterogeneous-space and homogeneous-space models. In this sense, homogeneous-space models can be considered as null models.

The most widely used population model is the so-called logistic model, defined as

$$\frac{dN(t)}{dt} = rN(t)\left(1 - \frac{N(t)}{K}\right). \quad (3.2)$$

The model describes the dynamics of the size (or density) of the population at time t, denoted by $N(t)$. The model has two parameters: r is the population's growth rate at

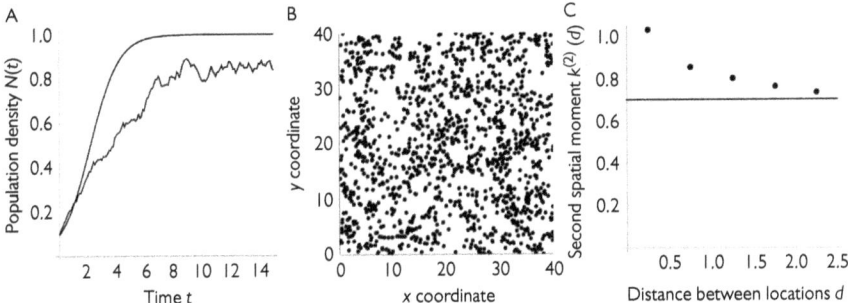

Figure 3.3 Simulations of the individual-based stochastic and spatial logistic model. (A) shows how the population density evolves in time. The smooth line depicts the prediction of the mean-field model (Eq. (3.2)) and the rugged line depicts the prediction of the individual-based model. (B) shows the spatial distribution of individuals at the final state of the simulation at time $t = 15$. The dots in (C) show the second spatial moment (Bolker and Pacala, 1997) measured from the spatial distribution in (B), and the line shows the prediction based on the assumption of complete spatial randomness. Parameter values $f = 2$, $m = 1$, initial population density 0.1, and initial distribution of individuals follows complete spatial randomness. Both the competition and dispersal kernels follow the density function of bivariate normal distribution with variance-covariance matrix $L^2\mathbf{I}$, where $\mathbf{I}$ is the two-dimensional identity matrix and the length scale parameter L is set to $L = 1$ for competition and $L = 0.75$ for dispersal. Both kernels thus integrate to unity over all space.

low density, and K is the carrying capacity of the environment. If starting from a low population density, the population will first grow exponentially with the rate r. As the population size increases, the growth rate starts to decrease, so that finally the population reaches an equilibrium state in which the population size equals the carrying capacity. This is illustrated by the smoothly increasing function in Figure 3.3A.

The logistic model is so simplistic that it can be considered as a cartoon rather than a realistic description of the dynamics of any real population (Law et al., 2003). To start with, the model implies that the population size is large enough so that the effects of demographic stochasticity can be ignored, and that dispersal is great enough so that the system can be treated as spatially well-mixed (Donalson and Nisbet, 1999), often referred as the mean-field assumption (e.g. Law et al., 2003). To make these assumptions explicit, we will start this section by constructing an individual-based model (IBM). The IBM makes assumptions and has parameters at the level of the individuals, and it is spatial and stochastic, so we will call it the stochastic and spatial logistic model. We will then simplify the IBM either by removing the effect of stochasticity, resulting in the deterministic spatial logistic model, or by removing the effect of space, resulting in the non-spatial stochastic logistic model, or by removing simultaneously the effects of stochasticity and space, resulting in the classical logistic model of Eq. (3.2). By doing so, we will show how the population-level parameters of the logistic model (r and K) are related to individual-level parameters such as mortality, fecundity, establishment, and competition.

3.2.1 Individual-based stochastic and spatial model

We construct the IBM in a continuous-space and continuous-time framework, and thus our model is technically a spatiotemporal point process (Table 1.1). We consider sessile organisms, such as plants, which are represented in the model as points in two-dimensional space. Like the basic random walk model is only a very rough caricature of the movement behaviour of any real organism, the spatial and stochastic logistic model that we will next describe is only a very rough caricature of the dynamics of any real population. This model is still a very useful starting point, on top of which additional complexity can be built.

In the spatial and stochastic logistic model, each individual is assumed to produce propagules (e.g. seeds) at a per capita fecundity rate f. We recall from the definition of a rate (see Appendix A.4) that this means that each individual will produce a propagule during a short time interval dt with probability fdt. If a propagule is produced, it will disperse (without any delay) into a location determined by a dispersal kernel that we denote here by $D(x, y)$ and which is centred around the location of the parent. Technically, the kernel describes the probability distribution of where the propagule will land. The propagule will establish (without any delay) with probability e, or alternatively it will die before establishment with probability $1 - e$. If it establishes, it becomes a new individual that starts (without any delay) to produce new propagules. Each individual has a background mortality rate of m, so that it dies during a short time interval dt with probability mdt. Additionally, competition among the individuals (for space or other resources) is assumed to increase the mortality rate. We assume that the mortality rate imposed by a competing individual to a focal individual depends on the distance between the individuals, and that the competing effects imposed by different individuals work in an additive manner. We denote the competition kernel by $C(x, y)$, so that the mortality rate of individual i is $m + \sum_{j \neq i} C(x_i - x_j, y_i - y_j)$, where the sum is taken over all other individuals j. We assume that both kernels are radially symmetric, meaning that they only depend on distance. Thus, we may write $D(x_i - x_j, y_i - y_j) = D(d_{ij})$ and $C(x_i - x_j, y_i - y_j) = C(d_{ij})$, where $d_{ij} = \sqrt{(x_i - x_j)^2 + (y_i - y_j)^2}$ is the distance between the locations.

The rugged line in Figure 3.3A shows a simulation of this model. The simulation suggests that population density increases from the initial density until it reaches a stationary state. At the stationary state, the population size remains approximately constant because the death rate on average equals the birth rate. Figure 3.3B shows the spatial distribution of individuals at the final time of the simulation. A visual inspection of this figure suggests that the individuals are not evenly distributed in space. This is because local dispersal generates an aggregated distribution, which is partly, but not entirely, removed by the self-thinning effect of local competition. The importance of this kind of self-emerging spatial structures has been increasingly recognized in studies of ecological dynamics of spatially structured populations (Bolker and Pacala, 1997; Murrell et al., 2004). To measure the spatial structure in a quantitative manner, we show in Figure 3.3C the second spatial moment $k^{(2)}(d)$, which measures the density of pairs of individuals that are at distance d from each other. The second spatial moment indicates that there is an excess of pairs of individuals at short distances, compared to the baseline expectation of squared population density (the line in Figure 3.3C), which is achieved at long distances. This corresponds to the visual observation that, in Figure 3.3B, the individuals form more tight clusters than what would be the case if they were randomly distributed. The clustering of individuals has consequences to the dynamics of the population. As the individuals have a higher density of neighbours than they would have in the case of complete spatial randomness, they experience a stronger competitive environment than they would otherwise do. This leads to an increased mortality rate, and consequently to a decreased population size. For this reason, the simulation of the IBM in Figure 3.3A falls below the smooth line which shows the prediction of the corresponding logistic model (Eq. (3.2)). Later in this section we will show that the logistic model is a particular limiting case of the IBM.

The IBM is relatively easy to simulate, but challenging to analyse. One approach for analysing IBMs is to describe the system with the help of an infinite hierarchy of spatial moments. The spatial moment $k^{(n)}$ of order n measures the density of n individuals at particular distances from each other. The first spatial moment $k^{(1)}$ measures population density, whereas the higher order moments, such as $k^{(2)}$ illustrated in Figure 3.3C, describe the spatial pattern. Readers interested in mathematical approaches for analysing this IBM and other spatiotemporal point processes are referred to the literature of spatial moment closures and pair-approximations (Bolker and Pacala, 1997; Ellner, 2001; Keeling et al., 2002; Brännström and Sumpter, 2005) and perturbation expansions around mean-field models (Ovaskainen and Cornell 2006a, b; Ovaskainen et al., 2014). But instead of attempting to analyse the IBM mathematically, we proceed here to derive simpler models by removing the influences of space, stochasticity, or both of these.

3.2.2 Simplifying the model: stochasticity without space

We start by removing the influence of space. To do so, we assume that the population inhabits a finite patch of area A. We denote the number of individuals that are present at time t by $n(t)$, which is connected to population density $N(t)$ by $N(t) = n(t)/A$. We assume that the population is well mixed, meaning that the propagules are dispersed with equal probability to any location within the patch, independently of the location of the parent producing the propagule. As each of the n individuals produces propagules at rate f and these establish with probability e, the transition from population size n to population size $n + 1$ takes place at rate fen. We further assume that the competitive effects within the patch are independent of the distances between the individuals or, equivalently, that the competitive effects that one individual imposes on other individuals are evenly distributed within the area A. With this assumption, the amount of competition imposed by one individual to another individual is c/A, where the constant c is the integral of the competition kernel C. As each of the remaining $n-1$ individuals competes with the focal individual, the per-capita mortality rate is $m + (c/A)(n-1)$. As any of the n individuals may die, the transition rate from population size n to population size $n-1$ is $mn + (c/A)(n-1)n$.

Combining these equations, the transition rates of the model are

$$\begin{cases} n \to n+1, & fen, \\ n \to n-1, & mn + \left(\dfrac{c}{A}\right)(n-1)n. \end{cases} \qquad (3.3)$$

This model has been called the stochastic logistic process (Norden, 1982), and it is technically a Markov process (Table 1.1). In this model, while the variable time is continuous, the population size changes by the discrete units of single individuals. The model is thus an appropriate description for a population with overlapping generations (Newman et al., 2004).

This Markov process is easy to simulate (see Appendix A.4), Figure 3.4 illustrating such simulations. In Figure 3.4A, we have assumed a patch of area $A = 40 \times 40 = 1600$, and thus this simulation is comparable to the simulation of the spatial version of the same model shown in Figure 3.3. We observe qualitatively the same behaviour, but now the model follows the baseline logistic model much more closely. This is because

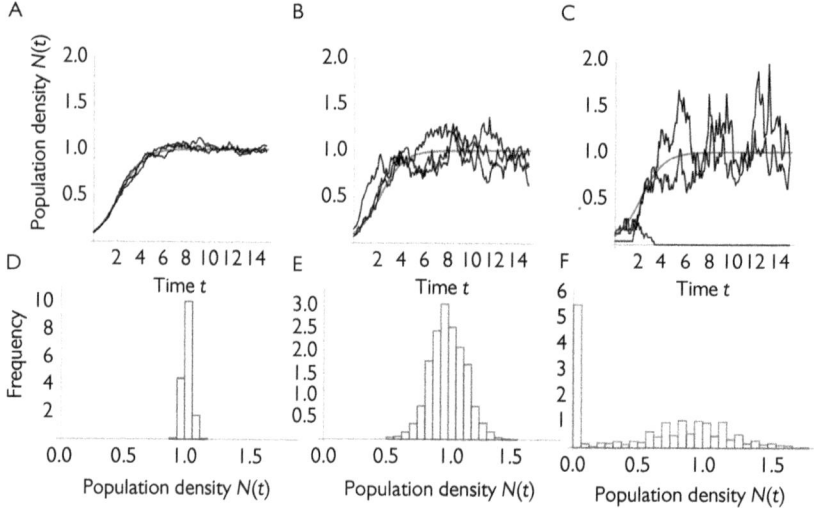

Figure 3.4 Simulations of the individual-based stochastic and non-spatial logistic model. Each of (A–C) shows three replicate simulations of population density evolving as a function of time t until $t = 15$. The simulations are conducted in patches of size 40×40, 10×10, and 5×5. As the models are non-spatial, patch size influences only the carrying capacity parameter. (D–F) show the distributions of populations size obtained by running the simulations until time $t = 100$, and dropping the data until $t = 15$. Parameter values as described in the legend to Figure 3.3.

the stochastic logistic model has one degree of complexity less than the spatial and stochastic logistic model: we have removed the spatial dimension, which causes, through generating aggregated distributions of individuals, additional variation in the dynamics. When decreasing the domain size ($A = 10 \times 10 = 100$ in Figure 3.4B, $A = 5 \times 5 = 25$ in Figure 3.4C), the number of individuals decreases and thus variation in the dynamics increases.

In Figure 3.4, all of the variation is generated by what is called demographic stochasticity, which is different to other kinds of stochasticity, like demographic heterogeneity and environmental stochasticity (Lande, 1993). Demographic stochasticity refers to the fact that births and deaths are random events, technically Bernoulli distributed random variables (see Appendix A.3). Thus, while e.g. the probability of an individual to produce an offspring during a short time interval dt is fdt, in a realization the individual either produces an offspring or it does not produce an offspring, rather than producing a fractional offspring of size fdt. As illustrated by Figures 3.4A–C, the relative role of demographic stochasticity increases as the population size decreases. This is to be expected, as in a large population the randomness related to individual birth and death events averages out at the population level (Donalson and Nisbet, 1999).

In Figure 3.4C, one of the three simulated realizations goes extinct, and this happens because of demographic stochasticity. In the initial phase the population is very small, and thus a couple of deaths taking place before the individuals have reproduced can drive the population to extinction. If the population goes extinct, it can never be recovered, because there are no parents to produce more offspring and because we assumed a closed population and thus no immigration from the outside world. Actually, in this model the population will eventually go always extinct with certainty, whatever the parameter values are. By visual inspection it may look like the populations simulated in Figure 3.4A would persist indefinitely. However, if we would run the simulation for long enough, they would go extinct with 100% probability. This is because in a finite population it is possible that all individuals die before any of them reproduces. The probability of such a scenario can be very small, but it is still greater than 0, which guarantees that eventual extinction will take place with certainty (Newman et al., 2004; Ovaskainen and Meerson, 2010). For this reason, the stationary state (see Appendix A.4) of the stochastic logistic model is concentrated to the extinction state, i.e. the absorbing state with $n = 0$. However, the expected time until extinction can be astronomically long, as it would be indeed the case with the parameter values used in the simulation of Figure 3.4A.

As a population may persist with high certainty over ecologically relevant timescales, it is interesting to go beyond the calculation of the mean extinction time and ask what kind of dynamics the model produces before the eventual extinction. The long-term dynamics before the eventual extinction are characterized by the so-called quasi-stationary distribution (see Appendix A.4), defined as the stationary distribution conditional on extinction haven't yet happened. Figures 3.4D–F show the distributions of population densities generated by continuing the simulations shown in Figures 3.4A–C from time $t = 15$ until time $t = 100$. Figures 3.4D and E correspond to quasi-stationary states. As one of the populations has gone extinct in Figure 3.4C, Figure 3.4F shows a mixture of the quasi-stationary state and the stationary state, the latter corresponding to the peak at 0. The width of the quasi-stationary state decreases as the area of the patch and thus the number of individuals within it increases, as expected from the discussion we had earlier about the influence of demographic stochasticity.

Earlier we analysed the behaviour of the stochastic logistic model with the help of simulations. This model and other single-step models (models that do not allow multiple simultaneous birth or death events) are actually simple enough to be analysed mathematically. For a review of such mathematical methods and some ecologically relevant results, we refer to Ovaskainen and Meerson (2010). One central result here is that the expected time to population extinction increases exponentially with system size (here, the area of the patch) if the underlying deterministic model (in this case, the logistic model) is above its extinction threshold, and thus in this case if the growth rate parameter r is positive. This again reflects the fact that in practice demographic stochasticity is important for small populations only.

3.2.3 Simplifying the model further: without stochasticity and space

Let us then simplify the stochastic logistic model further by removing the effect of demographic stochasticity. We can do it by assuming that the domain size A becomes infinitely large, in which case the number n of individuals also becomes infinitely large, and hence the randomness associated to demographic stochasticity completely averages out. To derive the deterministic version of the model (Eq. (3.2)) from the stochastic model (Eq. (3.3)), we again denote the current population density by $N = n/A$. The probability that the population will increase by one individual over a short time dt is $fendt = feNAdt$. If this happens, the population density will move from $N = n/A$ to $(n + 1)/A = N + 1/A$. The probability that the population will decrease by one individual, and thus that the population density will decrease by $1/A$, is $n(m + (c/A)(n - 1))dt = NA(m + cN - c/A)dt \approx NA(m + cN)dt$, where the approximation is justified as we consider the limit where $A \to \infty$ and thus the relative role of the term c/A becomes negligible. Combining these equations, the expected change in population density is

$$dN = feNAdt \times \left(\frac{1}{A}\right) - NA(m + cN)\,dt \times \left(\frac{1}{A}\right) = feNdt - N(m + cN)\,dt. \quad (3.4)$$

Dividing this equation by dt and rearranging the terms yields the differential equation

$$\frac{dN}{dt} = (fe - m)N\left(1 - \frac{N}{\left(\frac{fe-m}{c}\right)}\right). \quad (3.5)$$

This equation has exactly the same functional form as the classical logistic model (Eq. (3.2)), but the parameterization is different. Comparing the two equations shows that the population growth rate parameter $r = fe - m$, and thus it is the difference between the birth rate (which is the product of the fecundity rate and the establishment probability) and the density-independent mortality rate. The carrying capacity parameter $K = (fe - m)/c = r/c$, and thus it decreases with increasing competitive effects c, as expected. More interestingly, it increases with the population growth rate parameter r, meaning that environments which support a high population growth rate at low density (i.e. high levels of fecundity and establishment and low level of mortality) are also expected to support a high carrying capacity. This is not very surprising, but something that is hidden in the classical parameterization of the logistic model (Eq. (3.2)), which may lead one to think of r and K as two completely independent parameters.

While the dynamics of the deterministic (Eq. (3.5)) and stochastic (Eq. (3.3)) logistic models resemble each other (Figure 3.4), a major difference is that the deterministic model predicts that the population will persist indefinitely if $r > 0$ and that it will go extinct with certainty if $r < 0$. This is in apparent contrast with the stochastic version, in which the population will always go extinct with certainty, as noted earlier. However, in the part of the parameter space where the deterministic model predicts persistence ($r > 0$), the time to extinction in the stochastic model scales exponentially with the area of the patch, and thus also the stochastic model predicts that large populations will persist over ecological time-scales with very high probability.

3.2.4 Another way of simplifying the model: space without stochasticity

Another way to simplify the full spatial and stochastic individual-based model is to remove the effect of stochasticity, and thus derive a deterministic spatial model. This can be done by assuming that the local density of individuals is so high that each individual competes effectively with an infinite number of other individuals. In this case, there is no need to keep track of the locations of the individuals, as we may consider the population density $N(x, y, t)$ to be continuous over space. To derive an equation for population dynamics, we need to consider how the population density $N(x, y, t)$ at location (x, y) changes over a small time interval dt. Concerning births, an individual at a nearby location (x', y') may produce an offspring which disperses to (x, y). By the definition of the dispersal kernel, such a dispersal event takes place with probability density $D(x - x', y - y')$. As the population density at (x', y') is $N(x', y', t)$ and as the individuals located there produce propagules at rate f, and as those propagules establish with probability e, birth events take place at location (x, y) at the rate

$$\int_{x'=-\infty}^{\infty} \int_{y'=-\infty}^{\infty} feD(x - x', y - y') N(x', y', t) \, dx' dy' = fe(D \star N)(x, y, t). \quad (3.6)$$

Here $D \star N$ denotes the convolution (see Appendix A.2) between the dispersal kernel and the population density, and it means that these two spatial variables are combined by integrating them over all the space, as is done in the left-hand side of the equation. Analogously, the reader may attempt to derive that due to deaths the population density decreases at the rate $(m + (C \star N)(x, y, t))N(x, y, t)$. Here the convolution between the competition kernel and the density of individuals gives the average local density weighted by the competition kernel, which is the relevant variable determining in our model the density-dependent death rate.

Combining these considerations, we obtain the equation

$$\frac{\partial N}{\partial t} = fe(D \star N) - (m + (C \star N))N, \qquad (3.7)$$

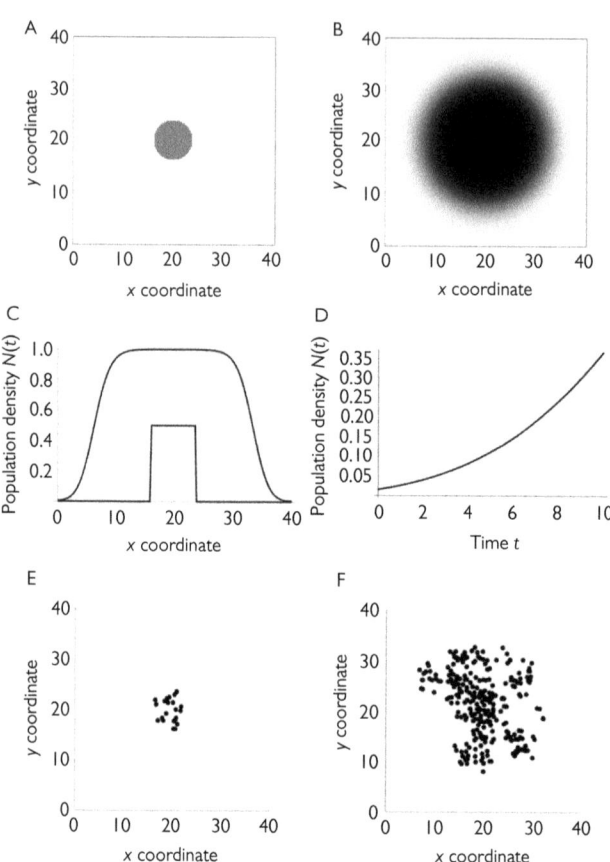

Figure 3.5 Simulations of the deterministic and spatial logistic model. (A) shows the initial state of the population, with population density being 0.5 in the grey area and 0 outside it. (B) shows the state of the population at time $t = 10$, with black colour corresponding to 1, which is the population density obtained at carrying capacity. (C) shows the population density along a 1-dimensional horizontal transect located in the middle of the domain ($y = 20$), the lower line corresponding to the initial condition of (A) and the upper line to the final condition of (B). (D) shows how population density (the number of individuals per unit area, averaged over the entire 40 × 40 simulation domain) increases in time. (E) and (F) show the initial and final states of a corresponding simulation of the stochastic and spatial logistic model. Parameter values as described in the legend to Figure 3.3.

where we have dropped for notational simplicity the fact that population density depends on space and time, $N = N(x, y, t)$, and that the two kernels depend on space, $D = D(x, y)$ and $C = C(x, y)$. As the equation involves convolutions and thus integrals over space, and a derivative with respect to time, it is mathematically an integro-differential equation (Table 1.1).

Integro-differential equations can be analysed mathematically (Bolker, 2003), but this is not very easy to do. For this reason, we illustrate the behaviour of Eq. (3.7) with the help of a numerical solution only. In the example of Figure 3.5, the population is initially (at time $t = 0$) present in the middle part of the domain at population density 0.5. By time $t = 10$, the population has expanded in a radially symmetric fashion and reached the carrying capacity $K = 1$ in the central part of its distribution. As the population grows in a radially symmetric manner, the total population size increases essentially as a quadratic function of time (Figure 3.5D). For comparison, Figures 3.5E and F show a simulation of the corresponding IBM, in which demographic stochasticity produces scatter in the realized pattern of population expansion.

To connect Eq. (3.7) to the baseline logistic model of Eq. (3.2), all we need to do is to assume that population density is constant over space, i.e. that $N(x, y, t) = N(t)$. In this case, convolving the population density with a kernel equals simply multiplying the population density by the integral of that kernel (see Appendix A). As the dispersal kernel integrates to 1 over all space, we have $D \star N = N$. As we denoted the integral of the competition kernel C by c, we have $C \star N = cN$. Inserting these simplifications into Eq. (3.7) results in Eq. (3.5) and thus in Eq. (3.2).

3.3 Population models in heterogeneous environments

Thus far, we have modelled population dynamics with the highly simplistic assumption that the environment is constant in both space and time. Real organisms are, however, faced with environmental variation, with strong influence to their dynamics and persistence (Wiens, 1976). As the vital rates that determine population growth rates, such as fecundity, mortality, and carrying capacity, are dependent on environmental conditions, environmental variation translates into variation in population dynamics. Additionally, as discussed in Sections 2.3–2.5, the movements of individuals are influenced by environmental heterogeneity. Movements determine where the individuals of the next generation will be, relative to the locations of the present individuals, and the environmental and population dynamical conditions faced by the next generation in turn influence the growth rate of the population.

In this section, we examine how spatial and temporal heterogeneity of the environment influences the distribution and dynamics of organisms. As with modelling in general, there is a large number of choices on how and at which detail environmental heterogeneity is included (Kareiva et al., 1990). For example, we could assume that the sources of heterogeneity vary in either space or time, or in both of these. Further, we could assume continuous or discrete variation in the environment, and either stochastic (e.g. unpredictable variation in weather) or deterministic (e.g. a seasonal environment) variation in the environment. To exemplify contrasting ecological situations and contrasting modelling frameworks, we consider three cases: (i) environmental stochasticity that varies over time, and spatial heterogeneity in (ii) continuous or (iii) discrete-space setting. In case (iii), we will also incorporate environmental stochasticity, and thus consider the interaction between spatial and temporal variation.

3.3.1 Environmental stochasticity

We first consider environmental stochasticity, i.e. random variation in the environmental conditions over time. To keep the discussion as simple as possible, we utilize here the non-spatial and deterministic mean-field model of population dynamics derived in the previous section (Eq. (3.5)). Among the many possible choices, we assume that environmental variation (say, in temperature) influences only the fecundity rate. Thus, we model population dynamics by Eq. (3.5), modified so that the fecundity $f = f(t)$ depends on time,

$$\frac{dN}{dt} = (f(t)e - m) N \left(1 - \frac{N}{\left(\frac{f(t)e-m}{c}\right)}\right). \tag{3.8}$$

We assume that fecundity fluctuates around its mean value in a stochastic manner. The stochasticity in the environment makes the species density fluctuate over time, even though the species itself follows the deterministic dynamics of Eq. (3.8). As shown by Figure 3.6, we consider three cases in which the magnitude of environmental fluctuation is equal, but its temporal scale of autocorrelation is different. In the uppermost panels, the environmental fluctuations are much faster than the population dynamics of the species. In this case, the effect of environmental stochasticity is essentially averaged out because the individuals face all kinds of conditions during their lifetimes, and thus their total reproductive output is relatively constant: periods of high fecundity are balanced by periods of low fecundity. In the middle panels, the temporal scale of autocorrelation is on the same order as the lifetimes of the individuals. In this case, the fluctuations in population size become pronounced, because favourable conditions last long enough so that the population has time to increase in size, and conversely adverse conditions last long enough so that the population has time to decline. In the lower panels the fluctuations in the environment are much slower than the lifetimes of the individuals, so that favourable or adverse conditions last over multiple generations. This makes the fluctuations in population size even more pronounced.

Figure 3.6 illustrates how the population density tracks environmental conditions with a time delay. Without the time delay generated by intrinsic population dynamics, the population would be at time t at the instantaneous carrying capacity $K(t)$ that corresponds to the prevailing environmental conditions,

$$K(t) = \frac{f(t)e - m}{c}. \tag{3.9}$$

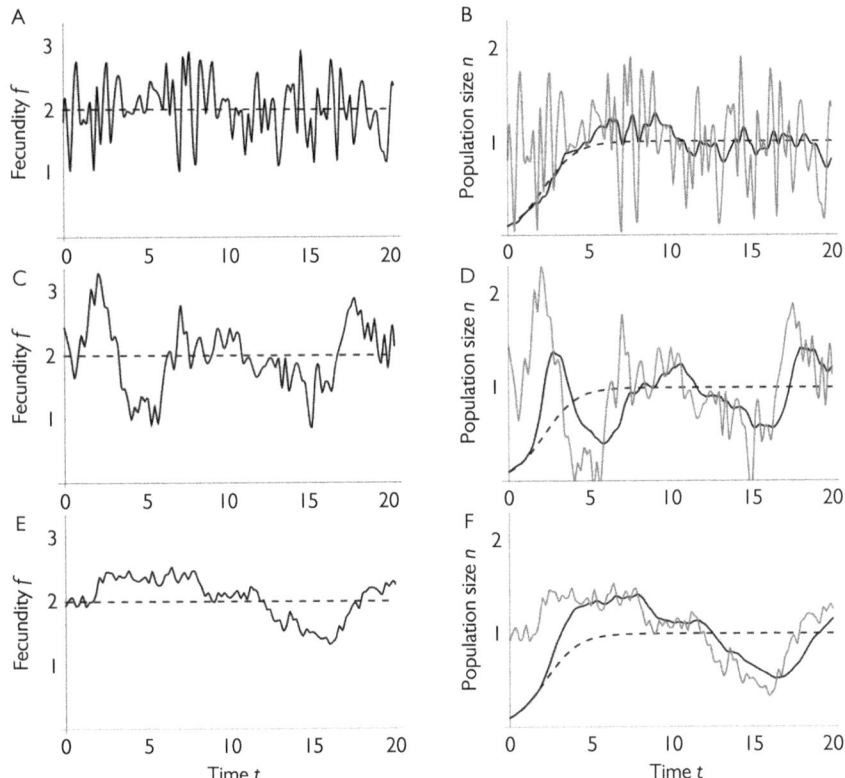

Figure 3.6 The influence of environmental stochasticity on population dynamics. Temporal variation in environmental conditions is assumed to influence population dynamics, as shown in (B, D, F), through its influence on fecundity f, as shown in (A, C, E). In all cases, f follows the normal distribution with mean 2 and standard deviation 0.5. (A, B) corresponds to a fast environment (temporal scale of autocorrelation $\tau = 0.1$), whereas (E, F) show a slow environment ($\tau = 10$), and (C, D) show an intermediate case ($\tau = 1$). The black continuous lines in (B, D, F) show population dynamics generated by Eq. (3.8), whereas the grey lines show the instantaneous carrying capacities computed by Eq. (3.9). The black dashed lines show the prediction of the model where fecundity is assumed to remain constant over time, with value $f = 2$.

In the case of the slowly varying environment (lower panels of Figure 3.6), the population is closely tracking the instantaneous carrying capacity, whereas in faster environments the population size is much influenced by the environmental conditions in earlier times too.

The influence of environmental stochasticity differs profoundly from the one caused by demographic stochasticity. This is because the former has an effect over all individuals simultaneously, whereas the latter influences each individual independently (Lande, 1993; Ovaskainen and Meerson, 2010). Fluctuations in population size induced by demographic stochasticity are of magnitude $\sqrt{K}$, and thus the relative magnitude of fluctuations, compared to population size, behaves as $1/\sqrt{K}$. This implies that fluctuations due to demographic stochasticity are density-dependent in the sense that they are important only for small populations (Figure 3.4). In contrast, fluctuations induced by environmental stochasticity are on the same order K as population size, and thus their relative magnitude is density-independent. Indeed, the influence of environmental stochasticity is visible in the simulations of Figure 3.6, where we have used the deterministic mean-field model, and thus technically assumed an infinitely large population size (see Section 3.2).

The magnitude of population fluctuations translates to extinction risk as it determines the likelihood that the population will move far away from the carrying capacity and in particular that it will hit 0 (Lande, 1993; Ovaskainen and Meerson, 2010). With demographic stochasticity only, the mean time to extinction T increases exponentially with carrying capacity, $T = C_1 \exp(aK)$. But how does the time to extinction scale with carrying capacity in the presence of environmental stochasticity? In the model defined by Eq. (3.9), the population actually never goes extinct, i.e. population size n never reaches exactly 0. This is because the model is a mean-field differential equation, in which population size is a continuous variable. Adverse environmental conditions may drive the population size to an arbitrarily low value, such as $n = 10^{-12}$. The value of $n = 10^{-12}$ would mean that the population does not even consist of a single individual, but only of a tiny fraction of a single individual. The fact that the mean-field model can predict such 'nano-individuals' (Wilson et al., 1998) is clearly not satisfactory, and it illustrates the limitations of models that do not consider the discreteness of individuals in general. Thus, to study the influence of environmental stochasticity on extinction times, we need to incorporate also demographic stochasticity. These two forms of stochasticity act together: environmental stochasticity can reduce the size of a large population to a small enough level so that the population eventually goes extinct due to demographic stochasticity. The combination of demographic and environmental stochasticities makes time to extinction scale with carrying capacity according to a power-law, $T = C_2 K^b$. Further, the exponent b decreases with the time-scale of temporal autocorrelation in environmental fluctuations (Ovaskainen and Meerson, 2010), meaning that even very large populations can have a substantial extinction risk in slowly varying environments, such as illustrated in the lowest row of Figure 3.6. This is intuitive, because in such environments the fluctuations in environmental conditions are not averaged out, and hence there is a risk of a long period of adverse conditions leading to a drastic population decline.

3.3.2 Spatial heterogeneity in continuous space: the plant population model

We next extend the individual-based stochastic and spatial model presented in Section 3.2 into heterogeneous space to build what we call here the plant population model. In the plant population model the individuals produce at a per capita fecundity rate f propagules that are dispersed according to a dispersal kernel D. In Section 3.2 the propagules were assumed to establish as adults with probability $e = 1$, but here we consider the case where some of the propagules will fail to establish, and thus $e < 1$. As in Section 3.2, the mortality rate of an established individual consists of the density-independent rate m and the competitive influences of all other individuals weighted by the competition kernel C.

In general, models can be extended to account for environmental heterogeneity by assuming that any or all of the model parameters vary in space and time. In the case of the present model, the model parameters are f, e, m, D, and C. For the sake of illustration, let us assume that there are two relevant environmental factors, which we call temperature and soil fertility. As our focus is now on spatial heterogeneity instead of temporal heterogeneity, temperature is not assumed to vary over time but over space—it can be e.g. the mean temperature at a given location. We assume that the fecundity f peaks at optimal temperature, whereas the establishment probability e increases with soil fertility. To simplify the visualization of the results, we assume that the temperature increases with the x coordinate, whereas soil fertility increases with the y coordinate, making fecundity and establishment functions of spatial location (Figure 3.7). We assume that the parameters m, D, and C are not influenced by environmental variation. We note that these are rather arbitrary assumptions on how the environment might influence the fundamental niche of the species, and we have selected these assumptions merely for the sake of illustrating modelling approaches.

Simulating the individual-based model of Section 3.2 with spatially varying fecundity and establishment parameters produces the distribution of individuals illustrated in the upper panels of Figure 3.8. The effects of the temperature and soil fertility are evident in this figure: the population density is highest in the part of the space where temperature is intermediate and soil fertility is the highest. The contour line in the figure differentiates between the conditions that do allow or do not allow for the persistence of the species, based on the corresponding mean-field model that is now specific to each spatial location (see Section 3.2 for the relationship between

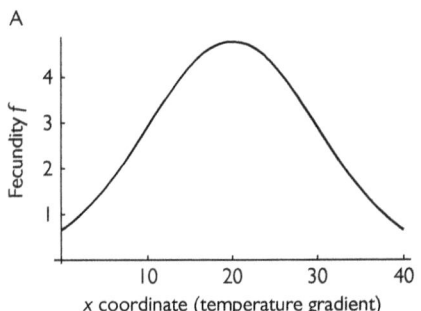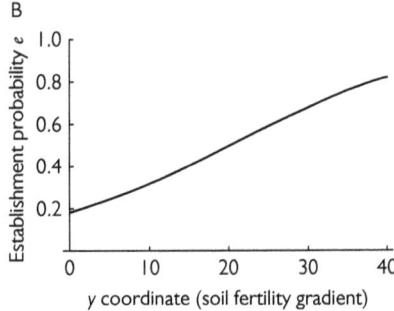

Figure 3.7 Spatial variation in fecundity and establishment assumed in the plant population model. The fecundity rate is assumed to peak at an intermediate temperature (A), whereas the establishment probability is assumed to be an increasing function of soil fertility (B). To enable the visualization of the results in geographical space (Figure 3.8), we assume that temperature equals with the x coordinate, whereas soil fertility equals with the y coordinate.

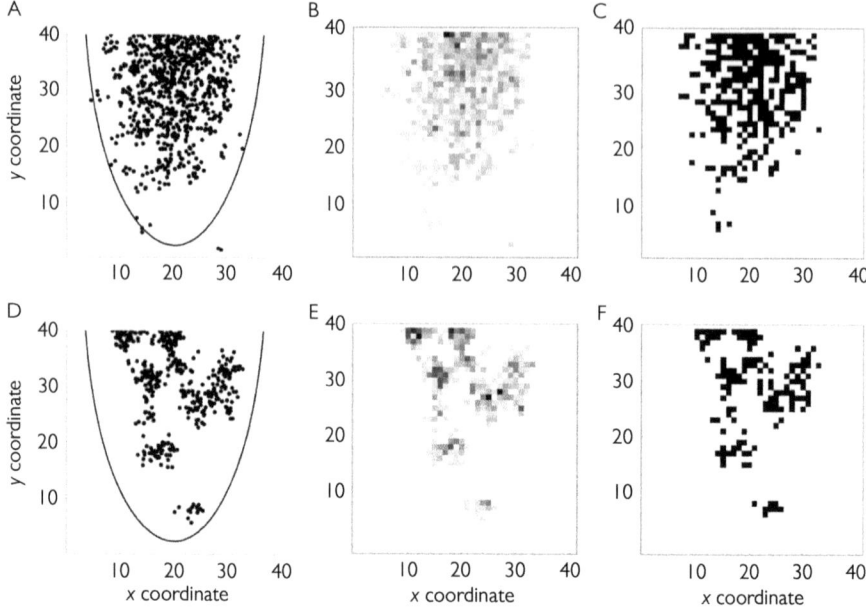

Figure 3.8 A simulated distribution of individuals at the stationary state of the plant population model. In the model variant considered in (A–C) we assume that density-independent mortality takes place independently among the individuals, whereas in (D–F) it takes place in a spatially correlated manner (parameterized as $L_\phi = 2$). (A, D) show the distributions of individuals, (B, E) the corresponding densities of individuals computed on a regular grid, and (C, F) presence–absence data that results from a simulated sampling process: we have assumed that each individual is detected with probability 0.5, and classified a grid cell as occupied if at least a single individual was detected. The data in (C, F) are analysed in Section 3.5 in the context of species distribution modelling. We have assumed spatial variation in fecundity and establishment as illustrated in Figure 3.7, and set the other parameters to the values assumed in the simulations of the homogeneous-space model of Chapter 3.2. In (A, D), the contour line corresponds to $fe = m$, and thus above the line the underlying mean-field model predicts that the species would persist, whereas below it the species is predicted to go extinct.

individual-based stochastic and spatial models and the corresponding mean-field models). The space above the line represents the fundamental niche of the species (see Section 3.1). In this model, the realized niche, i.e. the space actually occupied by the individuals, reflects very closely the fundamental niche. However, some individuals are found also from outside the fundamental niche, and conversely many areas within the fundamental niche have no individuals. This is partly due to the influence of demographic stochasticity, but it also illustrates a source-sink effect (Pulliam, 1988): a population may persist outside the fundamental niche if it receives a high enough flow of propagules produced within the fundamental niche.

In the lower panels of Figure 3.8 we have simulated otherwise the same model, but assumed that density-independent mortality takes place in a spatially correlated manner. To do so, we have assumed that 'catastrophes' arrive at a rate ϕ per unit area, that their centres are randomly located, and that they kill all individuals that are within a distance L_ϕ from their centre. We have set the parameter ϕ to $\phi = m/(\pi L_\phi^2)$ to ensure that the density-independent mortality rate of each individual is still the same (m) as in the baseline model which assumes independent deaths. This is an example of spatially correlated environmental stochasticity, or what has been called regional stochasticity (Hanski, 1991). As a result, the spatial pattern for the distribution of individuals is patchier than in the case of independent deaths, as can be expected. Further, with spatially correlated deaths the total number of individuals is smaller than in the case of independent deaths. This is because spatially correlated extinctions leave some areas completely unoccupied, and their re-colonization by local dispersal takes time.

3.3.3 Spatial heterogeneity in discrete space: the butterfly metapopulation model

While in the plant population model we incorporated space and time in a continuous setting, we now give a mathematically contrasting example by defining a discrete space and discrete time model. Even though the model is of general nature, it is partly motivated by empirical studies on the Glanville fritillary butterfly (see Box 3.1), and thus we call it the butterfly metapopulation model. The Glanville fritillary butterfly inhabits a fragmented network of habitat patches surrounded by an unsuitable matrix. It has annual generations, the flight season taking place during the summer, and the species spending the rest of the year as eggs, larvae, and pupae.

We assume that during the adult stage the individuals move accordingly to the diffusion model, with habitat-specific parameters, and thus that their movement tracks can be simulated as illustrated e.g. in Figure 2.14. In many butterfly species the males hatch before the females, and mating takes place soon after female emergence (a phenomenon called protandry; e.g. Wicklund and Fagerstrom, 1977). Thus, as a first approximation, we may assume that all females mate when they are born, and consequently we may ignore males in the population dynamical model. If a female is within a habitat patch, it is assumed to oviposit eggs at a constant rate f, whereas if it is in the matrix, it does not oviposit any egg. As a result, the number of eggs n_e oviposited in a habitat patch follows a Poisson distribution, with mean fT, where T is the total amount of time that all females spend in that patch. After the flight season is over, the eggs will hatch into larvae, and finally the larvae will hatch into adults of the next generation.

We assume that density dependence takes place during the larval stage, e.g. due to resource depletion. Among the many possible forms of density dependence, we will assume the much-studied Ricker model (Ricker, 1958). In the deterministic version, the number of adults n_a that hatch from the n_e eggs is

$$n_a = n_e \exp\left[r\left(1 - \frac{n_e}{K}\right)\right], \tag{3.10}$$

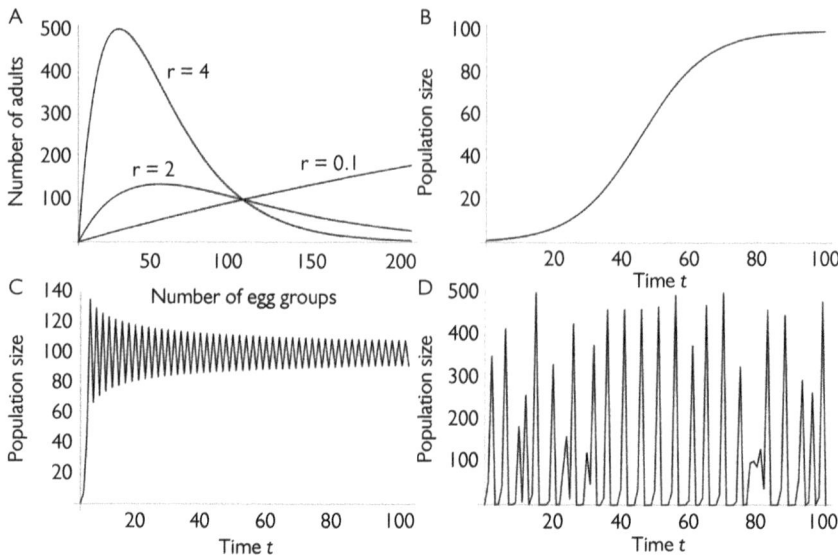

Figure 3.9 Dynamics of the Ricker model (Eq. (3.10)), which is used to model local population dynamics for the butterfly metapopulation model. (A) shows the number of adults that emerges from a given number of egg groups for three different values of the growth rate parameter ($r = 0.1, 2, 4$). (B) ($r = 0.1$), (C) ($r = 2$), and D ($r = 4$) show dynamics generated by the Ricker model, if assuming that all adults produce one egg group. In all panels, carrying capacity is set to $K = 100$.

whereas in the stochastic version the number of adults is Poisson distributed with mean given by Eq. (3.10). Here the parameters r relates to the number of adults that emerge from an egg group in the absence of density dependence, and K is the carrying capacity of the patch. As shown by Figure 3.9, the strength of density dependence, and thus the nature of population dynamics, depends critically on the parameter r. In the deterministic case, a low growth rate r leads to stable dynamics (Figure 3.9B), similar to those predicted by the continuous-time logistic models that underlies the plant population model. In contrast, an intermediate growth rate r leads to oscillatory dynamics (Figure 3.9C) and a high growth rate r to irregular dynamics (Figure 3.9D).

Let us next combine the population dynamical model of Eq. (3.10) with the movement model illustrated in Figure 2.14. To do so, we assume that the transition from eggs to adults within the patches follows the stochastic version of Eq. (3.10). We further assume that the growth rate r is the same for all patches, set to the low value leading to the stable dynamics shown in Figure 3.9B. We assume that the carrying capacity K_i of patch i is proportional to the area of the patch, irrespective of variation in patch quality that we incorporated in the movement model. After the adults emerge, they fly within the patch network according to the movement model, including the fact that edge-mediated behaviour makes them spend more time in

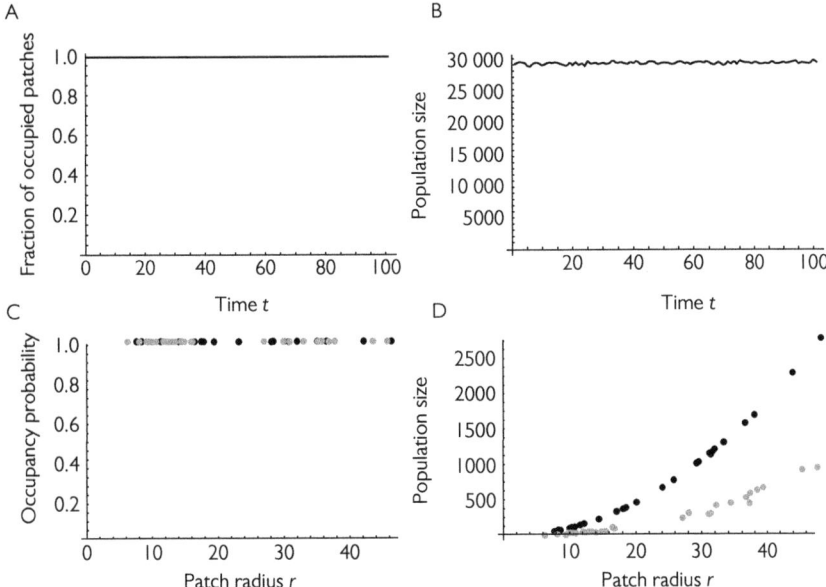

Figure 3.10 Simulation of the butterfly metapopulation model without environmental stochasticity. (A) shows the fraction of occupied patches as a function of time (here all patches are all the time occupied). (B) shows the total population size (number of adults) as a function of time. (C) shows patch-specific occupancy probabilities averaged over the simulation as a function of patch size. (D) illustrates the patch-specific population sizes, averaged over the years when the patch was occupied, as a function of patch size. In (C) and (D), the black and grey dots refer to high- and low-quality patches, respectively. Parameter values $r = 0.1$, $f = 0.25$, and $K_i = A_i/10$, where A_i is the area of patch i. The dynamics were simulated for 150 generations (out of which the first 50 were omitted as a transient) in the network of Figure 2.14, assuming the movement parameters described in the legend to that figure.

high-quality patches than low-quality patches. When they are within habitat patches, they lay eggs, which then form the adults of the next generation.

Population dynamics generated by the butterfly metapopulation model are illustrated in Figure 3.10. Figure 3.10D shows that population size increases with patch size, simply because carrying capacity scales with patch size. Population size is higher for high-quality patches than low-quality patches, because the individuals spend more time in high-quality patches (see Figure 2.14) and thus they oviposit more eggs there. But the butterfly metapopulation is not very dynamic at all in Figure 3.10! All the patches are constantly occupied, and the population size remains all the time close to 30,000 individuals. In contrast, real butterfly metapopulations can show great fluctuations both in terms of patch occupancy and in terms of population size (Box 3.1). The reason why the model fails to mimic real metapopulation dynamics is that we assumed a low growth rate, so the local population dynamics are stable. Further, the model involves only demographic stochasticity. As the total number of individuals is large, demographic stochasticity plays only a minor role. To generate colonization–extinction dynamics, we need more variability. In the context of the Glanville fritillary butterfly case study (Box 3.1), the main factor creating such variability is environmental stochasticity. So let us assume that the carrying capacities of the patches vary over the years. In Figure 3.11 we have randomized the carrying capacity K_i of each patch i for each year t from a log-normal distribution, where the mean has been set proportional to patch area. As a result, the local populations have now a substantial extinction risk, and the system persists in a balance between local extinctions and recolonizations. Furthermore, the total population size varies now greatly. As expected, the population sizes are greater in large patches and in high-quality patches (Figure 3.11D). Somewhat unexpectedly, the low-quality patches are more likely to be occupied than high-quality patches (Figure 3.11C). This is because the population sizes can peak to substantially higher numbers in the high-quality patches, which leads to over-consumption of the resources and thus the crash of the population. This example thus also serves to illustrate the sometime counterintuitive behaviour of models with non-linear dynamics, such as those described by Eq. (3.10).

Even in Figure 3.11 the butterfly metapopulation shows less pronounced fluctuations than is often the case in reality, such as in the example of Box 3.1. The reason

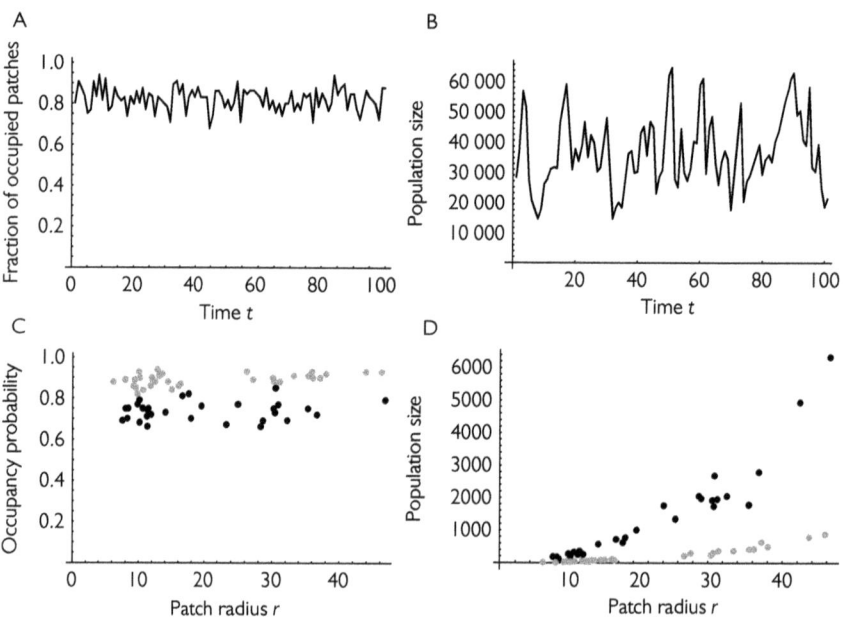

Figure 3.11 Simulation of the butterfly metapopulation model with spatially uncorrelated environmental stochasticity. The panels, symbols, and parameters are identical to those described in the legend to Figure 3.10, except that the carrying capacity for patch i and year t is set to $K_{it} = \exp(X_{it})(A_i/10)$, where the X_{it} are independent random variables sampled from the normal distribution $X_{it} \sim N(0, 3^2)$.

is that we assumed that the environmental conditions vary independently among the patches. Thus, when the conditions are adverse in one patch, they may be favourable in another patch, and in a large patch network the effects of adverse and favourable conditions average out in any given year. We next assume that temporal variation in the environment influences all patches simultaneously. This generates dynamics illustrated in Figure 3.12, with much variation both in patch occupancy and in total number of individuals. While in good years all patches are occupied, in bad years the whole system is almost empty. Indeed, the system is now prone to extinction—we needed to simulate the model 66 times before we obtained the simulation shown in Figure 3.12 in which the metapopulation persisted the full period of 100 generations.

3.3.4 The Levins metapopulation model and its spatially realistic versions

To study metapopulation dynamics analytically instead of performing simulations, let us move from the individual-based and spatially explicit metapopulation model to the simplest possible metapopulation model, which was formulated by

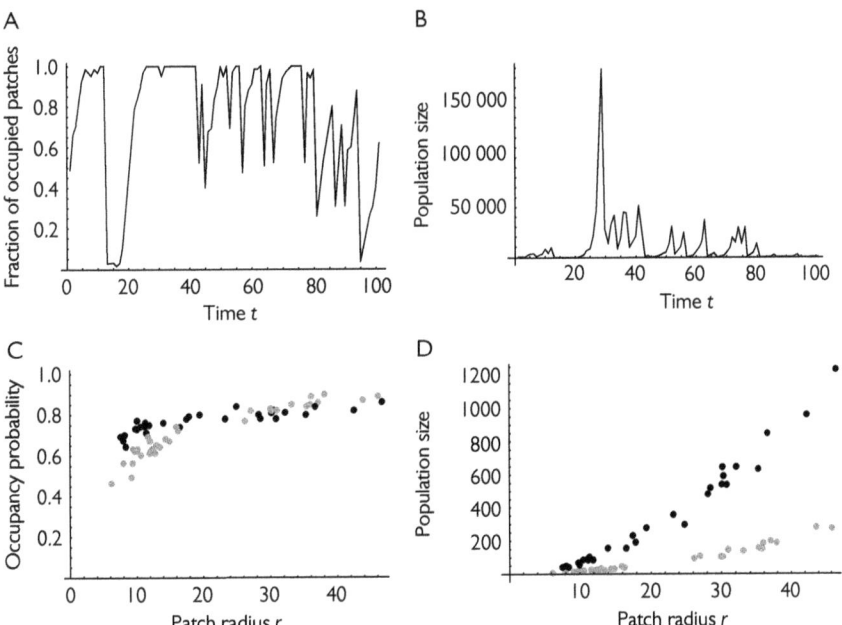

Figure 3.12 Simulation of the butterfly metapopulation model with spatially correlated environmental stochasticity. The panels, symbols, and parameters are identical to those described in the legend to Figure 3.10, except that the carrying capacity for patch i and year t is set to $K_{it} = \exp(X_t)(A_i/10)$, where the X_t are independent random variables sampled from the normal distribution $X_t \sim N(0, 3^2)$.

Levins (1969, 1970). In this model, the landscape consists of an infinite number of identical and equally connected patches. The patches are scored simply as occupied or empty, so that their local population sizes are not accounted for. A patch occupied by a local population is assumed to go extinct with the extinction rate e. Empty patches are assumed to become occupied with the colonization rate cp, where c is a colonization rate parameter, and p is the fraction of patches currently occupied. The colonization rate is assumed to be proportional to the fraction of occupied patches because the occupied patches are those that send out the migrants, i.e. the potential colonizers. The dynamical variable of the model is the fraction of occupied patches at time t, denoted here simply by p, though the reader should keep in mind that it depends on time, and thus $p = p(t)$. As the fraction of patches that can potentially go extinct is p, and as the fraction of patches that can potentially become colonized is $1 - p$, the dynamics of the model are described by the differential equation.

$$\frac{dp}{dt} = cp(1-p) - ep. \qquad (3.11)$$

Setting $dp/dt = 0$ shows that this model has two equilibrium states: one corresponding to metapopulation extinction ($p^* = 0$), and another one given by $p^* = 1 - e/c$. The latter one is positive and thus ecologically meaningful only if $c > e$. This condition gives the extinction threshold of the model: the metapopulation will persist if and only if the colonization rate parameter is greater than the extinction rate parameter. If this is not the case, the system will always decline to the equilibrium state $p^* = 0$ corresponding to metapopulation extinction.

Equation (3.11) is deterministic because it describes a system with infinitely many habitat patches. In the deterministic case, the model prediction is clear-cut: either the system goes extinct or then it persists forever. We have already discussed the relationship between deterministic and stochastic models at some length in the context of homogeneous-space population models (Section 3.2). Let us complement that discussion by constructing a stochastic variant of Levins model, and by comparing its behaviour to that of the deterministic version. In the stochastic model, there is a finite number n of habitat patches, each of which can be occupied or empty. Let us denote the number of occupied patches by m, so that the fraction of occupied patches is $p = m/n$. As in the deterministic model, the extinction rate of an occupied patch is e, and the colonization rate of an empty patch is $cm/n = cp$. This system is a Markov process, in which the dynamical variable is the number of occupied patches m at any given time. Extinction of an occupied patch corresponds to the transition $m \to m - 1$, which takes place for any of the m occupied patches at rate em. Colonization of an empty patch corresponds to the transition $m \to m + 1$, which takes place for any of the $n - m$ empty patches at rate $c(m/n)(n - m)$. This model has three parameters: the extinction and colonization rate parameters e and c, which are identical to those of the deterministic model. The additional parameter is the total number of patches n.

In Figure 3.13 we have simulated the dynamics of this Markov process over time. The simulations resemble the dynamics predicted by the deterministic model of

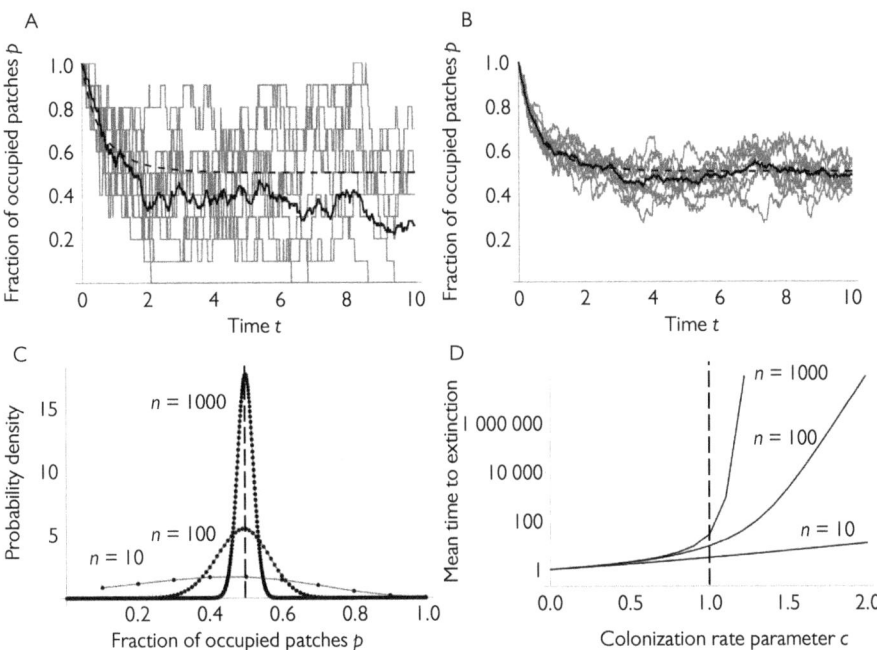

Figure 3.13 Metapopulation dynamics predicted by stochastic and deterministic variants of the Levins metapopulation model. (A) and (B) show how the fraction of occupied patches evolves over time in a system with 10 patches (A) and in a system with 100 patches (B). The grey lines show 10 replicate simulations of the stochastic Markov process, the continuous black line their average, and the dashed line the prediction of the deterministic model (Eq. (3.11)). (C) shows the quasi-stationary distribution of the fraction of occupied patches for metapopulations with $n = 10$, $n = 100$, and $n = 1000$ patches. The dashed line shows the equilibrium state $p^* = 1 - e/c$ of the deterministic mean-field model, which corresponds to the limit of $n \to \infty$. (D) shows how the mean time to extinction behaves as a function of the colonization rate for metapopulations with $n = 10$, $n = 100$, and $n = 1000$ patches. In all panels, the extinction rate parameter is set to $e = 1$, and in (A–C) the colonization rate parameter is set to $c = 2$. In (A) and (B), the system is initiated with all patches occupied, whereas in (D) the mean time to extinction is computed by assuming that the system is initially at the quasi-stationary state.

Eq. (3.11), but they show stochastic fluctuations around the deterministic expectation. The match between the stochastic and the deterministic models improves as the number of patches increases (Figures 3.13A, B), both in the sense of the mean prediction of the stochastic model matching with the prediction of the deterministic model and in the sense of the magnitude of the fluctuations decreasing.

In addition to running simulations, it is possible to analyse the behaviour of the Markov process by constructing a transition matrix (see Appendix A.4). With the help of the transition matrix, we may e.g. derive the distribution of population sizes that is reached after a transient period. This is called the quasi-stationary distribution of the Markov process, defined as the stationary distribution conditional on the system having not yet gone extinct. Figure 3.13C shows the quasi-stationary distribution for the fraction of occupied patches, assuming that there are either $n = 10$, 100, or 1000 patches. Reflecting the discussion of the previous paragraph, with increasing number of patches the fraction of occupied patches is more tightly concentrated around the prediction of the deterministic model. Figure 3.13D shows how the mean time to population extinction behaves in the stochastic model as the function of the colonization rate parameter c and the number of patches n. As we have fixed the extinction rate parameter to $e = 1$, the deterministic model would predict that the system goes extinct if $c < 1$ and that it persists if $c > 1$. In the stochastic model, the metapopulation goes eventually extinct with certainty (i.e. with probability 1), whatever are the parameter values. The time to extinction increases both with increasing value of the colonization rate parameter c and with increasing system size n. With a large number of patches n, the stochastic model makes essentially the same prediction as the deterministic model: it predicts a short time to extinction if $c < 1$, but such a long time to extinction for $c > 1$ (note the logarithmic scale in Figure 3.13D) that, in practice, the metapopulation does not go extinct over ecologically relevant timescales. It is possible to show mathematically that, above the deterministic extinction threshold, the time to extinction increases exponentially with the number of habitat patches (Ovaskainen and Meerson 2010 and references therein).

One limitation of the Levins model is that it assumes identical and equally connected patches, whereas real metapopulations show variation in the sizes, qualities, and connectivities of the habitat patches. Extensions of the Levins model to heterogeneous networks have been termed spatially realistic metapopulation models (Moilanen and Hanski, 1995; Hanski, 2001). As one simple example, we may assume that the extinction rate E_i of patch i is inversely related to its area A_i, so that $E_i = e/A_i$. The justification here is that large patches have large populations, and that extinction rate decreases with increasing population size. We may further assume that the colonization rate C_i of patch i is a sum of contributions from the occupied patches, the contributions of which increase with increasing areas of the source patches (because larger patches send more migrants) and with decreasing distance d_{ij} between the source patch j and the target patch i (because migrants are more likely to disperse to a nearby patch). Assuming an exponential dispersal kernel, we may model the colonization rate as

$$C_i = \sum_{j \neq i} A_j \exp(-\alpha\, d_{ij}) O_j, \tag{3.12}$$

where $O_j \in \{0, 1\}$ denotes the occupancy state of the source patch, and the parameter α sets the spatial scale of dispersal.

The qualitative behaviour of this spatially realistic metapopulation model is similar to the behaviour of the spatially implicit Levins model. Most importantly, in the deterministic version of this model, the metapopulation will persist if and only if a threshold condition is satisfied. The threshold condition is given by $\lambda_M > e/c$, where λ_M is called the metapopulation capacity of the patch network (Hanski and Ovaskainen, 2000). The metapopulation capacity λ_M measures both how much habitat there is in total, and how connected the habitat patches are to each other. Mathematically, it is given by the leading eigenvalue of the matrix **M** with elements $M_{ij} = A_i A_j \exp(-\alpha\, d_{ij})$.

3.4 The persistence of populations under habitat loss and fragmentation

The progressive destruction of habitats has been recognized as a major threat to biological diversity and considered to be a primary cause behind the current species extinction crisis (World Resources Institute, 2005, and references therein). Figure 3.14 illustrates the concepts of habitat loss and fragmentation. By habitat loss we mean that the amount of habitat is decreased, which generally relates to the reduction of habitat area. However, habitat loss can also take place through reduction of habitat quality, e.g. due to decreased availability of resources per unit area. By habitat fragmentation per se we refer to a situation in which large continuous habitats are

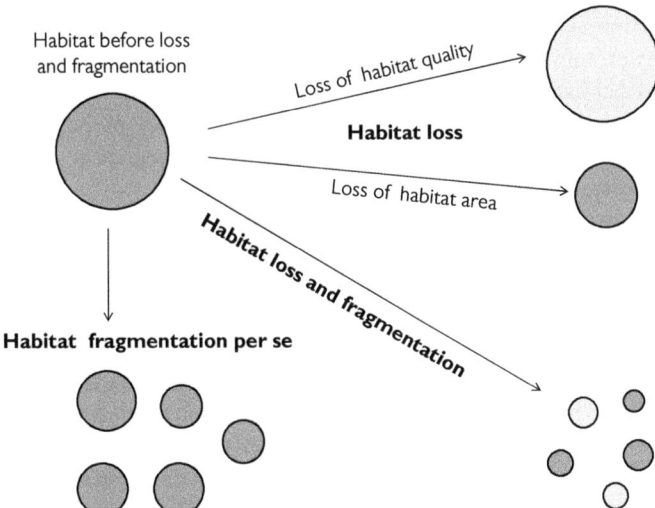

Figure 3.14 Concepts of habitat loss and fragmentation. Habitat before loss and fragmentation may represent e.g. a large and spatially contiguous natural forest. Many species specialized to natural forests utilize dead wood as their resource. From the viewpoint of such a species, the forest may experience habitat loss in terms of habitat quality, e.g. if forest management reduces the amount of dead wood. The forest may experience habitat loss also in terms of habitat amount, e.g. due to conversion of part of the forest into cultivated fields. The concept of habitat fragmentation per se refers to a situation where the amount of habitat does not change but the habitat is present in smaller blocks. In practice habitat loss and fragmentation appear together, e.g. some remnants of the original forest remaining in a landscape dominated by cultivated fields.

replaced by many small habitat fragments, but there is no habitat loss and thus the total area and quality of the habitats are kept the same. While habitat fragmentation per se does not appear alone (Didham et al., 2012), it is interesting to compare population dynamics in landscapes with the same amount of habitat but with different levels of fragmentation. In reality habitat loss and fragmentation act together, converting continuous landscapes into networks of patches isolated from each other by a matrix of inhospitable habitats. Obviously, increasing the level of habitat fragmentation makes the habitat patches increasingly isolated from each other (Bender et al., 2003). This is expected to lead to less movement among the patches (Andreassen and Ims, 2001; Section 2.4), and thus to lowered colonization probabilities of empty patches. Isolated populations are also threatened by inbreeding, which can lower their viability and increase their extinction risk (Saccheri et al., 1998; Schmitt and Seitz, 2002). Habitat fragmentation leads to an increased amount of edges between different habitat types (Ewers and Didham, 2006), which can provide different conditions from the interior habitats e.g. due to changed microclimatic conditions. While edges are for many species of lower quality than the interior habitats, many other species actually are favoured by edges (Ewers and Didham, 2006, and references therein).

In this section, we illustrate the ecological consequences of habitat loss and fragmentation by examining how the dynamics of population dynamical models respond to decreased habitat availability. To be able to compare the responses of different model types, we apply habitat fragmentation to all three models considered in Section 3.3: the *plant population model*, the *butterfly metapopulation model*, and the *Levins metapopulation model*.

3.4.1 Habitat loss and fragmentation in the plant population model

We assume that initially (at time $t = 0$) all the landscape consists of suitable habitat, and thus we start from the homogeneous-space model considered in Section 3.2. We then assume that habitat loss takes place instantaneously at time $t = 10$, decreasing the amount of habitat area to 20% of its original extent, so that 80% of the habitat is lost. In Scenario A we assume that the remaining habitat remains in one contiguous block, and thus there is no fragmentation (Figure 3.15A). In Scenario B we assume the remaining habitat is present in a fragmented manner in 25 small patches. We assume that the species is not able to establish in the matrix, i.e. the lost habitat has become unsuitable for it. Moreover, we assume that the conversion of habitat from suitable to unsuitable does induce an immediate death of individuals located there. Thus, after habitat loss, the individuals located in the matrix die out one by one, whereas in the remaining habitats also new individuals continue to be established.

As 20% of the original habitat remains after habitat loss, the carrying capacity of the landscape drops to 20% of its original extent. Consequently, the most straightforward prediction, which is the prediction of the mean-field model discussed in Section 3.2, is that also the population size drops to 20% of its initial number. This prediction is a good approximation in the case of Scenario A where, after a

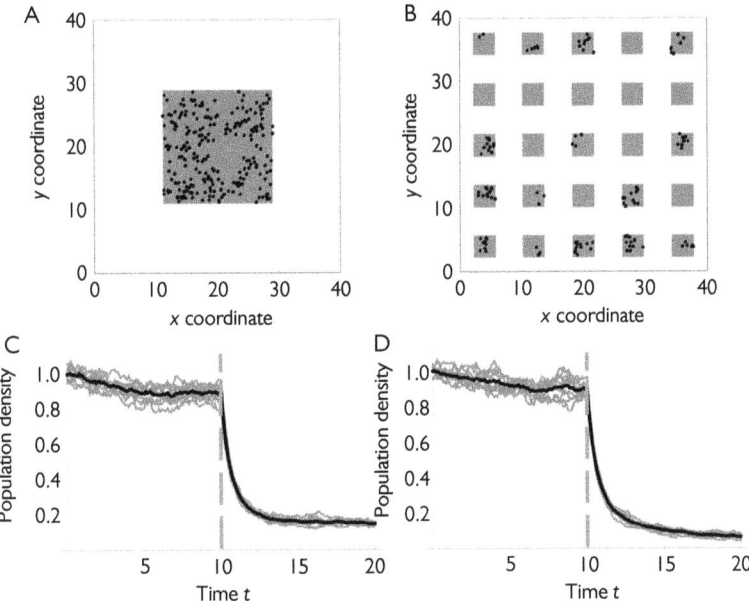

Figure 3.15 The influences of habitat loss and fragmentation on the dynamics of the plant population model. Initially at $t = 0$ the entire domain consists of suitable habitat. At time $t = 10$ we assume that 80% of the habitat is lost, with the grey areas in (A) and (B) showing the distributions of the remaining habitats for Scenarios A and B (see text), respectively. The dots in (A) and (B) show the distributions of individuals that are present at the final time $t = 20$ of the simulation. (C) and (D) show the time evolution of population density in Scenarios 1 and 2, respectively. The grey lines show 10 replicate simulations, and the black line their average. The vertical dashed line shows the time $t = 10$ at which time habitat loss takes place. Parameter values of the model are equal to those described in the legend to Figure 3.3.

transient period, population density stabilizes to approximately 20% of its original value (Figure 3.15C). This is so because the remaining block of habitat is so large that the dynamics of the species are essentially uninfluenced within it. As the local population density within the habitat remains essentially constant, the global population density, i.e. the number of individuals divided by the size of the entire landscape, drops proportionally to the amount of habitat loss.

While the behaviour of Scenario B (Figure 3.15D) resembles that of Scenario A (Figure 3.15C), it is not identical: population density drops to a lower value in Scenario B than it does in Scenario A. This is because in Scenario B there is an effect of habitat fragmentation on the top of that of habitat loss. Let us consider the set of individuals that are present in one patch as a local population. As the local populations are small in the fragmented landscape, they are prone to go extinct (Levins, 1970; Keymer et al., 2000; Molofsky and Ferdy, 2005; Altermatt and Ebert, 2010), as has happened for some of the patches in Figure 3.15B. Once a local population

100 • *Population ecology*

is extinct, the recolonization of the patch requires that a propagule produced in another patch lands into the focal patch. As the distances between the patches are large compared to the typical dispersal distances, such events are rare. As a consequence, the local populations gradually go extinct one by one, the result of which is seen in Figure 3.15D not only as a lower occupancy than in Figure 3.15C, but also as a longer transient time.

3.4.2 Habitat loss and fragmentation in the butterfly metapopulation model

We will next consider habitat loss in the context of the butterfly metapopulation model. In Section 3.3, we simulated the dynamics of the butterfly metapopulation model in the network shown in Figure 2.14, with three model variants: no environmental stochasticity (illustrated in Figure 3.10), spatially uncorrelated environmental stochasticity that influences the habitat patches independently (Figure 3.11), and spatially correlated environmental stochasticity that influences all patches simultaneously (Figure 3.12). Let us now assume that the network experiences habitat loss. As discussed in the beginning of this section, this could happen in a multitude of ways: there could be loss of entire habitat patches, either randomly or in an aggregated manner, or loss of patch area or loss of patch quality. Among these, we assume that no entire patches are lost, but that all patches lose part of their area. In Figure 3.16, we have simulated the dynamics of the model in the original network (left-hand panels), in a network in which the areas of all patches have been reduced to 10% of their original extent (middle panels), and in a network in which the areas of all patches have been reduced to 5% of their original extent (right-hand panels).

Figure 3.16 illustrates that in an increasingly fragmented network, the fraction of occupied patches decreases. This is because of several factors. First, the individuals will emigrate faster from smaller patches (Kuussaari et al., 1996), thus having less time for laying eggs into their natal patch, leading to a slower growth rate at low density, and thus increasing the risk of local extinction. Second, in smaller patches the carrying capacities will be smaller, and thus the local populations will be restricted to smaller population sizes (Hanski, 1999), again increasing their risk of local extinction. Third, smaller local populations will send out fewer migrants (Andreassen and Ims, 2001; Ewers and Didham, 2006), decreasing the colonization rate of empty patches. And fourth, smaller patches will be less likely to be found by dispersing individuals (see Section 2.4), which further decreases the colonization rate of empty patches.

3.4.3 Habitat loss and fragmentation in the Levins metapopulation model

To study the influence of habitat loss and fragmentation on metapopulations more analytically, let us finally apply Levins model (Eq. (3.11)) in the context of habitat loss and fragmentation. To do so, we follow Lande (1987), who extended the Levins model by assuming that only a fraction h of the patches is suitable for occupancy.

3.4 The persistence of populations under habitat loss and fragmentation • 101

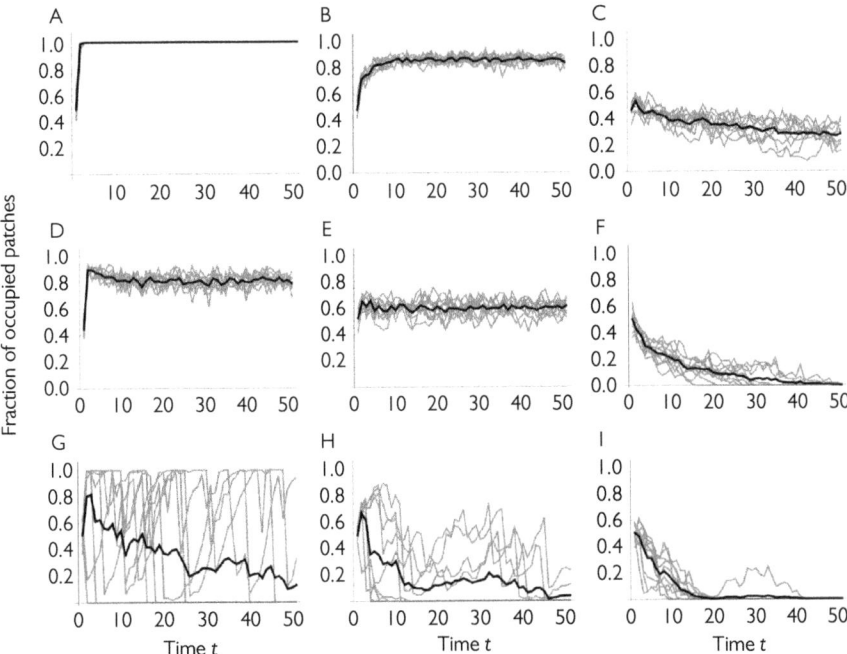

Figure 3.16 The influences of habitat loss and fragmentation for the dynamics of the butterfly metapopulation model. All panels show how the fraction of occupied patches evolves over time, when starting the system so that initially each patch is empty with probability 0.5, and occupied at its carrying capacity with probability 0.5. The grey lines show 10 replicate simulations, and the black line their average. (A–C) corresponds to a model without environmental stochasticity, (D–F) to a model with spatially uncorrelated environmental stochasticity, and (G–I) to a model with spatially correlated environmental stochasticity. The left-hand column (A, D, G) show simulations in the patch network of Figure 2.14. In the middle column (B, E, H) the areas of the habitat patches have been decreased to 10% of their original size, whereas in the right-hand column (C, F, I) their areas have been decreased to 5% of their original size. The parameter values are identical to those used in Figures 3.10–3.12.

In the extended model the migrating individuals may end up in unsuitable patches, reducing the colonization rate from cp to chp. The dynamical equation of Lande's model is thus

$$\frac{dp}{dt} = chp(1-p) - ep. \tag{3.13}$$

The extinction threshold in Lande's model is $ch > e$, which can be rewritten as $h > e/c$. Thus, the model predicts that the metapopulation will persist if the amount of habitat is above a certain threshold value, whereas below that threshold the metapopulation will go extinct. This is what we observed with the simulations of the butterfly metapopulation model: decreasing the amount of habitat availability first decreases the fraction of occupied patches, and eventually it makes the entire system go extinct, as is the case in Figures 3.16F and I. The key insight here is that a metapopulation can go deterministically extinct even if some perfectly suitable habitat remains, simply because there is too little of that.

As discussed in Section 3.3, a spatially realistic version of the Levins–Lande model predicts that the population will survive if and only if $\lambda_M > e/c$, where λ_M is the metapopulation capacity of the patch network. The metapopulation capacity accounts not only for the amount of habitat, but also for the fragmentation pattern. The spatially realistic model can be used to examine how the extinction threshold depends on the scenario of habitat loss: e.g. random loss of patches, systematic loss of patches, or reduction in patch area (Hanski and Ovaskainen, 2000; Ovaskainen, 2002; Ovaskainen and Hanski, 2003b).

3.5 Statistical approaches to analysing population ecological data

In this section, we discuss a selection of statistical approaches that can be applied to population ecological data. To connect the statistical approaches to the mathematical models presented earlier in this chapter, we fit the statistical models to data generated by those mathematical models. As we did in Section 2.5 when analysing movement data, we pretend that we have acquired the data through some survey methods, and we apply the statistical analyses with the aim of gaining insights into the ecological processes that produced the data.

We start by attempting to identify signals of demographic stochasticity, environmental stochasticity, and density-dependence from time-series data, first considering a solely statistical approach, and then fitting the stochastic logistic model with Bayesian inference. We then illustrate species distribution modelling with data generated by the plant population model. Finally, we construct a stochastic patch occupancy model and fit it to data generated by the butterfly metapopulation model.

3.5.1 Time-series analyses of population abundance

Time-series analysis of ecological data is a broad topic which is in the focus of several books, such as *Ecological Time Series* by Powell and Steele (1995), *Population Cycles: The Case for Trophic Interactions* by Berryman (2002), *Complex Population Dynamics* by Turchin (2003), and *Introductory Time Series with R* by Metcalfe and Cowpertwait (2009). In this section, we illustrate time-series analysis by applying autoregressive models to the data that we generated in Sections 3.2 and 3.3 with the non-spatial stochastic logistic model.

One of the simplest time-series models is the autoregressive model of order 1, which is denoted by AR(1). The AR(1) model is defined by

$$n_{t+1} - n_t = \alpha + \beta n_t + \varepsilon_t, \tag{3.14}$$

3.5 Statistical approaches to analysing population ecological data • 103

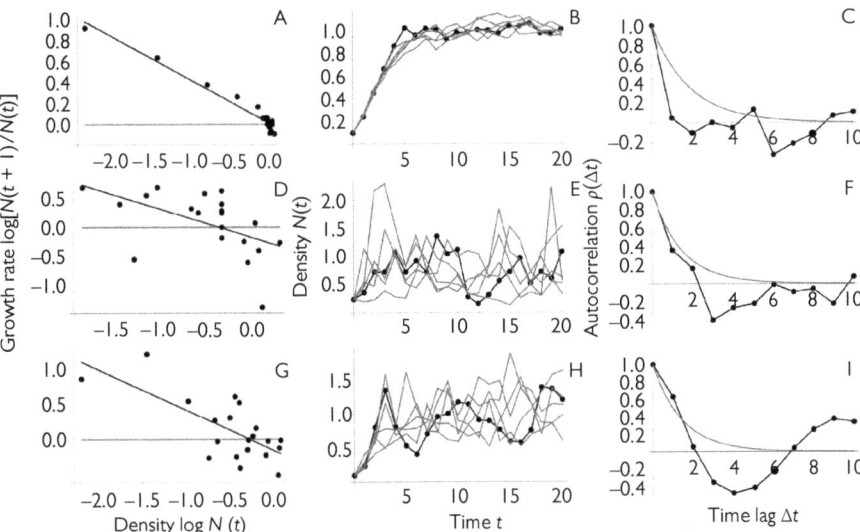

Figure 3.17 Time-series analyses of population data. The data, shown by the dots, were obtained from the time-series of Figure 3.4A (here the upper row of panels), Figure 3.4C (here the middle row of panels), and Figure 3.6D (here the bottom row of panels). We have sampled these data to intervals of 1 time unit and used the first 20 time units. In (A, D, G) the dots show the data used to fit the AR(1) model of Eq. (3.14), the black lines show the model fits, and the grey lines show zero lines. In (B, E, H) the black lines show the data, and the grey lines show five simulation replicates based on fitted AR(1) models. In (C, F, I) the dots show the autocorrelation functions (ACFs) of the data (excluding the first 5 time units which represent a transient), whereas the grey lines show the theoretical ACFs predicted by the AR(1) models. For parameter values of the AR(1) models, see Table 3.1.

where $\varepsilon_t \sim N(0, \sigma^2)$ is normally distributed noise, n_t is the state of the population at time t, and α and β are parameters to be estimated. The response variable of the model is the difference $n_{t+1} - n_t$, i.e. the growth rate of the population from time t to time $t + 1$.

In the upper row of panels in Figure 3.17 we apply the AR(1) model to data generated by the stochastic logistic model without environmental stochasticity and with a large population size, i.e. with a small amount of demographic stochasticity. Before starting the analysis we have log-transformed the data, and thus defined $n_t = \log N_t$, where N_t stands for the density (number of individuals per unit area) of the species. Equation (3.14) is simply a linear model (see Appendix B.1), where the intercept α relates to overall population size, and the slope β models how the growth rate depends on current population density. As shown by Figure 3.17A, the slope is negative, meaning that the growth rate decreases as the current population density increases. This creates a density-dependent effect: at low density the growth rate is positive, whereas at high enough density the growth rate becomes negative. The growth rate is 0 at the point where the grey and black lines cross in Figure 3.17A, giving an estimate of the carrying capacity. Figure 3.17B compares the data to the predictions of the fitted models, obtained by using Eq. (3.14) to generate five simulated time-series for the log-transformed population density, and the back-transforming by $N_t = \exp(n_t)$. The model fits well to the data, as it generates a similar pattern of a transient population increase followed by eventual fluctuations around the carrying capacity.

In the middle row of panels in Figure 3.17 we apply the same AR(1) analysis to data generated by otherwise the same model, but with a smaller population size and thus a higher level of demographic stochasticity. This results in a larger amount of unexplained variation σ^2 (Figure 3.17D and Table 3.1), which is reflected in the simulations of the fitted model as a higher amount of fluctuation around the stationary state (Figure 3.17E). In the lower row of panels in Figure 3.17 we consider data generated by a model with infinitely large population size (i.e. no demographic stochasticity), but now with environmental stochasticity. The model fits for data with either demographic (Figure 3.17D) or environmental stochasticity (Figure 3.17G) look rather similar, making the point that it is generally difficult to induce the source of stochasticity from time-series data only. However, as discussed in depth by Lande et al. (2003), with sufficient data this is possible, since these two forms of stochasticity scale differently as a function of population size (Lande et al., 2003). In brief, while fluctuations in population size due to environmental stochasticity are on the same order N as the population size, with demographic stochasticity they are on the order $\sqrt{N}$.

Figures 3.17C,F,I show the autocorrelation functions (ACFs) of the data as well as the theoretical ACF based on the fitted models. The ACF for lag Δt is defined as the correlation of the time-series between times t and $t + \Delta t$. For the AR(1) the theoretical expectation is $(1 + \beta)^{\Delta t}$, where β is the density-dependence parameter of Eq. (3.14). In the two upper panels of Figure 3.17, where the data were generated without environmental stochasticity, the exponentially decaying ACF generated by the AR(1) is at least qualitatively in line with the data. The ACF decreases a bit slower in Figure 3.17C than in Figure 3.17F, reflecting the fact that the population

Table 3.1 Parameter estimates of AR(1) models (Eq. (3.14)).

Model	α	β	Variance σ^2
U	0.02	−0.41**	0.0027
M	−0.17	−0.50*	0.22
B	−0.04	−0.50**	0.10

Models U, M, and B correspond to models fitted to the data shown in the upper, middle, and bottom rows of panels in Figure 3.17. Model fitting was performed by standard linear model analysis with maximum likelihood estimation. Statistical significances shown for the parameter β are based on the t-statistic, with $0.01 < p < 0.05$ denoted by * and $p < 0.01$ denoted by **.

size fluctuates faster in Figure 3.17E than in Figure 3.17B. In these models temporal autocorrelation is generated simply by intrinsic population dynamics: if demographic stochasticity perturbs the population size to a low density, it takes some time before the population grows back to the carrying capacity.

With data generated with environmental stochasticity there is a clear mismatch between the theoretical and empirical ACFs. This is because this time-series was taken from a model where environmental conditions varied slowly in time, i.e. with temporal autocorrelation. In Figure 3.17I, the data show negative autocorrelation for time-lags 3–6, corresponding to the cyclic nature of the time-series. In the AR(1) model, the ACF always decreases exponentially to 0, and it is thus not able to reproduce cyclic dynamics. To improve the model fit, two alternative routes could be taken. First, if data on the temporally varying environmental condition would be available, one could utilize it as a predictor in the model. Second, one could increase the order of the autoregressive model by considering an AR(2) model, or more generally an AR(n) model. We do not consider such extensions here, but we will return to the AR(2) model in Section 4.5 where we analyse cyclic population dynamical data generated by a predator-prey model.

3.5.2 Fitting Bayesian state-space models to time-series data

The time-series analyses in Section 3.5.1 yielded information on population-level parameters r and K, the contributions of various sources of stochasticity, and patterns of temporal autocorrelation. In this subsection we take a more mechanistic approach by fitting to data the individual-based stochastic logistic model rather than the solely statistical AR(1) model. This subsection also serves to illustrate Bayesian inference, for which reason we discuss here the process of model fitting in some detail. A more technical introduction to Bayesian inference is found from Appendix B.2.

We start by recalling from Section 3.2 that the stochastic logistic model is technically a Markov process (see Appendix A.4), and its state is described by the number of individuals $n(t)$ at time t. The population size can take any integer value, $n = 0, 1, 2, \ldots$. The present individuals produce new individuals with per-capita fecundity rate f, and thus the population level birth rate is fn. The per capita death rate is $m + c(n - 1)$, where m is the density-independent background mortality rate and the parameter c describes the additional death rate imposed by competitive effects of the $n - 1$ other individuals to the focal individual. Thus, the population level death rate is $[m + c(n - 1)]n$. The mean-field version of this model is (see Section 3.2)

$$\frac{dN}{dt} = (f - m)n - cn^2 = rn\left(1 - \frac{n}{K}\right), \quad (3.15)$$

where $r = f - m$ is the per capita growth rate of the population at low density and $K = r/c$ is the carrying capacity. We reparameterize the model so that instead of (f, m, c) we consider as the primary model parameters $\theta = (f, m, K)$. This is because the carrying capacity K is easier to interpret at the population level than the competition parameter c. Here we have combined all the model parameters into the vector θ.

To simulate the model for any parameter values θ, the carrying capacity K can be converted to the corresponding competition parameter as $c = \max[(f - m)/K, 0]$, where the maximum with 0 is needed to ensure that the competition parameter does not go negative even if the growth rate parameter r is negative. Note that with a negative growth rate $r<0$ there is no need to limit the growth with density-dependent mortality as the population will have a tendency to decrease even when at low density.

The black line in Figure 3.18A shows a single simulation of this model, initiated with five individuals at time $t = 0$ and run until time $t = 10$, with the parameter values $f = 3$, $m = 1$, and $K = 50$. Thus we have set $\theta^{TRUE} = (3, 1, 50)$, where the superscript TRUE indicates that these values are the underlying true values which were used to simulate the dynamics, and which we will try to estimate back. As the model is a Markov process, the births and deaths take place in continuous time instead of at discrete time steps. Further, as the model has a discrete state-space ($n = 0, 1, 2, \ldots$), the population size jumps in Figure 3.18A abruptly between integer values, with births increasing the population size by one ($n \to n + 1$) and deaths decreasing it by one ($n \to n - 1$).

Bayesian state-space models generally combine a process model with an observation model, as we did in the analyses of capture-mark-recapture data in Section 2.5. Let us assume that this time-series comes from a laboratory experiment in which the

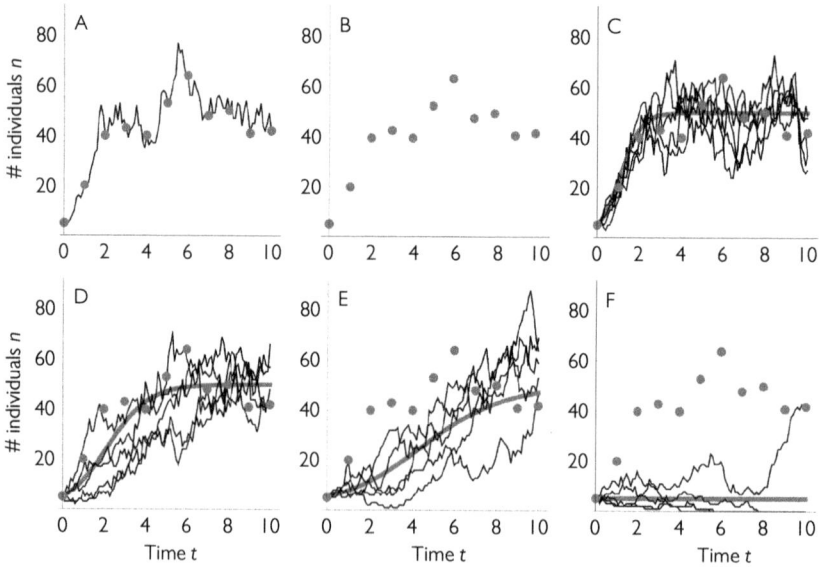

Figure 3.18 Simulations of the stochastic logistic model. The black lines show individual-based simulations generated by the Gillespie algorithm (see Appendix A.4), whereas the thick grey lines show the time-evolution of the mean-field model (Eq. (3.2)). The grey points show data sampled from the simulation of (A). Parameter values $f = 3$, $m = 1$, and $K = 50$, expect $f = 2$ (D), $f = 1.5$ (E), and $f = 1$ (F).

experimenter initiated the population with the 5 individuals in the morning of day 0, and then counted the number of individuals from morning 1 to 10. To simplify the treatment, we assume that there is no observation error, so that the observed number of individuals equals the actual number of individuals. The data collected by the experimenter is showed by the grey dots in Figure 3.18A: the numbers of individuals seen in days 1–10 were 20, 40, 43, 40, 53, 64, 48, 50, 41, and 42.

Now, let us move to the process of parameter estimation. To do so, we pretend that we do not know the underlying continuous-time process that is visible in Figure 3.18A, but only the observed data, i.e. the 10 numbers listed at the end of the previous paragraph and illustrated in Figure 3.18B. In reality, we would not know the correct model structure, i.e. that the population follows the stochastic logistic model. However, we ignore here the issue of model selection, and thus assume that we would know that the dynamics follow the stochastic logistic model. Thus, the question we wish to address is how informative the data of Figure 3.18B are about the parameter values generating these data.

Before approaching this question with formal statistical inference, let us first attempt to gain an intuitive understanding on the concept of likelihood by simulating the model with different parameter values, and comparing those simulations visually with the data. Figure 3.18C shows five replicate simulations of the model, done by exactly the same parameter values used to generate the original data. Even though the model and the parameters are identical to those used to generate the data, each realization is somewhat different. This is simply because the model is stochastic. As a consequence, it is not to be expected that any of these realizations would match with the data exactly, and indeed none of the five replicate simulations do so. But otherwise the overall characteristics of the time-series showed in Figure 3.18C match well with the variation in the data, and thus it looks plausible that the data could have been generated by the parameter values used in Figure 3.18C, as they indeed were.

In Figure 3.18D we compare the data to simulations conducted by parameter values that do not coincide with the true values. Here we have kept the parameters $m = 1$ and $K = 50$ at their true values, but decreased the fecundity rate from $f = 3$ to $f = 2$. This is seen as a slower growth rate both in the underlying mean-field model (shown by thick grey line) and the five stochastic realizations. But yet some of these realizations match the data reasonably well. So just by comparing Figures 3.18C and D we could not be certain that the true parameter value is $f = 3$ instead of $f = 2$. Decreasing the fecundity rate further ($f = 1.5$ in Figure 3.18E and $f = 1$ in Figure 3.18F) makes the simulations deviate more from the data, and thus we would conclude that especially the value $f = 1$ seem very unlikely.

The essence of likelihood-based parameter estimation is captured by this discussion: the question is which parameters or parameter combinations could have plausibly produced the observed data. To turn these considerations into a statistical framework, the key is to define the word 'plausibly' more precisely. In likelihood-based inference (Appendix B.2), this is done by computing the likelihood of observing the data, conditional on the parameter values. To illustrate, if the parameter vector is set to θ^{TRUE} and the population is initiated with five individuals in the morning of day 0, the probability of observing exactly 20 individuals in the morning of day 1 (as is the case with the data we generated) is $p_1 = 0.03815$. How we computed this probability is not important here, but an interested reader may have a look at Appendix A.4, where we discuss how the time-evolution of a Markov process can be solved with the help of a linear system of differential equations. We may next compute the probability of the second data point by restarting the model at time $t = 1$ with 20 individuals, and computing that the probability of observing exactly 40 individuals at time $t = 2$ is $p_2 = 0.043095$. With the same logic we may compute the remaining probabilities $p_3, \ldots, p_{10}$, where e.g. $p_{10} = 0.0368283$ is the probability of the population consisting of 42 individuals at time $t = 10$ if there were 41 individuals at time $t = 9$.

Let us put the data from all days into the vector $y = (20, 40, 43, 40, 53, 64, 48, 50, 41, 42)$. We aim to compute the likelihood of the data, denoted by $p(y|\theta^{\text{TRUE}})$, i.e. the probability of observing the data y given the parameter vector θ^{TRUE} and knowing that the initial population size was 5 individuals. This probability is the product of the single-step probabilities, giving $p(y|\theta^{\text{TRUE}}) = p_1 p_2 \cdots p_{10} = 0.00000000000000198 = 1.98 \cdot 10^{-15}$. Thus, it is extremely unlikely to generate exactly the same data even with exactly the same parameter values.

But even though the likelihood $p(y|\theta^{\text{TRUE}})$ is very low, it can still be high compared to the likelihood $p(y|\theta)$ of generating the same data with some other set of parameter values θ. Let us denote the parameter values used in Figures 3.18D, E, and F as θ^D, θ^E, and θ^F, respectively. By calculating the likelihood of the data with these parameter values gives $p(y|\theta^D) = 6.82 \cdot 10^{-16}$, $p(y|\theta^E) = 5.34 \cdot 10^{-17}$, and $p(y|\theta^F) = 1.27 \cdot 10^{-19}$. Thus, with the true values θ^{TRUE}, the probability of observing the data is approximately 3, 40, or 15,000 times higher than with the parameter values θ^D, θ^E, or θ^F, respectively.

We may repeat this exercise for all possible values of the fecundity parameter f. This is done in Figure 3.19A, where the continuous line shows the likelihood $p(y|\theta)$ of observing the data as a function of the fecundity parameter f, while keeping the other parameters m and K at their true values, $m = 1$ and $K = 50$. Values near to $f^{\text{TRUE}} = 3$ yield a higher likelihood of observing the data than values far away from $f^{\text{TRUE}} = 3$. The peak of the function gives the maximum likelihood (ML) estimate. The ML estimate is $f \approx 2.8$, which is near to the true value but not exactly at that value.

This discussion implies that if we would know the true values of the parameters m and K, we would be able to estimate the fecundity parameter f. But we don't know those other parameters, as this model (like most models) contains simultaneously several unknown parameters. So we next relax the assumption that we know the true values of the two other parameters m and K. Let us start by keeping $K = 50$ at its true value, but guessing that the value of the parameter m would be $m = 0.5$ instead of $m = 1$. The dashed line in Figure 3.19A shows how the likelihood of observing the data now behaves as a function of the fecundity parameter f. In this case, the ML estimate (the peak of the curve) is $f \approx 2.2$. Thus, our inference of f changed, i.e. it is conditional on what we think about the value of the parameter m. This illustrates

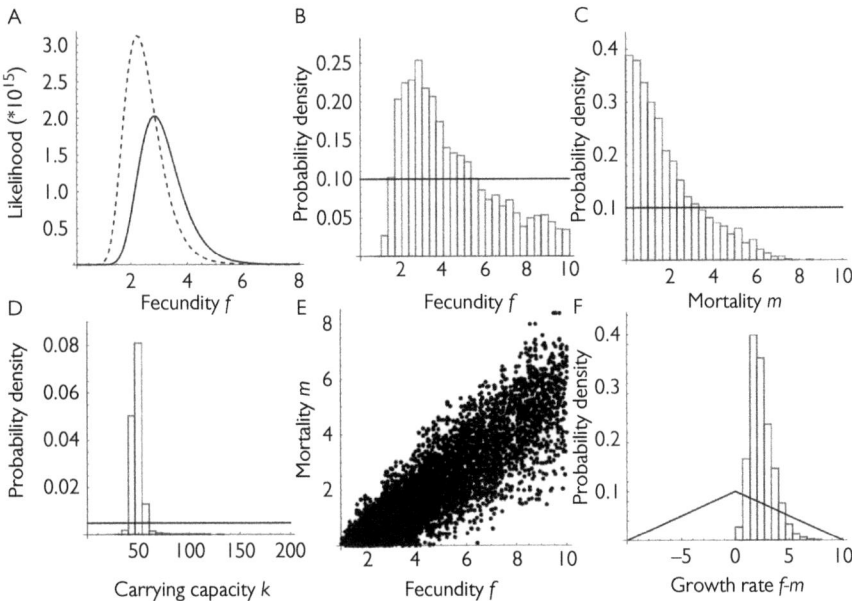

Figure 3.19 Bayesian inference applied to the stochastic logistic model. The figure shows likelihood profiles and posterior distributions for the data shown by the grey dots in Figure 3.18. (A) shows the likelihood profile as a function of the parameter f, assuming that $m = 1$ and $K = 50$ (continuous line) or that $m = 0.5$ and $K = 50$ (dashed line). (B–D) and (F) show the marginal prior (lines) and posterior (histograms) distributions for the parameters $f, m, K,$ and $f - m$, respectively. (E) shows 1,000 samples from the joint posterior distribution of the parameters f and m. The posteriors were sampled with the Metropolis–Hastings algorithm (Appendix B.2).

why estimation cannot be done separately for each of the parameters, but for their combination $\theta = (f, m, K)$. Thus, we wish to know which parameter combinations θ are 'plausible' in the sense that they would generate the observed data with high likelihood.

When conducting ML estimation, we simply search for the parameter combination θ^{ML} that maximizes the likelihood of observing the data, i.e. a vector θ^{ML} for which the value of $p(y|\theta^{ML})$ is higher than $p(y|\theta)$ for any other choice of the parameter vector θ. A numerical search shows that with the data of Figure 3.18, the ML estimate is $\theta^{ML} = (1.85, 0.02, 47.85)$, with $p(y|\theta^{ML}) = 5.7 \cdot 10^{-15}$. The ML parameters are not very close to the true parameter values. For example, while the true value of the mortality parameter is $m = 1$, the ML estimate is $m = 0.02$. If continuing the statistical inference with the ML approach, the next step would be to compute the confidence intervals (CIs) of the ML parameter values. For example, the 95% CI of the ML estimate is the interval which would include the true value in 95% of the cases, if the experiment would be replicated. For readers interested in how to derive CIs in the context of ML estimation, we recommend e.g. Royall (1986) or Bolker (2007).

Instead of continuing with the ML approach, we proceed with Bayesian inference. To do so, we need to define a prior distribution $p(\theta)$, which describes what we knew about the parameter values before seeing the data. In some cases we may truly have prior information. For example, we could have done another experiment, in which we had followed singly grown individuals throughout their lifetimes, from which data we could have estimated the density-independent death rate m directly. However, let us assume that we have no such prior information available. But let us assume that a researcher with a lot of experience with the focal species tells that neither of the parameters f and m should be greater than 10, and that the parameter K should not be greater than 200. Naturally, all these three parameters need to be positive as well. Then one option is to define the prior distribution to correspond to the assumption that $0 \leq f, m \leq 10$, and $0 \leq K \leq 200$, but within these bounds any parameter values are equally likely. This assumption is an example of a uniform prior. We note that there are many other possible choices for the prior than the uniform prior. However, we will not discuss here the issue of the prior distribution in more detail, but refer the interested reader to e.g. Kass and Wassermann (1996). Mathematically, the uniform prior means that $p(\theta) = 1/20000$ if $\theta = (f, m, K)$ is within the above-mentioned bounds, and that $p(\theta) = 0$ if θ is outside those bounds. But where did the number 1/20000 come from? It is the size of the box $[0, 10] \times [0, 10] \times [0, 200]$ which contains all the parameter combinations that we considered possible. Thus, setting $p(\theta) = 1/20000$ and $p(\theta) = 0$ for θ, respectively, being inside and outside the above-mentioned bounds ensures that the integral of $p(\theta)$ over the whole parameter space is 1. This is necessary, as $p(\theta)$ is a probability distribution.

After defining the model and the prior, we are ready to estimate the parameters. In the context of Bayesian inference, the parameter estimates are obtained by computing the posterior distribution (Appendix B.2). The posterior distribution contains the full information about parameter uncertainty, from which summaries such as mean estimates or credibility intervals of parameters can be computed (Appendix B.2). We briefly note that the Bayesian credibility intervals are not identical to the CIs that can be derived for ML estimate: the 95% credibility interval is the range of values to which the true parameter value belongs to with 95% certainty. The process of sampling the posterior distribution is solely a technical step: all biologically relevant considerations are included in the prior $p(\theta)$ and in the likelihood of the data $p(y|\theta)$. For the example considered here, we sampled the posterior using the Metropolis–Hastings algorithm (see Appendix B.2). The simplest way to look at the posterior distribution is to plot it separately for each parameter, as we have done in Figures 3.19B, C, and D for the parameters f, m, and K, respectively. These distributions are called marginal posterior distributions, referring to the fact that the influences of the other parameters have been integrated out when considering each parameter in turn. In these panels, the histograms show the posterior distributions and the lines show the prior distributions. Both of these are probability distributions, and thus their integrals (the areas under the curves) equal 1.

In case of the carrying capacity K, the marginal posterior distribution is much more concentrated around the true value than the prior distribution, and thus the data were informative about this parameter. In contrast, the marginal posterior distributions for the parameters f and m are very wide, meaning that there remains a lot of uncertainty in these parameters. One reason why we failed to estimate these parameters more accurately is that the population dynamics mainly depend on the overall growth rate $r = f - m$ rather than on the individual values of these two parameters. For this reason, the parameters f and m are positively correlated in the posterior distribution, as illustrated by the joint posterior distribution of these two parameters shown in Figure 3.19E. The positive correlation between these two parameters indicates either that both f and m are likely to have a high value or that both of them are likely to have a low value. Thus, by plotting the joint distribution of these two parameters we see that the uncertainty in these parameters is not as high as it appears when plotting them marginally.

One advantage of the Bayesian framework is that it is straightforward to compute posterior distributions not only for the original model parameters, but also for any other parameters that can be derived from the original model parameters. Figure 3.19F shows the posterior distribution of the growth rate parameter r, generated by sampling a parameter combination θ from the posterior, computing the value of $r = f - m$, and resampling new values of θ to generate the distribution of r. This procedure accounts for the correlations among the parameters as it is based on sampling parameter combinations θ, not individual parameters separately. The posterior distribution of r is more tightly concentrated around its true value ($r = 2$) than are the fecundity and mortality parameters around their true values. This is to be expected, as the data are more informative about r than separately about f and m. This discussion also explains why the ML estimate was not successful in estimating the parameters f and m accurately: the likelihood of the data varies essentially with the growth rate r rather than separately with the parameters f and m.

The reader may wonder how it was possible to estimate f and m separately at all, as the underlying deterministic dynamics (the grey lines in Figure 3.18) only depend on the value of the parameter r, and not on f and m separately. The reason here is that the stochastic dynamics depend to some extent also on the separate values of f and m. To see this, consider e.g. the case with $m = 0$ and $r = f$. With these parameter values, there is no mortality if the population is at low density. Thus, if the population fluctuates up and down already at low density rather than increasing monotonously, the data will tell that such a parameter combination is not possible.

Yet, the value of r is much less accurately estimated than the carrying capacity K. This is because actually only the first two data points ($t = 1, 2$) in Figure 3.18B are informative about the growth rate, whereas the remaining eight data points essentially show fluctuations around the carrying capacity. Thus, to obtain more accurate estimates of r, one should acquire more data from a population that is in low density, e.g. by replicating the initial phase of the experiment several times.

The reader may also wonder how Figure 3.19F shows a prior distribution for the parameter r, even if we did not specify one directly. This is because the priors for the model parameters f and m implicitly define a prior also for the derived parameter r. The reader may utilize the material for sums of random variables presented in Appendix A.3 to deduce that the prior of r indeed has the triangle shape shown by the line in Figure 3.19F. This is the distribution of a random variable defined as the difference between two random variables, which both have a uniform distribution in the range from 0 to 10.

3.5.3 Species distribution models

During the past decades there has been a great increase in the use of species distribution models (SDMs), also referred to as ecological niche models (ENMs) or habitat suitability models (HSMs; e.g. Hirzel and Le Lay, 2008). These models aim at explaining and predicting variation in the occurrence or abundance of a species as a function of environmental variables, and thus at describing the species niche in a statistical manner. There is a large array of methods for SDMs, including many parametric and non-parametric alternatives for the functional relationships between environmental and species variables, such as generalized linear models (GLMs), generalized additive models (GAMs), and regression tree models (e.g. Guisan and Zimmermann, 2000; Guisan et al., 2002; Guisan and Thuiller, 2005; Elith et al., 2006; Elith and Leathwick, 2009). The purposes of applying SDMs are usually to identify those environmental variables that relate to species occurrence, and to quantify how those variables are linked to the species occurrence.

In this section we illustrate the use of SDMs by applying a GLM-based approach to data generated by the plant population model of Section 3.3. As illustrated in Figure 3.20, we have divided the study area into a set of squares which have been sampled for the presence–absence of the species. We assume here that we know that the relevant environmental variables are soil fertility and temperature, and thus we omit here the step of model selection during which the environmental variables would be identified. As a starting point for the rich literature on model selection in the context of SDMs, we refer to the literature cited in the previous paragraph.

As a specific type of a SDM, we use probit regression to link the patch occupancy data to the predictor data, thus modelling the presence or absence $y_i \in \{0, 1\}$ of the species at site i as

$$\Pr(y_i = 1) = \Phi(\beta_0 + \beta_1 S_i + \beta_2 T_i + \beta_3 T_i^2), \qquad (3.16)$$

where S_i and T_i are the measured soil fertility and temperature covariates, and β_0, β_1, β_2, and β_3 are the regression coefficients to be estimated. The function Φ is the cumulative distribution function of the standard normal distribution $N(0, 1)$, and it works as the link function of probit regression (see Appendix B.1). We have assumed a linear response to soil fertility and a squared response to temperature to allow the model to capture the occurrence patterns visible in the data: occurrence probability is maximized at an intermediate temperature and it increases monotonically with soil fertility, reflecting the assumptions which we made when generating the data.

The parameter estimates of the fitted models are shown in Table 3.2 and their predictions are illustrated in Figures 3.20C and D. From these panels we see that the models fit the data in the sense that they predict high occurrence probabilities

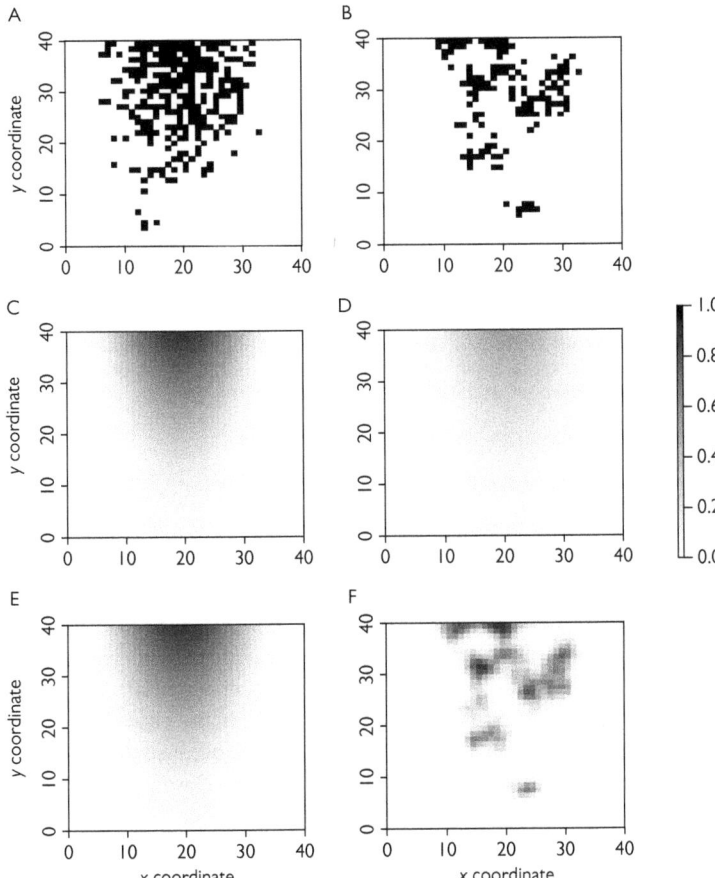

Figure 3.20 Fitting species distribution models without and with a spatially structured residual. (A, B) show the presence–absence data (presence shown in black, absence in white) to which the models are fitted. The other panels show the predicted occupancy probabilities based on the probit regression model of Eq. (3.16) without (C, D) and with (E, F) a spatially correlated residual (with Matérn covariance structure). The data are the same as in Figure 3.8 and have thus been obtained by sampling the plant population model. (A, C, E) correspond to the plant population model with independent mortality, whereas (B, D, F) to the plant population model with mortality caused by spatially correlated disturbances. The models were fitted using Inla-R (Rue et al., 2009).

in regions where the species are present and low occurrence probabilities in regions where the species are absent. One way of quantifying the explanatory power of binary data model is to use Tjur's (2009) R^2, computed as the mean model prediction for those sampling units where the species occurs, minus the mean model prediction for those sampling units where the species does not occur. This gives $R^2 = 0.395$ for

Table 3.2 Parameter estimates of the species distribution models (SDMs) illustrated in Fig. 3.10.

Model	β_1	β_2	β_3
1	0.078 (0.068 ... 0.089)	0.48 (0.41 ... 0.56)	−0.013 (−0.015 ... −0.011)
1*	0.078 (0.068 ... 0.091)	0.48 (0.41 ... 0.57)	−0.013 (−0.015 ... −0.011)
2	0.053 (0.042 ... 0.064)	0.39 (0.3 ... 0.48)	−0.009 (−0.012 ... −0.007)
2*	0.166 (0.084 ... 0.288)	0.90 (0.47 ... 1.47)	−0.026 (−0.041 ... −0.014)

The values shown give the parameter estimates (posterior mean and the 95% central credibility interval) for the parameters of Eq. (3.16), i.e. the effects of soil fertility (β_1), temperature (β_2), and square of temperature (β_3). Models 1 and 2 refer to data generated without and with spatially correlated disturbances, thus corresponding to the left-hand and right-hand panels of Figure 3.20. The models with the asterisk (*) include a spatial autocorrelation structure.

Figure 3.20C and $R^2 = 0.174$ for Figure 3.20D. This is in line with the fact that the data in Figure 3.20A follow a clearer pattern than the data in Figure 3.20B. We recall that the data in Figure 3.20B are noisier because they are influenced by spatially correlated disturbances, which the statistical model of Eq. (3.16) does not account for. As with this example, SDMs applied to real data are likely to include only a subset of those variables that have influenced the distribution and abundance of species (Dormann et al., 2007; Hirzel and Le Lay, 2008; de Knegt et al., 2010), the influences of the missing variables thus remaining unexplained. In many cases it can be difficult to improve the model by adding more variables, either because the underlying processes are not known or because measurements of the relevant covariate (e.g. here the locations and times of the disturbances) are not available.

One way for accounting for unmodelled processes is to use a random effect, which may have spatial or other such structure (see Appendix B.1; McIntire and Fajardo, 2009). In Figures 3.20E and F we have included a spatially structured random effect, more precisely by assuming the Matérn covariance structure (e.g. Minasny and McBratney, 2005). For the data not influenced by spatially correlated disturbances this leads to almost no difference: the prediction shown in Figure 3.20E is almost identical to that shown in Figure 3.20C, and the new Tjur's $R^2 = 0.397$ improves the previous one only little. In contrast, for the data generated with the spatially correlated disturbance, the effect of the spatial term is clearly visible in the prediction (Figure 3.20F vs Figure 3.20D), and the new value of Tjur's $R^2 = 0.499$ is substantially higher than that without the spatial term. Thus, the inclusion of the spatially structured random effect indicated that the original model missed some relevant variables. In other words, the fact that inclusion of the spatial term greatly improves the model fit hints to the statistical modeller that, on top of the modelled influences of soil fertility and temperature, there are also some other processes that have influenced the data. The parameters of the Matérn covariance structure quantify that the spatial scale of random variation is 2.84 spatial units, which is on the same order as the spatial scale of the disturbance we assumed.

Importantly, the inclusion of the spatially correlated residual also changed the inference about the effects of the covariates (temperature and soil fertility) by adding more uncertainty to their estimates (Table 3.2). This is because with spatial autocorrelation the data contains less information about the effects of the covariates than would be the case with independent data points, i.e. the effective number of data points is now smaller. Thus, without inclusion of the spatially correlated residuals our inference about the effects of the covariates was not valid: the data violated the assumption of independence of residuals (see Appendix B.1).

3.5.4 Metapopulation models

In the previous section, we modelled the distribution of a species by dividing a continuous landscape into a regular grid and by modelling the presence–absences of the species on that grid. In highly fragmented landscapes consisting of patches surrounded by unsuitable matrix, it is natural to model patch occupancy directly on the patches themselves rather than dividing the landscape into a computational grid (Hanski, 1999). In this section we fit a stochastic patch occupancy model (SPOM; Hanski, 1994, 1997; Moilanen, 1999) to data generated by the butterfly metapopulation model of Section 3.3. As the name implies, SPOMs do not model variation in the sizes of the local populations, just score the patches as being empty or occupied. While thus far we have modelled either temporal data or spatial data, with the SPOM we consider spatiotemporal data. We thus denote by y_{it} whether patch i is occupied ($y_{it} = 1$) in year t or not ($y_{it} = 0$) and model the dynamics of patch occupancy through two components: colonizations and extinctions. Figure 3.21A illustrates the temporal aspect of the data by showing how the fraction of occupied patches fluctuates over time.

A key feature of spatially realistic SPOMs is that they connect the processes of colonization and extinction to the structure of the landscape, i.e. patch areas and connectivities (Hanski, 1994, 1999; Moilanen, 1999). Because larger patches tend to have larger populations, and because larger populations have a lower probability of going extinct (Hanski, 1999; Lande et al., 2003; Ovaskainen and Meerson, 2010), it is reasonable to assume that the probability of local extinction depends on patch area. We thus model the extinction probability as

$$\mathrm{logit}\big[\Pr(y_{it+1} = 0 | y_{it} = 1)\big] = \beta_1 + \beta_2 A_i^{\gamma_{ex}}, \qquad (3.17)$$

where A_i is the area of patch i, and β_1, β_2, and γ_{ex} are the parameters to be estimated. Note that while with the SDMs we applied probit regression, here we apply logistic regression. As discussed in Appendix B.1, these choices are rather arbitrary.

Empty patches become colonized because migrants originating from the occupied populations move to the empty patch and settle there. Thus, colonization probability of a focal patch is expected to increase with increasing number of occupied patches, especially if those other patches are large (they have large populations), if they are near the focal patch (the migrants are likely to disperse to the focal patch), and if the focal patch is large (the migrants are more likely to find it and to settle there).

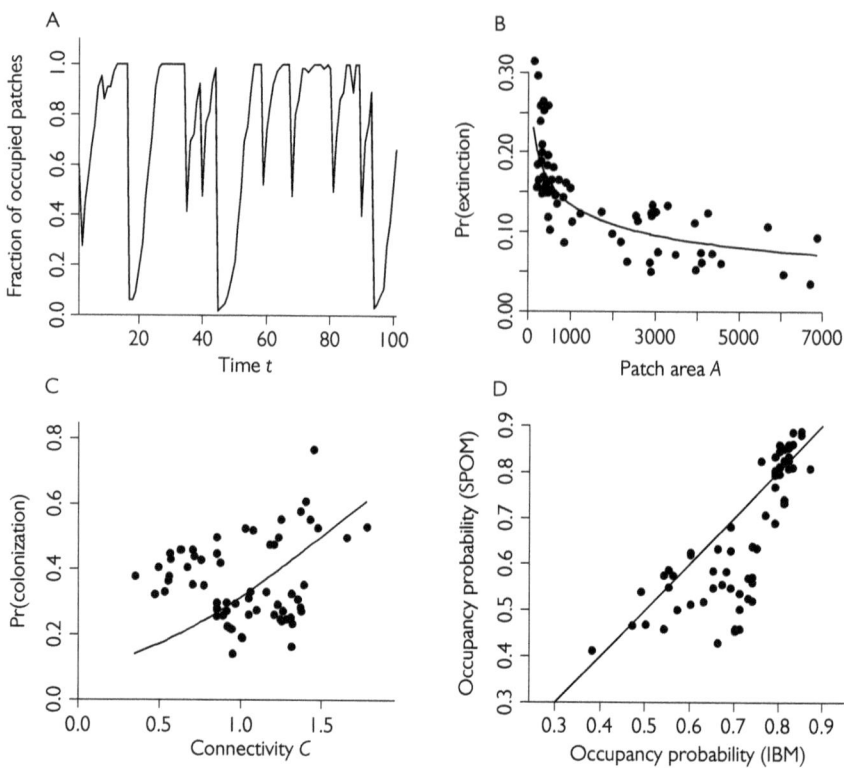

Figure 3.21 A stochastic patch occupancy model (SPOM) fitted to the time-series of patch occupancy data simulated in Figure 3.12A (redrawn in A). The lines in (B, C) show the posterior mean predictions for extinction probability of an occupied patch as a function of patch area (Eq. (3.17)) and the colonization probability of an empty patch as a function of connectivity (Eq. (3.18)). The dots in these panels illustrate the raw data, shown as time-averaged values for each patch, and connectivity set to its posterior mean value. (D) compares the occupancy probabilities of the original model to those predicted by the SPOM, the line depicting the identity $y = x$.

To capture these influences statistically, we model the colonization probability of an empty patch i as

$$\text{logit}\left[\Pr(y_{it+1} = 1 \mid y_{it} = 0)\right] = \beta_3 + \beta_4 C_{it}, \quad (3.18)$$

where we define the population dynamical connectivity of patch i at time t as

$$C_{it} = A_i^{\gamma_{im}} \sum_{j \neq i} A_j^{\gamma_{em}} e^{-\alpha d_{ij}} y_{jt}. \quad (3.19)$$

Here d_{ij} is the distance between the target patch i and the source patch j, and the parameter α models the influence of distance on migration success. The parameters γ_{im} and γ_{em} measure how patch area influences immigration and emigration, respectively. Note that we have included in the sum the term y_{jt} to account for the fact that only occupied patches can send migrants.

We estimated the parameter vector $\theta = (\beta_1, \beta_2, \beta_3, \beta_4, \gamma_{ex}, \gamma_{em}, \gamma_{im}, \alpha)$ using Bayesian inference, sampling the posterior distribution with a Metropolis–Hastings algorithm (see Appendix B.2). The posterior mean estimates (and their 95% credibility intervals) of $\beta_2 = -1.34$ ($-2.9\ldots-0.28$) and $\gamma_{ex} = 0.13$ ($0.07\ldots0.23$) quantify how extinction probability decreases with patch area (Figure 3.21B). The positive estimate of the parameter $\beta_4 = 1.58$ ($1.28\ldots1.87$) indicates that colonization probability increases with population dynamic connectivity (Figure 3.21C). Connectivity increases with the size of the target patch, $\gamma_{im} = 0.43$ ($0.32\ldots0.54$) and its scale is $1/\alpha = 500$ ($250\ldots1250$) spatial units. But connectivity rather decreases than increases with the size of the source population, $\gamma_{em} = -0.84$ ($-1.48\ldots0.22$), which may sound counter-intuitive. The reason here is that while large patches have more potential migrants, an individual migrant is more likely to emigrate from a small patch than from a large patch. With the parameters we assumed in the simulation with which we generated the data, the negative effect of patch area on emigration probability was higher than the positive effect of patch area on population size.

The patch-specific occupancy probabilities by the SPOM correlate with those of the original metapopulation (Figure 3.21D). The match is, however, not perfect. This is to be expected, as the SPOM is only a statistical approximation of the process that generated the data (here, the individual-based butterfly metapopulation model). For example, the SPOM does not account for quality variation among the habitat patches, and it does not account for the spatially correlated nature of environmental stochasticity that we assumed when generating the data, i.e. that all patches were simultaneously influenced by beneficial or adverse conditions.

3.6 Perspectives

In this chapter, we have introduced mathematical and statistical approaches in the context of single-species population ecology. Like in the movement ecology chapter, also here we have focused just on a small subset of the wide range of possible approaches. In particular, we rooted most of the discussion around two models, which we called the plant population model and the butterfly metapopulation model. We derived the plant population model from the logistic population model, whereas the butterfly metapopulation model was derived by combining the Ricker population model with diffusive movements in a patch network. After introducing the baseline models, we showed how they can be gradually extended to incorporate more biological realism, so that they become useful in addressing specific questions, such as how populations respond to habitat loss and fragmentation. We hope to have also demonstrated that one can address the very same question with different model types, and that by doing so one gets a more comprehensive picture of the ecological phenomenon than by applying only one model type.

3.6.1 The invisible choices made during a modelling process

Models are always simplifications of reality. The simplifications made during the modelling process can and should be listed as assumptions that can be expressed verbally. However, often many of the underlying assumptions remain invisible in the sense that they are not explicitly spelled out. Thus, the reader of any modelling study should think critically about which kinds of hidden assumptions the presented model may involve. This is also the case in the present chapter, where we have made a number of explicit and implicit assumptions that may not hold for natural population.

As a first example, in the plant population model we assumed that density dependence acts on mortality, but it can equally well act on fecundity or establishment. North et al. (2011) constructed 12 variants of the heterogeneous-space plant model by assuming that density dependence acts on mortality, fecundity, or establishment, and that habitat quality affects mortality, fecundity, establishment, or carrying capacity. As a second example, in all of our models we have assumed that all individuals are identical, even though demographic heterogeneity can make a big difference (Brännström and Sumpter, 2005). One modelling approach that focuses particularly on the effect of demographic population structure is that of matrix population models (Caswell, 2001). As a third example, the growth rate in all of our models responds instantaneously to changes in density. This assumption can be relaxed, e.g. with the help of delay differential equations (e.g. Kuang, 1993) or by building intermediate states into the model (e.g. propagules first establish as seedlings which later become reproductive adults). As a fourth example, we have simply assumed a predefined carrying capacity, rather than modelling density dependence more mechanistically through the limitation of resources required for survival and reproduction. We will return to this issue in Chapter 4 where we consider multispecies models. As a fifth example, in the butterfly metapopulation model we simplified the landscape to a set of habitat patches embedded in a uniform matrix. Matrix heterogeneity can have important effects on population dynamics, e.g. if the risk of mortality or the movement behaviour of the species differs among different matrix types (Fahrig, 2007). As a sixth example, in the butterfly metapopulation model we considered two extremes of environmental stochasticity: independent dynamics among the patches or globally correlated dynamics. In reality, the type of stochasticity is likely to fall between these two extremes, where correlation in environmental stochasticity experienced by two populations decreases with increasing distance, as was the case with the variant of the plant population model where we assumed that mortality is generated by spatially localized catastrophes.

As tempting as it would be to add these (and many other) extensions into population models, one should always consider whether such increased complexity is necessary to address the questions in mind. Quite often, the simplest models provide the most important insights, as they purify the causal links from mechanisms to their consequences. For example, the butterfly metapopulation model presented in this section is more realistic than the Levins model. But as the Levins model is simpler, it provides clearer insights to those questions that can be addressed by it.

Another kind of a 'hidden' assumption relates to the choice of the mathematical framework that is used to construct the model. We attempted to make such links visible by modelling the same ecological situation with many kinds of models, and by linking those models to each other. In particular, as a starting point for the plant population model, we defined an individual-based spatial and stochastic logistic model in continuous time and continuous space, the model of which was technically a spatiotemporal point process. We then showed how the model can be simplified by excluding the effects of space or stochasticity, or both. The resulting simplified models belonged to other mathematical frameworks, such as Markov processes or differential equations. But also here we could have considered alternatives. For example, our spatial and deterministic version of the model (Eq. (3.7)) was technically an integro-differential equation, where the 'integro'-part refers to the fact that the model involves a convolution and thus an integral over space. Convolution terms model interactions (competition, dispersal) that extend over some finite area of space. Integro-differential equations can be further simplified by replacing convolutions by completely local interactions, which can be modelled by spatial derivatives. If we would have followed this path, we could have simplified Eq. (3.7) into the partial differential equation

$$\frac{\partial N}{\partial t} = rN\left(1 - \frac{K}{N}\right) + D\left(\frac{\partial^2 N}{\partial x^2} + \frac{\partial^2 N}{\partial y^2}\right), \quad (3.20)$$

where $N = N(x, y, t)$ is the population density as a function of space and time. For a comprehensive treatment of spatial ecology with partial differential equations, including models of heterogeneous space, we refer to Cantrell and Cosner (2003).

3.6.2 Some key insights derived from population models

One key insight that we derived from the most simplistic population models is that a local population is expected to go extinct if the death rate exceeds the birth rate, and that a metapopulation is expected to go extinct if the rate of local extinctions exceeds that of colonizations. In deterministic models, the (meta)population either persists forever or goes extinct, whereas in stochastic models the (meta)population either persists for a long but finite time, or goes extinct quickly. From the ecological point of view, the take home message is the same, whether the model is deterministic or stochastic. But where stochasticity and its type can really make a difference is how population dynamical fluctuations, and in particular extinction risk, scales with population size. While demographic stochasticity matters only for small populations, environmental stochasticity is equally influential for small and large populations.

Much of the emphasis in this chapter was in the influence of environmental heterogeneity, which can be of spatial, temporal, or spatiotemporal nature. We assumed that model parameters vary in space or time to examine how variation in environmental conditions influences the dynamics and persistence of populations. In brief, environmental variability matters because it translates into variability in the four processes that determine the growth rate of a local population: births, deaths, immigration, and emigration. As a first approximation, populations are found under conditions that fall within their fundamental niche. However, it is not only the current environmental conditions in the present location that matter, but also the past environmental conditions, as well as the conditions in the nearby locations. This is because populations track environmental variation with a transient, and because individuals move around in space.

As a particular application of heterogeneous-space models, we discussed how habitat loss and fragmentation can increase the rates of deaths and local extinctions, and decrease the rates of births and recolonizations, thus hampering the balance between these, and eventually leading to (meta)population extinction. We also illustrated that populations are not expected to follow changes in environmental conditions immediately, but with time delay, as was the case in the response of the plant population model to fragmentation (Figure 3.15). Further analyses with similar models have shown how the duration of the transient period depends both on the properties of the species and on the magnitude and type of habitat loss and fragmentation (Ovaskainen and Hanski, 2002; Hastings, 2010). One key trait that determines the response of species to fragmentation is their dispersal capability (Henle et al., 2004). In the plant population model, we assumed passive propagule dispersal, implemented through a dispersal kernel. In a highly fragmented landscape, the majority of passively dispersed propagules will land on the unsuitable matrix and thus become wasted. In the butterfly metapopulation model, we assumed that the species utilizes local knowledge of the landscape structure, implemented in the form of edge-mediated behaviour. Active habitat selection helps the individuals to spend a substantial fraction of their time within the habitat patches even in highly fragmented landscapes, and it is thus likely to render the metapopulation less vulnerable to fragmentation effects than is the case with passive propagule dispersal.

3.6.3 The many approaches to analysing population data

In our final section, we illustrated some commonly used statistical approaches to population ecology, including the analysis of temporal, spatial, and spatiotemporal data. Applying the statistical tools allowed us to identify the presence of density dependence and link parameters related to growth rate and carrying capacity to environmental variation.

As is the case with mathematical models, statistical approaches have also many more or less explicit choices and options among which one must choose. For example, in the case of the time-series analysis, instead of fitting the autoregressive models we could have applied autocorrelation functions, partial autocorrelation functions, autoregressive moving average models, and autoregressive integrated moving average models (Powell and Steele, 1995; Metcalfe and Cowpertwait, 2009).

3.6 Perspectives

Further, we could have explored in more detail the lag structure with the help of partial autocorrelation functions (PACF), which can be used to estimate the order of an AR process (Ranta et al., 2006). Similarly, with the species distribution models we could have applied many other kinds of model structures instead of the generalized linear models. As all models are only caricatures of reality, there is often no 'correct' model, and the question is rather whether the model applied is sufficient in extracting ecologically relevant information from the data.

Among the approaches considered here, SDMs are an especially large and rapidly extending field (e.g. Guisan and Zimmermann, 2000; Guisan and Thuiller, 2005; Elith et al., 2006; Elith and Leathwick, 2009; Wisz et al., 2013). SDMs applied to presence–absence data can be viewed as models of the species niche, as they identify environmental conditions under which the species is or is not able to persist, and consequently the conditions under which it is or is not present (Hirzel and Le Lay, 2008). However, for a number of reasons, such as source-sink dynamics (Hanski, 1999; Hirzel and Le Lay, 2008; Stevens and Baguette, 2008), one cannot draw an equality between the species presence and a positive growth rate parameter r. This was exemplified by the analysis of the data generated by a model with localized catastrophes, the data of which involved the species absence in many areas within the fundamental niche. Incorporating a spatially structured residual was helpful in identifying the existence of some spatial process that was not explained by the covariates included in the model.

There has been a recent increase in modelling a species' niche not only phenomenologically using correlation techniques, but also more mechanistically within the framework of biophysical or energy budget modelling (e.g. Kooijman, 2000; Kearney and Porter, 2006; Buckley et al., 2010; Kearney et al., 2010, 2013; Dormann et al., 2012). Such models extend more traditional SDMs by incorporating explicit links between functional traits of organisms and their environments (Kearney and Porter, 2009), e.g. in terms of the exchange of heat, water, and nutrients between the organisms and their environment. As biophysical approaches enable an assessment of the constraints related to survival, development, growth, and reproduction (Kearney et al., 2010), they hold much potential for linking data on population growth rates and species distributions on the underlying ecological mechanisms.

4

Community ecology

4.1 Community assembly shaped by environmental filtering and biotic interactions

In Chapter 2 we discussed why, where, when, and how individual organisms move, and described mathematical and statistical tools for analysing movement processes and patterns. For example, an elephant needs to move in the savannah in order to find resources, such as fresh grass to eat. Yet, individuals are not isolated in nature but they interact with conspecifics. Grass is required not only for our focal elephant but also for its conspecifics, and the competition for the shared resources creates density-dependent effects. We devoted Chapter 3 to scale up from the individual level to study the dynamics of populations. Grass is, in addition, required not only for elephants but for many other African herbivores, too. An assemblage of two or more interacting species that share the same environment at the same time is referred to as an ecological community (Begon et al., 1996). The elephant competes with zebras and many other herbivores for grass and other resources. The overlap in resource use between elephants and other herbivores determines the strength of interspecific competition, making elephants compete less e.g. with giraffes that feed on upper parts of the canopy than with zebras. The grass eaten by the herbivores needs to grow again, and so its own population dynamics are influenced by the herbivore pressure, feeding back to the dynamics of the herbivores. When moving around to find food, mates, and other resources, our focal elephant encounters natural enemies, which may range from large predators to invisible parasites and pathogens. While some of the invisible organisms may parasite or infect the elephant, others may establish neutral or beneficial interactions with it—such as the bacterial communities that take part in grass digestion. More generally, in natural communities, organisms face spatiotemporal variation not only in terms of abiotic conditions, but also in terms of biotic conditions (Wisz et al., 2013): they compete for limiting resources, eat each other, and are connected in many other ways in mutually harmful, beneficial, or neutral relationships (Agrawal et al., 2007).

As illustrated in Figure 4.1, community ecology seeks to identify the underlying mechanisms that create and maintain the compositions and dynamics of species assemblages at different spatial and temporal scales (Vellend, 2010). Abiotic environmental factors, e.g. temperature or rainfall, play an important role in determining

Quantitative Ecology and Evolutionary Biology. Otso Ovaskainen, Henrik Johan de Knegt & Maria del Mar Delgado. © Otso Ovaskainen, Henrik Johan de Knegt & Maria del Mar Delgado 2016. Published 2016 by Oxford University Press. DOI 10.1093/acprof:oso/ 9780198714866.001.0001

4.1 Community assembly shaped by environmental filtering and biotic interactions • 123

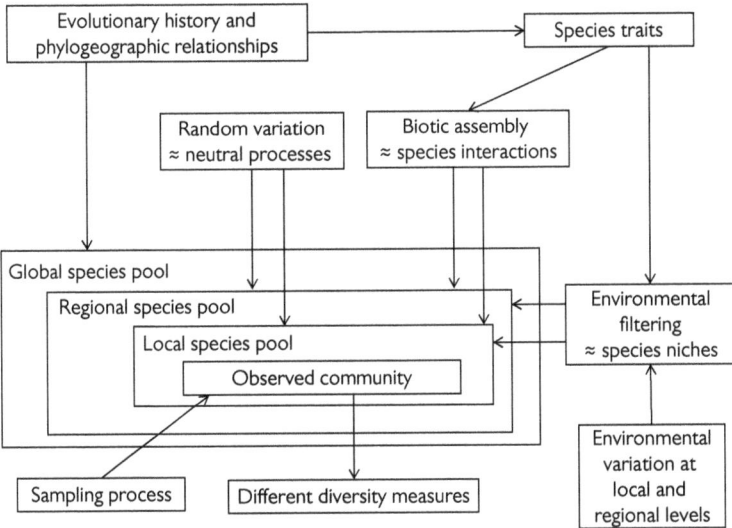

Figure 4.1 The combined effects of biotic and abiotic factors operating at different temporal and spatial scales determine the composition and dynamics of ecological communities. The responses of the species to these factors depend on their traits and other life-history characteristics, which in turn are ultimately shaped by evolutionary history and therefore constrained by phylogenetic relationships.

where species may establish or not, thus constraining their geographic distributions and influencing their local densities (Wright, 2002; Tuomisto et al., 2003; Jabot and Etienne, 2008). But also biotic (intra- and interspecific) interactions are fundamental in influencing species assemblages at a given time and location (Götzenberger et al., 2012). Abiotic factors can modify biotic interactions, e.g. the influence of competition being often more pronounced under harsh environmental conditions. Thus, it is the joint actions of biotic and abiotic factors which ultimately determine the composition and dynamics of ecological communities (e.g. Van de Koppel et al., 2006; Bee et al., 2009). These factors act through species traits related to reproduction and survival, which in turn are shaped by evolutionary history and thus constrained by phylogenetic relationships.

In environments that face repeatable disturbances, such as forests undergoing fires, species composition often evolves in a predictable manner, a phenomenon that is called ecological succession. Communities are also shaped by ecological drift, i.e. solely stochastic events associated to dispersal and demography, as emphasized by the neutral theory of biogeography (Hubbell, 2001; Leibold and Mcpeek, 2006). Due to ecological drift, dissimilar communities can be found in similar environments.

Commonly used characterizations of community structure include various aspects of species diversity, such as species richness (the number of species) and measures based on the relative species abundances, such as the Shannon (H) and Simpson (λ)

indexes (Lande, 1996; Jost, 2007; Anderson et al., 2011; Chiarucci et al., 2011). Measures of species diversity can be computed at various spatial scales, such as within and among communities (Whittaker, 1960; Tuomisto 2010a, b, c). The total species diversity in a landscape (γ diversity) can be partitioned into components measuring the mean species diversity at a local scale (α diversity) and the differentiation among local communities (β diversity). Other common studied aspects of ecological communities involve the spatiotemporal dynamics of species abundances (see Chiarucci et al., 2011, and references therein), the organization of species into trophic networks (e.g. Bascompte and Jordano, 2013), and the identification and analyses of phylogenetic relationships within and among communities (e.g. Ives and Helmus, 2010; Wisz et al., 2013).

4.1.1 Ecological interactions

A long-term interaction between two species is called symbiosis (Douglas, 1994). For example, plants provide carbon for mycorrhizal fungi, and in return the fungi provide phosphorous and nitrogen for the plants. This is an example of a mutualistic interaction, in which the species offer to each other resources that are inexpensive for them to produce, but that would be more expensive or even impossible to produce for the other partner. In parasitic interactions, the parasite benefits whereas the host is harmed. Also in predator–prey interaction the predator benefits and the prey is harmed, but unlike parasitism, this is a short-term interaction. An example of mutually antagonistic interaction is competition, in which two species compete for the same limited resource such as food, shelter, physical space, or sunlight. A further type of interaction is commensalism, in which one of the species benefits from the interaction but its influence is apparently neutral for the other species.

Predator–prey and host–parasite interactions can have a strong influence on the nature of population dynamical fluctuations (Lambin et al., 2002; Wisz et al., 2013). For example, population cycles are commonly thought to arise from both parasite-host (May and Hassell, 1981; Colin and Dam, 2007) and predator–prey relationships (Korpimaki and Norrdahl, 1989; Gilg et al., 2006), because both of these can introduce a time delay in the regulation of population growth (Korpela et al., 2014). Functional response describes how the rate at which a predator consumes prey depends on prey density, whereas numerical response describes how the growth rate of a predator population depends on prey density (Solomon, 1949; Holling, 1959). The forms of functional and numerical responses depend on the behaviour of the predators and the prey, as predators may e.g. adjust their search activity as a function of prey density (Zemek and Nachman, 1999). They further depend on the presence of alternative prey species, on the predator's preferences for the different prey species, and on the predator's interactions with other predators (Abrams and Ginzburg, 2000). Numerical and functional responses can also depend on plasticity in development, as predators may e.g. grow faster at high prey densities (Wang et al., 2009). The combined influences of such factors on predator's response to prey density or a parasite's response to host density has been called the total response (e.g. Gilg et al., 2006, and references therein). Knowledge about the shapes of numerical and functional responses is essential for understanding the dynamics and persistence of interacting populations. However, their measurement is challenging, because making direct observations can be difficult and because measurements in laboratory systems may not translate to natural conditions (De Roos and Persson, 2001; Hjelm and Persson, 2001; Bergström et al., 2006).

In Chapter 3, we constructed models with density-dependence generated by intraspecific competition. Also interspecific competition can lead to density regulation: if the density of a species increases, it imposes a stronger competitive effect on the species with which it competes, whose population density consequently decreases. The outcome of such competitive interactions depend on the relative strengths of intra- and interspecific competitive abilities of the two species. In the case of neutral dynamics, the two species are assumed to be identical, and consequently only their total density is regulated by competition. If the two species are otherwise identical but one has a greater competitive ability, the weaker competitor will eventually be excluded from the system, a phenomenon called competitive exclusion (Hardin, 1959; Armstrong and McGehee, 1980). But as different species generally have dissimilar ecological niches, interspecific competition tends to be less strong than intraspecific competition, allowing the species to coexist. From an evolutionary point of view, competition is a central force driving niche separation, specialization, and speciation (but see Tobias et al., 2014 for further discussions).

4.1.2 Fundamental and realized niches and environmental filtering

The concept of a niche refers to all factors required for a species to exist. It describes the abiotic (e.g. temperature, humidity, pH, soil, sunlight) and biotic (e.g. food resources) environment that the individuals need for their growth, survival, and reproduction (Colwell and Rangel, 2009). Individuals, however, are not found from all parts of the environment that fall within the species' niche, and sometimes individuals may be found from locations outside the niche. The concept of fundamental niche refers to those conditions under which a species is in principle capable of persisting, whereas the concept of realized niche refers to those conditions under which populations of the species are actually found. The mismatch between the fundamental and realized niches can be e.g. due to dispersal limitation (individuals not arriving to areas that are within the fundamental niche), source-sink dynamics (individuals produced by a source population dispersing to areas that are outside the fundamental niche), or antagonistic biotic interactions (e.g. a population going extinct due to predation from an area where it could persist in the absence of a predator).

The match between the abiotic environmental conditions and species niche can be considered as the first filter that determines which species may be present in a certain location. In a spatially heterogeneous environment, environmental filtering creates spatial variation in community composition (Leibold, 1995; Kraft et al., 2015). However, as discussed earlier, the realized species communities found at any given locality are also shaped by biotic interactions (Gotelli, 1999; Wisz et al., 2013), dispersal, and pure chance events (Figure 4.1). For example, many invasive species have rapidly colonized large areas in other continents. Thus, the conditions within those newly colonized areas are within their niche, but the niche was not realized before the colonization.

While identical species cannot coexist in the long-term, species that originally share a common niche (e.g. after a speciation event) can evolve to coexist through a process termed niche differentiation (Leibold and Mcpeek, 2006). During niche differentiation, each species becomes specialized to some of the resources within its original niche, and thus a smaller niche is formed. The evolutionary process that creates niche differentiation among closely related species has been called character displacement (but see discussions in Tobias et al., 2014), and it relates to adaptive radiations during which an ancestral species develops into a set of related species that specialize to use different kinds of resources.

4.1.3 Organizational frameworks for metacommunity ecology

Local communities are not closed and isolated from each other, but they are connected by dispersal. Just as the collection of interacting local populations is called a metapopulation, the collection of interacting local communities is called a metacommunity (Leibold et al., 2004). A metacommunity is more than a collection of independent local communities, as interactions among nearby communities alter their local dynamics, and conversely the dynamics of the local communities influence the properties of regional communities.

As reviewed by Leibold et al. (2004) and Logue et al. (2011), processes generating metacommunity dynamics can be classified into four paradigms, called the species-sorting, mass-effect, patch-dynamics, and neutral paradigms. The species-sorting paradigm assumes a heterogeneous environment and the dominance of environmental filtering and niche differentiation. In this case, species compositions and relative species abundances within local communities are simply determined by the matches between the species niches and the environmental conditions (Shmida and Wilson, 1985; Leibold, 1998; Cottenie et al., 2003). Thus, this paradigm assumes a high enough rate of dispersal to allow the species to reach those localities that fall within their niches, but a low enough rate of dispersal so that the local communities are not much influenced by immigrants arriving from elsewhere. In contrast, the mass-effect paradigm (Mouquet and Loreau, 2002, 2003) assumes a high level of dispersal such that the local abundances of the species are much influenced also by the environmental conditions in other localities, as is the case with source-sink dynamics. While the species-sorting and mass-effect paradigms focus on heterogeneous environments, the patch-dynamics perspective views the landscape as a collection of identical habitat patches that undergo perturbations. In such a case, regional coexistence of species can be obtained e.g. by a trade-off between competitive ability and colonization ability (Levins and Culver, 1971; Hastings, 1980; Tilman, 1994), which can lead to predictable successional dynamics. Finally, the neutral paradigm assumes that all species have identical niches (Hubbell, 2001; Rosindell et al., 2012), and consequently that

4.1 Community assembly shaped by environmental filtering and biotic interactions • **127**

metacommunities are structured by ecological drift only. While many ecologists may not consider the neutral model as a realistic description of real metacommunities, it serves as a useful null model. For example, some community-level patterns, such as the species–area relationship and species–abundance relationship, can be generated by the neutral model (He and Legendre, 2002; Rosindell and Cornell, 2009), suggesting that such patterns may not carry a strong signal of the underlying niche-based processes.

4.1.4 The outline of this chapter

In this chapter, we extend the single-species perspective of Chapter 3 to ecological phenomena involving two or more interacting species. As an empirical example that has stimulated the development of some of the models of this chapter, we refer to David Tilman's classical experiments on plant communities, illustrated in Box 4.1 and in Figure 4.2.

Following the structure of the previous chapter, we first assume in Section 4.2 that the ecological community inhabits a homogeneous environment. We model

---

Box 4.1 An empirical example of community ecological research

David Tilman and colleagues conducted a large number of empirical (e.g. Wedin and Tilman, 1993; Tilman et al., 1996; Fargione et al., 2003) and theoretical (Tilman et al., 1994; Tilman, 1994) studies on competitive communities, focusing e.g. on the links between biodiversity and ecosystem processes (e.g. Tilman et al., 1996). To illustrate competitive mechanisms, we review here the experimental study of Fargione et al. (2003), which examined the roles of neutral versus niche-based processes in community assembly. In 1994, Fargione et al. (2003) established a set of experimental plots by sowing seeds of 24 species of perennial grassland plants, the number of species planted to each plot varying from 1 to 24. The actual experiment started 3 years later in 1997, when seeds of 27 other perennial grassland plant species were introduced to the plots. After 2 more years, the plots were surveyed to examine whether the newly introduced species had established.

The results of this experiment demonstrated that the success of the newly introduced species (whether measured by species diversity, biomass, or vegetation cover) declined with the number of species originally sowed to the plots (Figure 4.2). This was the case because the resident species communities had depleted resources (e.g. soil nitrate and water, and availability of bare ground and light) to a level that prevented the establishment of many of the newly introduced species. To link the results to species traits, Fargione et al. (2003) classified the species to four groups called C3 grasses, C4 grasses, forbs, and legumes. These groups differ in their ability to utilize resources (e.g. Wedin and Tilman, 1993). For example, while the C4 grasses are generally the strongest competitors for nitrogen, C3 grasses can access nitrogen early in the season, forbs from deeper soil layers, and legumes from the air. The experimental results demonstrated that these niche-based differences were important for community assembly: the establishment success of the newly introduced species was poorest if other species of the same group were present already in the resident community (Fargione et al., 2003).

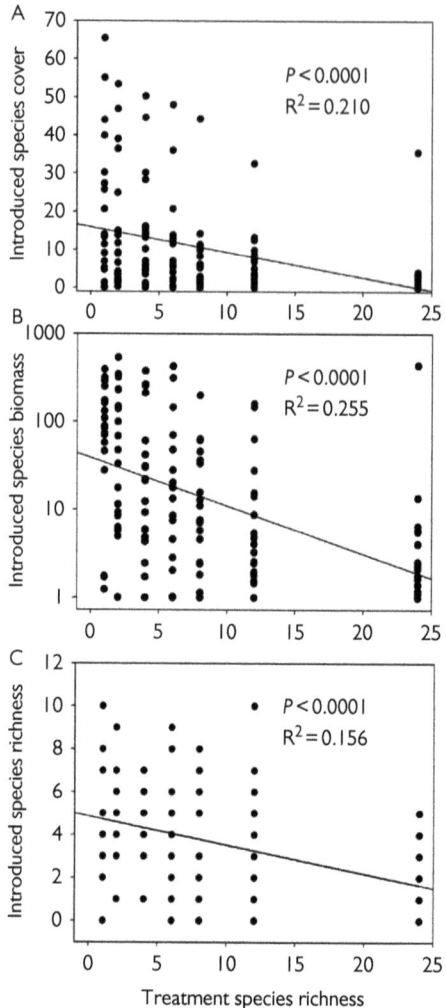

Figure 4.2 Experimental results by Fargione et al. (2003). In all panels, the x-axis shows species richness in the resident community, which varies from 1 to 24 according to how many plant species were originally sowed to the plot. The y-axis shows the invasion success of 27 other plant species that were introduced to the plots 4 years later, measured either by cover (A, unit percent), biomass (B, unit g/m^2), or species richness (C). The lines show linear regressions. Reproduced with permission from US National Academy of Sciences.

the dynamics of three types of ecological interactions, namely competitive, resource–consumer, and predator–prey interactions. We then move to heterogeneous environments in Section 4.3, where we extend the plant population model of Section 3.3 into a plant community model, in which different plant species compete for shared resources. We thus consider how environmental filtering and species interactions jointly influence the assembly and dynamics of communities. In Section 4.4 we utilize the plant community model to analyse how species communities respond to habitat loss, thus extending the single-species viewpoint of Section 3.4 again. Finally, in Section 4.5 we discuss statistical tools for the analysis of various kinds of data that can be acquired for species communities, and illustrate the interplay between theory and data by applying the statistical tools to data generated by the models introduced in Sections 4.2–4.4.

While in this chapter we move a step forward by considering how ecological interactions influence the dynamics and persistence of interacting species, we still ignore evolutionary processes—this is because we will study them in the final Chapter 5.

4.2 Community models in homogeneous environments

In Section 3.2, we considered the single-species model of logistic population growth, with model parameters relating to fecundity, establishment, and mortality. As discussed in Section 4.1, these parameters are influenced by the interactions between the focal species and other species (Vellend, 2010). Here we will modify the single-species model of Section 3.2 to study ecological phenomena involving two or more species.

There are three main types of interactions. (i) Predator–prey, host–parasite, and consumer–resource interactions increase the growth rate of one species and decrease that of the other species. (ii) Competitive interactions decrease the growth rates of both species. (iii) Mutualistic interactions increase the growth rates of both species. Out of these interaction types, we will construct a competition model, a resource–consumer model, and a predator–prey model. Instead of building the four model variants considered in Section 3.2 (stochastic and spatial individual-based model, non-spatial stochastic model, deterministic spatial model, and non-spatial and deterministic mean-field model), we consider here the two extremes only, i.e. the individual-based model (IBM) and the mean-field model (MFM). Further, we simply write down the corresponding MFMs without detailing how they were derived from the IBMs. A reader interested in deriving the mean-field models may follow the recipe introduced in Section 3.2.

4.2.1 Competitive interactions

Species with overlapping ecological niches compete for the resources that they need for growth, survival, and reproduction, e.g. nutrients, habitats, or territories. Competition can inhibit population growth in many ways, and the effect of this ecological interaction will depend on the competitive abilities of the two species (see Box 4.1 and Figure 4.2 for an empirical example). The principle of competitive exclusion (Hardin, 1959; Armstrong and McGehee, 1980) states that a stronger competitor, i.e. a species whose members are more efficient at finding or exploiting resources, will drive a weaker competitor to extinction. While simple models indeed make such a prediction, more detailed models have pointed out circumstances in which competition decreases population sizes but the species are able to coexist (Schoener, 1973, 1976; Amarasekare, 2003).

We start by considering a two-species competition model in a homogeneous environment, with the aim of exploring conditions that allow or prohibit species to coexist. To do so, we extend the single-species plant population model of Section 3.2 into a model of two competing species, called species 1 and species 2. The model involves both intra-specific (within-species) competitive interactions and inter-specific (between species) competitive interactions. In the general case, both species may have their own fecundity (f_1 and f_2), establishment (e_1 and e_2), and mortality (m_1 and m_2) parameters. As with the single-species model, we assume that the death rates of both species increase linearly with the local density of their conspecifics, parameterized by the competition kernel $C_{11}(d)$ for species 1 and $C_{22}(d)$ for species 2. Additionally, the dynamics of the two species are coupled by assuming that the species also impose competitive effects to each other: the effect of an individual of species 1 on the death rate of an individual of species 2 is parameterized by the competition kernel $C_{12}(d)$, and conversely $C_{21}(d)$ measures the effect of species 2 on species 1.

With the addition of interspecific competition, the single-species mean-field model of Section 3.2 is modified to the two-species model

$$\begin{cases} \dfrac{dN_1}{dt} = f_1 e_1 N_1 - N_1 (m_1 + c_{11} N_1 + c_{21} N_2), \\ \dfrac{dN_2}{dt} = f_2 e_2 N_2 - N_2 (m_2 + c_{12} N_1 + c_{22} N_2), \end{cases} \quad (4.1)$$

where the competition parameters c_{11}, c_{12}, c_{21}, and c_{22} are integrals over all space of the corresponding competition kernels C_{11}, C_{12}, C_{21}, and C_{22}. As discussed in Appendix A.2, Eq. (4.1) is a system of differential equations. Due to the second-order terms (N_1^2, N_2^2 and N_1, N_2), it is a non-linear system of differential equations, and thus the time-dependent solution cannot be found analytically using the methods of Appendix A. However, as illustrated later, it is possible to analyse the qualitative behaviour of this model analytically, as well as to simulate it numerically.

We will use the two-species competition model described in the previous paragraph to exemplify some key aspects of competitive interactions. Rather than examining the full parameter space, we simplify by assuming that the two species are identical in the sense that they have equal fecundity f, establishment e, and mortality m parameters. We further assume that the within-species competition rates (c_{11} and c_{22}) are identical and denote them by c.

Let us first assume asymmetric competition. We assume that the species 1 is a stronger competitor than species 2, so that the competitive effect of species 1 on species 2 (c_{12}) is larger than the effect of species 2 on species 1 (c_{21}). The dynamics generated under this assumption are shown in Figure 4.3. Even though we have assumed that species 2 has initially higher density than species 1, eventually species 1 takes over due to its better competitive ability (Figure 4.3A). Figure 4.3B shows the dynamics as a phase-space plot. In this panel, the arrows show in which direction

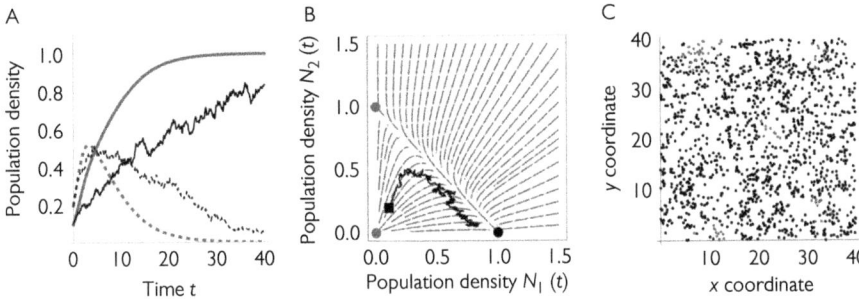

Figure 4.3 Population dynamics of two competing species. (A) shows how the densities of the two species evolve over time, either based on the mean-field model (grey lines) or on a single realization of the individual-based stochastic and spatial model (black lines). The two species are assumed to be otherwise identical except that species 1 (shown by continuous lines) is a stronger competitor than species 2 (shown by dashed lines). (B) shows the same dynamics in the phase-space, where the density of species 1 is shown on the x-axis and the density of species 2 is shown on the y-axis. The arrows show the dynamics and the dots the equilibria predicted by the mean-field model. The black dot shows the stable equilibrium in which only species 1 is present, whereas the grey dots show two unstable equilibria. The black line shows the trajectory of the individual-based simulation of (A), the black square showing the initial point. (C) shows the distribution of individuals at the end of the simulation, with species 1 shown by black and species 2 by grey dots. Parameter values for both species are the same as in the simulation of the single-species model shown in Figure 3.3, except that the initial density of the species 2 is set to 0.2. The competition parameter are $c_{11} = c_{22} = 1$, $c_{12} = 1.2$, $c_{21} = 0.8$.

the mean-field model evolves if starting from a given initial condition. Following the flow of the arrows shows that the system will eventually reach the equilibrium shown by the black dot, in which species 2 is absent and species 1 is at its equilibrium density $N_1^* = 1$. Thus, in this model species 1 always excludes species 2. The black line in Figure 4.3B depicts the simulated trajectory of the IBM in the phase-space. As expected, the dynamics of the IBM are approximately but not exactly captured by the MFM, for the same reasons as discussed in the single-species context in Section 3.2. At the end of the simulation, species 2 is present only in a few remnant populations (Figure 4.3C), whereas species 1 is abundant.

Let us then assume that also between-species competition parameters are identical. We denote by α the ratio of between-species to within-species competition, so that $c_{11} = c_{22} = c$ and $c_{12} = c_{21} = \alpha c$. With this simplification, the model reads as

$$\begin{cases} \dfrac{dN_1}{dt} = feN_1 - N_1\left[m + c\left(N_1 + \alpha N_2\right)\right], \\ \dfrac{dN_2}{dt} = feN_2 - N_2\left[m + c\left(N_2 + \alpha N_1\right)\right]. \end{cases} \quad (4.2)$$

This model is neutral (Hubbell, 2001; Rosindell and Cornell, 2009; Haegeman and Loreau, 2011; Rosindell et al., 2012) in the sense that both species have identical

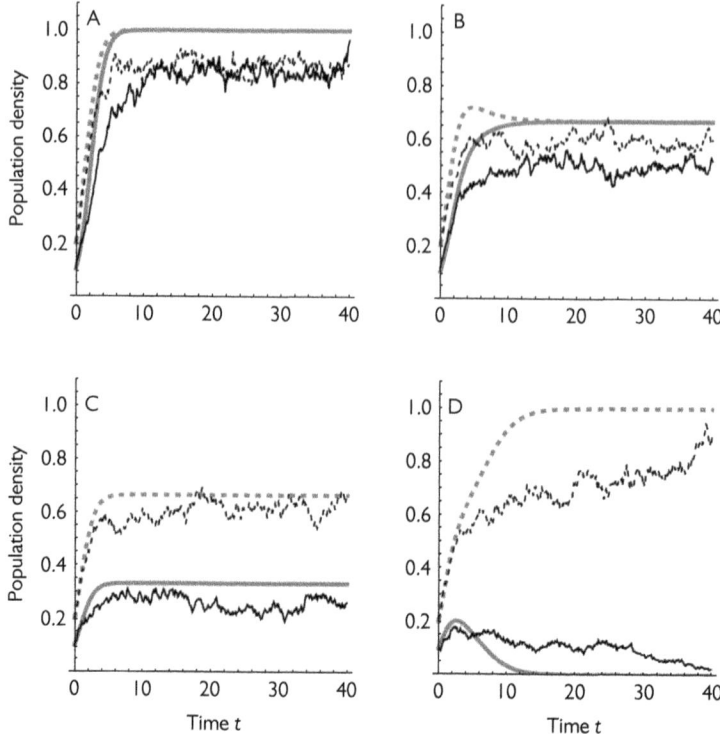

Figure 4.4 Time-series of population dynamics of two identical species, with different levels of interspecific competition. The symbols and parameter values are the same as in Figure 4.3A, except that the competition rates are $c_{11} = c_{22} = 1$, $c_{12} = c_{21} = \alpha$, where $\alpha = 0$ (A), $\alpha = 0.5$ (B), $\alpha = 1$ (C), or $\alpha = 1.5$ (D).

parameter values, but it is not neutral in the sense that the individuals still distinguish between conspecifics and heterospecifics. If $\alpha < 1$, the individuals compete more with their conspecifics than with their heterospecifics, while for $\alpha > 1$ they compete more with their heterospecifics than with their conspecifics. Figure. 4.4 illustrates how the dynamics of this model evolve over time, Figure 4.5 shows the same dynamics in the phase-space, and Figure 4.6 displays the spatial distribution of individuals in the end of the simulation. In all three figures, the four panels correspond to different levels of interspecific competition: $\alpha = 0$ (A), $\alpha = 0.5$ (B), to $\alpha = 1$ (C), and $\alpha = 1.5$ (D).

If $\alpha = 0$, the species follow their dynamics independently of each other, and thus the outcome of this model is identical to running the separately two realizations of the single-species model. We recall from Section 3.2 that the single-species model leads to the equilibrium density $N^* = (fe - m)/c$. Thus, the two-species model has as its equilibrium state $(N_1^*, N_2^*) = (N^*, N^*)$, as illustrated in Figures 4.4A and 4.5A. In Figure 4.6A the spatial distributions of the individuals are aggregated within the species (as is the case in Figure 3.3B), but they are independent of each other.

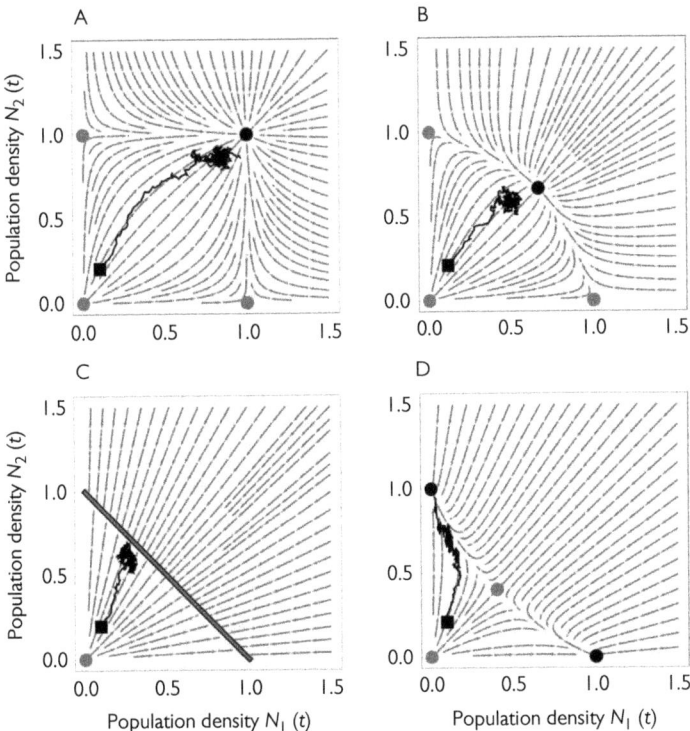

Figure 4.5 Phase-space plots of population dynamics of two identical species, with different levels of interspecific competition. The parameter values in (A–D) equal those in Figures 4.4A–D. The symbols are as described in the legend to Figure 4.3B. In (C), the thick grey line shows a continuum of equilibria. These equilibria are stable in the sense that all trajectories approach the line (hence the black outer line), but unstable in the sense that within the line the dynamics are neutral (hence the grey filling).

Adding a competitive effect between the species increases the mortality rates and thus lowers the equilibrium densities of both species. If $\alpha < 1$, the level of interspecific competition is lower than that of intraspecific competition, and thus the two species still coexist. The equilibrium density is reduced to $N_1^* = N_2^* = (fe - m)/c(1 + \alpha)$, as exemplified by comparing panels A and B of Figures 4.4–4.6. The spatial distribution of the individuals shows a negative correlation, so that individuals of species 1 are at low density in locations where individuals of species 2 are at high density, and vice versa (Figure 4.6B).

If the level of interspecific competition is higher than that of intraspecific competition ($\alpha > 1$), the two species cannot coexist (Schoener, 1973, 1976). Consequently, in Figure 4.5D the equilibrium corresponding to coexistence is unstable, and the system drifts to a situation in which either of the species becomes extinct. After the extinction of one species, the competitive pressure on the other species is released, and

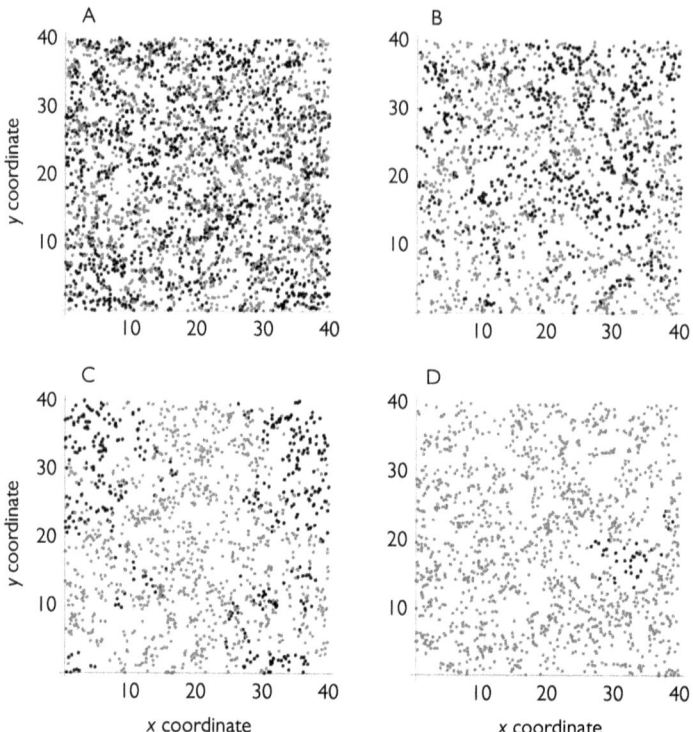

Figure 4.6 Spatial distributions of two identical species, with different levels of interspecific competition. The parameter values in (A–D) equal those in Figures 4.4A–D. The symbols are as described in the legend to Figure 4.3C.

thus the system converges to the equilibrium state of the single-species model. In the deterministic mean-field model the winning species is determined solely by which species is more abundant initially. In a stochastic model, the species that is initially less abundant is more likely to go extinct, but due to demographic stochasticity also the species which is initially more abundant may go extinct.

A peculiar special case is obtained with the parameter value $\alpha = 1$, in which case the individuals compete equally much with their conspecifics and heterospecifics, and thus the model is fully neutral (Hubbell, 2001; Haegeman and Loreau, 2011). In this case, we may sum the two equations of Eq. (4.2) together, yielding

$$\frac{dN}{dt} = feN - N(m + cN), \qquad (4.3)$$

where $N = N_1 + N_2$ is the total number of individuals. Equation (4.3) coincides with the mean-field model of Section 3.2, and thus now the collection of all individuals follows single-species dynamics. Consequently, the equilibrium state for the total number of individuals is $N^* = (fe - m)/c$. In the two-species model, there is a continuum of equilibria, shown by the line in Figure 4.5C. Here the densities of both species can be anything in the range $(0, N^*)$, as long as their sum is $N_1^* + N_2^* = N^*$. In the mean-field model, there is no push to any direction along the continuum of equilibria, and so the balance between N_1^* and N_2^* is neutral. In the stochastic model, demographic stochasticity makes the proportions of individuals belonging to each species perform a random walk.

4.2.2 Resource–consumer interactions

The single-species model of Section 3.2, as well as the competition model described earlier, assume that the mortality rate of a focal individual increases with the density of the other individuals. However, these models do not explain why this is the case. As discussed in Section 4.1, density-dependence may be due to direct interference competition among conspecifics, in which case it is natural to assume (as we did) that local density per se influences death rate. But density-dependence may equally well be an indirect effect due to interactions between the species and its abiotic or biotic environment, mediated through agents such as parasites or predators, or the availability of food resources or vacant territories (Volkov et al., 2005; Chisholm and Muller-Landau, 2011; Johnson et al., 2014).

Consumer–resource interactions are central to ecological food web dynamics, and they include e.g. prey–predator, host–parasite, and plant–herbivore systems (Murdoch et al., 2003). We start with a simple model in which we assume that new resources are introduced in the system at a constant rate rather than being produced by the present resource population. Thus, this model is most easily interpreted in terms of abiotic resources rather than biotic resources that follow their own population dynamics.

We assume that new resources appear in the system at birth rate b, whereas existing resources are subject to decay at death rate d. Thus, without the presence of the consumer, the mean-field model for the resources (the density of which we denote by R) reads $dR(t)/dt = b - dR(t)$. The reader can easily verify that in this model the resources reach the equilibrium density $R^* = b/d$. In the corresponding IBM, the resource units are discrete entities. Their expected density (number per unit area) in the stationary state is R^*, and their stationary distribution follows complete spatial randomness.

Let us then add a consumer into the system, and denote its density by N. To keep the IBM simple, we assume that the consumer is immobile, but that its ability to consume a resource unit depends on its distance to it. We denote the rate at which a consumer consumes a resource unit located at distance d from it by $\Lambda(d)$, and call this the consumption kernel. In the general case, consumers consume resources because they need them for their survival and reproduction. To make the model as simple as possible, we assume here that resources are needed only for reproduction. We make further the highly simplifying assumption that a consumer can produce an offspring only if it consumes a resource unit. We denote by e the conversion efficiency of resources, i.e. the probability by which a consumer produces a new consumer following a consumption of a single resource unit. The newly produced offspring is then dispersed according to a dispersal kernel $D(d)$ to a location where it establishes as a new consumer. Unlike in the single-species model of Section 3.2, we don't assume any explicit form of density-dependence, and thus the mortality rate m of the consumers is independent of their density.

The MFM that corresponds to the IBM described in the previous paragraphs is given by

$$\begin{cases} \dfrac{dR(t)}{dt} = b - dR(t) - \alpha R(t)N(t), \\ \dfrac{dN(t)}{dt} = e\alpha R(t)N(t) - mN(t). \end{cases} \quad (4.4)$$

In this equation, the consumption rate α is defined as the integral of the consumption kernel Λ over all space. Like was the case with the single-species model of Section 3.2, the dispersal kernel does not appear in the equation. This is because in the MFM the spatial location is irrelevant. As the dispersal kernel is assumed to integrate to unity over all space, it influences only where new individuals are born, not the rate at which they are born.

Due to the presence of the product term $R(t)N(t)$, Eq. (4.4) is a non-linear system of differential equations, making it again difficult to analyse its time-dependent solutions analytically. However, it is possible to find equilibrium solutions (R^*, N^*) by setting the derivatives $dR(t)/dt$ and $dN(t)/dt$ to 0. It is easy to verify that there are two equilibrium states. One equilibrium state is the consumer-free equilibrium with $R^* = b/d$ and $N^* = 0$, in which the resources stabilize to their equilibrium level set by the balance between their renewal and decay rates. The more interesting equilibrium state is

$$\begin{cases} R^* = \dfrac{m}{\alpha e}, \\ N^* = \dfrac{be}{m} - \dfrac{d}{\alpha}, \end{cases} \quad (4.5)$$

where the consumer reaches an equilibrium density set by the balance between its death and fecundity rates, the latter controlled by resource availability. As can be expected, consumer density increases with the consumption rate α, the conversion efficiency e and the renewal rate of the resources b, whereas it decreases with the decay rate of the resources d and the mortality rate of the consumers m.

In particular, Eq. (4.5) implies that the consumer can persist ($N^* > 0$) if and only if

$$\dfrac{be}{m} > \dfrac{d}{\alpha}. \quad (4.6)$$

If the threshold condition of Eq. (4.6) is not satisfied, the system has only one ecologically relevant equilibrium point, which corresponds to the extinction of the consumer. This is the case in the simulations shown by the upmost row of panels in Figure 4.7. As we have initiated the resources at a high density, the consumer population increases initially. However, it soon starts to decline, because the rate of resource renewal is not high enough to sustain the consumer population.

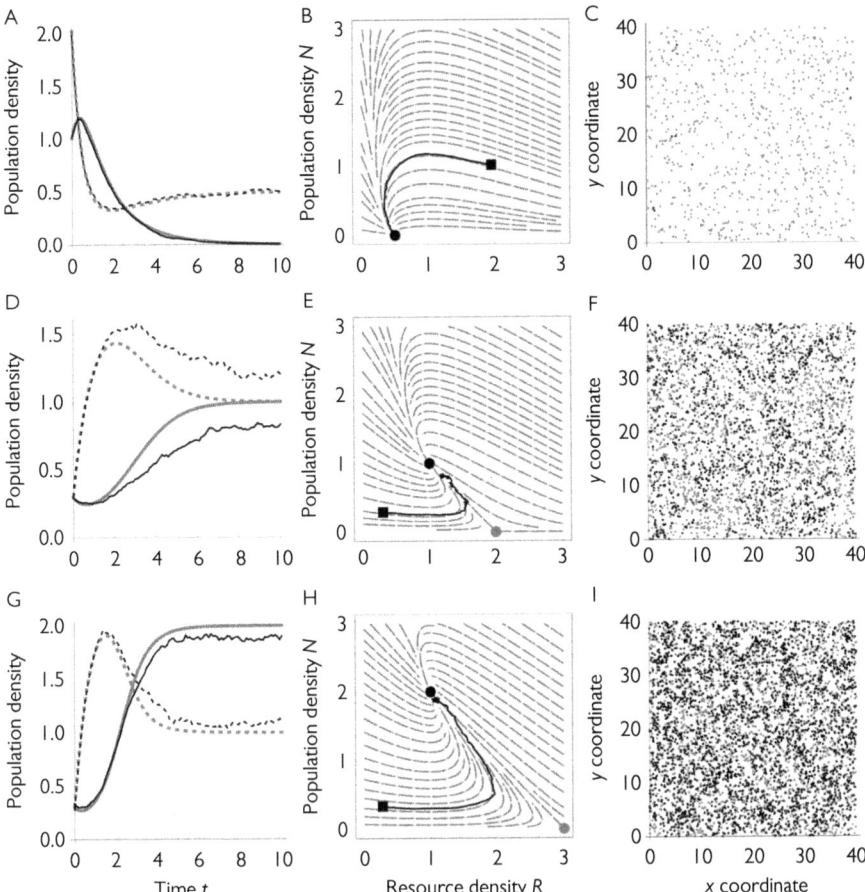

Figure 4.7 Resource–consumer dynamics with different levels of resource renewal rate. The left-hand panels (A, D, G) show how the densities of resources (dashed lines) and consumers (continuous lines) evolve over time, based on either the mean-field model (grey lines) or a single realization of the individual-based stochastic and spatial model (black lines). The middle panels (B, E, H) show the same dynamics in the phase-space, where the density of the resource is shown on the x-axis and the density of the consumer on the y-axis. The symbols are as described in the legend to Figure 4.3B. The right-hand panels (C, F, I) show the distributions of resource units (grey dots) and consumers (black dots) at the end of the simulations. The resource renewal rate is set to $b = 0.5$ (A–C), $b = 2$ (D–F), or $b = 3$ (G–I). Initial densities of both resources and consumers are set to 0.2, except in the upper panels (A–C), where the initial density of resources is 2 and the initial density of consumers is 1. Other parameter values set to $d = \alpha = e = m = 1$.

The threshold condition for coexistence (Eq. (4.6)) holds in the middle row of panels in Figure 4.7. As we have initiated the system at a low density of resources and consumers, and as the birth rate of the consumers depends on the product between these two densities, it is initially low. The low density of consumers enables the resources to grow at a fast rate. Consequently, the resources reach a high abundance during the transient phase, before they start to be depleted by the growing consumer population.

In the lowest row of panels in Figure 4.7 we have further increased the resource renewal rate, allowing the consumer to reach a higher abundance. Interestingly, in this model the equilibrium density of resources does not depend on the renewal or decay rates of the resources (Eq. (4.5)), which is the same in the middle and lowest rows of panels in Figure 4.7. Thus, the increased resource renewal rate is translated into a higher density of consumers, not of resources. By Eq. (4.5), the equilibrium density of resources decreases with factors that make the consumer more abundant and more effective: increasing consumption rate, increasing establishment probability, and decreasing mortality rate.

Note that we have not assumed explicitly any density-dependence for the consumer, but still its density does not increase exponentially, as would be the case for the single-species model without density-dependence. This is because the reproduction rate of the consumer depends on the availability of resources, as illustrated by the phase-space plots (Figures 4.7B, E, H). If the density of consumers is low and that of resources is high, the consumers starts to grow exponentially. As the density of consumers increases, they deplete resources at an increasing rate, which then lowers the birth rate of the consumers. Thus, in this model density-dependence operates through the birth rate, which decreases with increasing consumer density due to the interactions between the consumers and the resources. Equally well, we could have assumed that resources are needed for survival, in which case density-dependence would have operated through the consumer death rate.

4.2.3 Predator–prey interactions

The predator–prey model that we next construct is identical to the resource–consumer model, except that now also the prey (i.e. the resource R) follows its own population dynamics. This modifies the mean-field model as

$$\begin{cases} \dfrac{dR(t)}{dt} = bR(t) - dR(t) - \alpha R(t)N(t), \\ \dfrac{dN(t)}{dt} = \alpha e R(t)N(t) - mN(t), \end{cases} \quad (4.7)$$

where the only difference to Eq. (4.4) is that the birth rate of the prey is $bR(t)$ instead of the constant birth rate b. Without the term $-dR(t)$ which models the death rate of the prey in the absence of the predator, Eq. (4.7) would be the classical Lotka–Volterra model (Volterra, 1926; Lotka, 1932).

The system of differential equations given by Eq. (4.7) produces oscillatory dynamics (Figure 4.8). Assuming that initially the number of predators is small, the

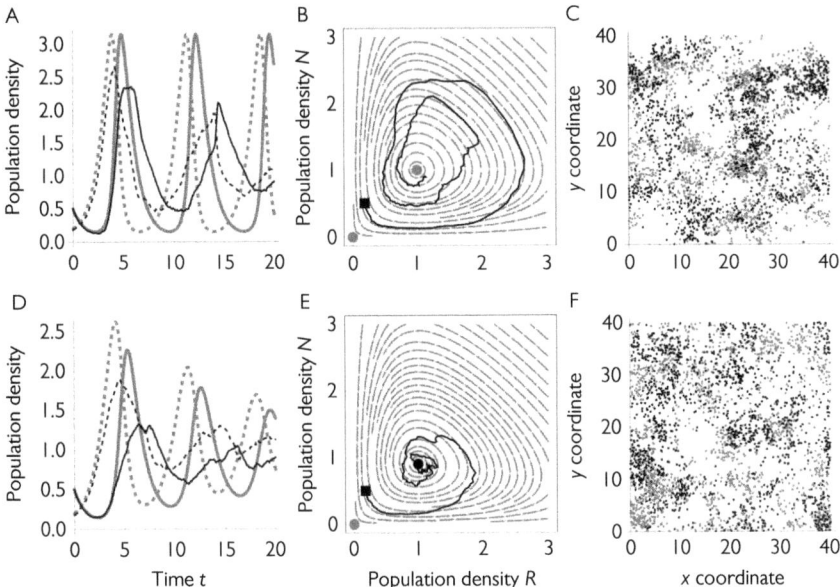

Figure 4.8 Predator–prey dynamics without and with internal density regulation of the prey population. Symbols are as described in the legend to Figure 4.7, if considering the prey as the resource and the predator as the consumer. Parameter values are $b = 2$, $d = 1$, $\alpha = 1$, $e = 1$, and $m = 1$. In the upper row (A–C) the prey does not have internal density regulation and thus $c = 0$, whereas in the lower row (D–F) $c = 1/10$. Initial density was set to 0.2 for the prey and 0.5 for the predators.

availability of prey per predator is high, making the predator population increase in number. The growth of the predator population increases the death rate of the prey, thus making them decline. The reduced availability of prey then decreases the predation rate, making the predators decline, which completes the population cycle. Mathematically, this model shows periodic oscillations.

The dynamics illustrated in the upper row of panels in Figure 4.8 consists of many closed loops (Figure 4.8B) that do not converge to a common limiting cycle. This means that the initial condition is important. When starting the mean-field model with a given number of predators and prey, it will return after one cycle back exactly to those same numbers. Thus if we start the model from a different initial density, it will make a different loop, and return again back to that initial condition. As the stochastic model shows random variation around its deterministic skeleton, its inherent stochasticity makes its alternate between the loops in a random manner (Figure 4.8B).

The sensitivity of the MFM to initial conditions is not very satisfactory from an ecological point of view, as one would not expect that the initial conditions matter still after a very long time. If releasing a number of rabbits and foxes to a large enclosure, and another number of rabbits and foxes to another identical enclosure, one would expect that eventually the two would follow similar dynamics, even if the initial numbers of rabbits and foxes were different. One feature that makes the differential equations of Eq. (4.7) so sensitive to the initial conditions is that it does not have any other density regulation for the prey than that provided by the predators. In the resource–consumer model the resources reach a finite density in the absence of the consumer, but in the predator–prey model the prey grows exponentially in the absence of the predator. This is clearly unrealistic, as the growth of the prey population must be eventually controlled by its own resources.

One possibility to resolve this problem would be to construct a model of three trophic levels by extending the resource–consumer model so that the consumer would be preyed upon by a predator. As the dynamics of the consumer would be controlled by the availability of the resources, they would not grow unlimited even in the absence of the predator. As a less-mechanistic but qualitatively similar alternative, we may assume that the death rate of the prey increases with prey density, as we did in the single-species model. This modifies the mean-field model into

$$\begin{cases} \dfrac{dR(t)}{dt} = bR(t) - dR(t) - \alpha R(t)N(t) - cR(t)^2, \\ \dfrac{dN(t)}{dt} = \alpha e R(t)N(t) - mN(t). \end{cases} \quad (4.8)$$

The addition of the density-dependence for the prey stabilizes the dynamics, as illustrated by the lower row of panels in Figure 4.8. Now the oscillations become damped, and the system approaches eventually a stable equilibrium point. However, we note that many real predator–prey systems show continuous oscillations (Kendall et al., 1999; Wang et al., 2009), whereas the deterministic model of Eq. (4.8) does not. Some additional mechanisms, such as the inclusion of a handling time (Li et al., 2013), are needed to generate sustained limit cycles.

4.3 Community models in heterogeneous environments

In Section 4.2 we discussed how ecological interactions influence the dynamics and persistence of simple two-species communities. We assumed homogeneous environmental conditions, and thus ignored the process of environmental filtering (see Section 4.1). When discussing single-species population ecology in Chapter 3, we extended homogeneous-space models to heterogeneous-space models by assuming that the dynamics of the species are influenced by the match between its ecological niche and the environmental conditions. The concept of environmental filtering relates to the fact that species have different ecological niches, and thus environmental variation makes different species thrive in different parts of the space (Leibold, 1995; Cottenie, 2005; Kraft et al., 2015). Now we extend the homogeneous-space community models of Section 4.2 into heterogeneous space, and thus examine how environmental filtering and ecological interactions jointly influence the dynamics and distributions of species.

4.3.1 The case of two competing species

We start by combining the heterogeneous-space plant population model of Section 3.3 with the homogeneous-space two-species competition model of Section 4.2. As a result, we have a heterogeneous-space model of two competing plant species. As shown by Figure 4.9A, we assume that the species differ in their thermal optima, which is lower for species 1 than for species 2. The species further differ in terms of their establishment rates: while both species have a higher establishment rate in fertile-rich soils than in fertile-poor soils, the overall establishment probability is higher for species 2 (Figure 4.9B). As a trade-off, we assume that species 1 is a stronger competitor than species 2: the mortality rate of species 2 increases in the presence of species 1, whereas species 1 does not experience any competition from species 2. Otherwise, the two species have identical parameter values. This model can be considered to be an example of the colonization–competition trade-off, which has been studied using a variety of model types (Higgins and Cain, 2002; Levine and Rees, 2002; Cadotte, 2007). Here species 1 is a superior competitor but inferior colonizer, whereas species 2 is an inferior competitor but superior colonizer.

Figure 4.10 shows a simulated distribution of individuals in this model. As species 1 (shown by the black dots in Figure 4.10A) is not influenced by the presence of species 2, its distribution follows its fundamental niche (shown by the black line) in the same way as in the single-species model of Section 3.3. Thus, its density is highest in areas that match its temperature optimum, and in areas with high soil fertility. As species 2 (shown by the grey dots in Figure 4.10A) is suppressed by competition from species 1, its realized niche covers only a part of its fundamental niche. In particular, species 2 is rare in the part of the parameter space where the soils are fertile and the temperature is intermediate. This is because species 1 is abundant under those conditions, and thus species 2 is outcompeted, even if the conditions there would otherwise be suitable it. Consequently, while the distribution of individuals

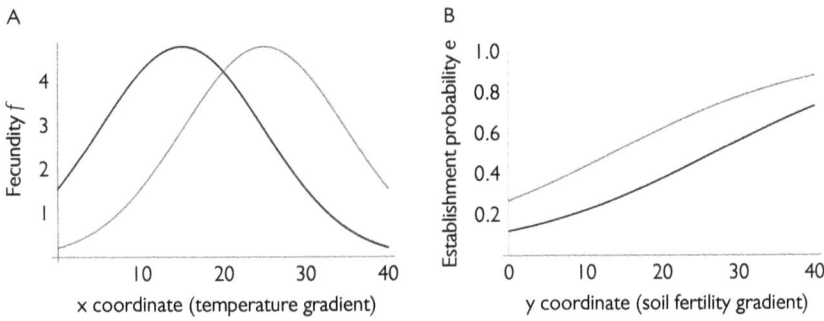

Figure 4.9 Spatial variation in fecundity and establishment assumed in the two-species plant population model. (A) shows that the fecundity rate peaks at a lower temperature for species 1 (black line) than for species 2 (grey line). (B) shows that the establishment probability increases as a function of soil fertility for both species, but the overall level is higher for species 2.

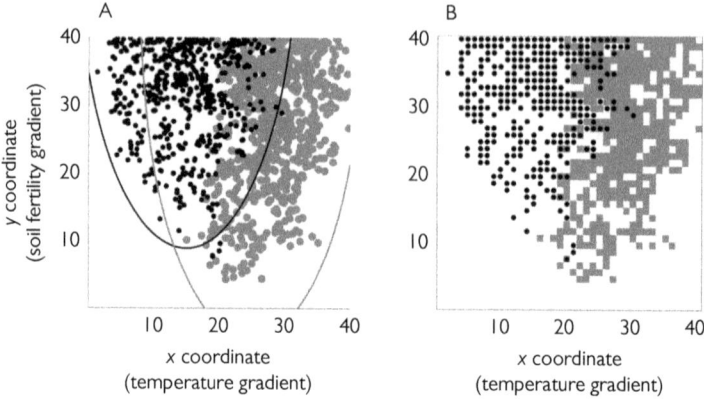

Figure 4.10 A simulated distribution of individuals at the stationary state of the two-species plant population model. (A) shows the distribution of individuals, with black dots corresponding to species 1 and grey dots to species 2. The lines show the equality $fe = m$, and thus they depict parameter combinations above which the species are predicted to persist or go extinct according to the mean-field model, assuming no interspecific competition. (B) shows presence–absence data sampled from the distribution of individuals shown in (A). Here we have assumed that each individual is detected with probability 0.2, and that a grid cell is classified as occupied if at least a single individual was detected. The occurrences of species 2 are shown by grey squares, whereas those of species 1 with black dots, so that cells in which both species are present are visible. We have assumed spatial variation in fecundity and establishment as illustrated in Figure 4.9. All other parameter values are set equal to those of the homogeneous-space model of Chapter 3.2, except that the interspecific competition kernels are set to $C_{12} = 2C$, and $C_{21} = 0$, where C is the intraspecific competition kernel that is assumed to be equal for both species.

clearly indicates that species 1 prefers soils with high fertility, for species 2 such a pattern is not equally evident. This is not because species 2 would not prefer soils with high fertility, but because it is outcompeted from such areas. As competition makes species 2 rare in areas where species 1 is abundant, one would expect a negative co-occurrence pattern among them. In other words, one would expect that the number of grid cells in Figure 4.10B where the two species occur together would be smaller than expected by chance. We will test this hypothesis in Section 4.5 by applying a statistical species community model to these data.

4.3.2 The case of many competing species

Let us then consider a community with a large number n of species, with $n = 100$ in the example to be considered. We extend the two-species model to a multispecies model that we call the plant community model. To do so, we assume that each species has its own response to temperature, with both the optimal temperature and the sensitivity to temperature varying among the species (Figure 4.11A). According to

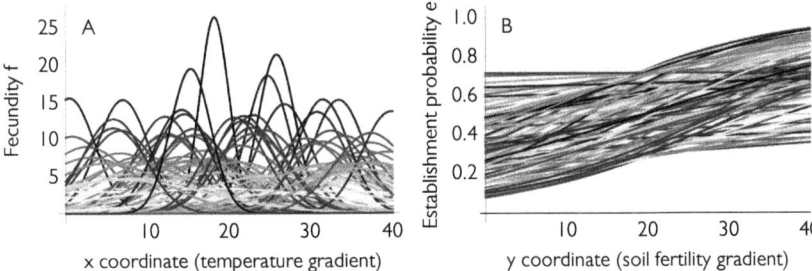

Figure 4.11 Spatial variation in fecundity and establishment assumed in the plant community model. To generate variation in fecundity (A), we sampled the optimal temperature for each species from a uniform distribution covering the whole temperature range present in the environment. The variance of the normally distributed fitness function was randomized from a log-normal distribution. To generate variation in establishment probability as a function of soil fertility (B), we sampled both the intercept and the slope of the logit-linear model from uniform distributions, restricted to positive values for the slope. The lines have been coloured according to variance in temperature response, so that the most specialized species is shown in black, and the most generalist species is shown in light grey.

a very broad definition, generalist species perform many tasks poorly, whereas specialist species perform few tasks well (Begon et al., 1996). Here we have assumed a trade-off that makes some species generalists and others specialists: species with wide thermal tolerance have a lower fecundity at their optimal temperature than species which have a narrow niche for thermal tolerance. The species differ also in terms of their establishment probability: we have assumed variation among the species both in their overall establishment level and in their sensitivity to soil fertility (Figure 4.11B). However, for all species we have again assumed that the probability of establishment increases with increasing soil fertility. As with the two-species example, we assume that the species are identical in their other parameters values, except for interspecific competition, which we describe in the following.

Figure 4.12 illustrates a community simulated from this model. In the upper panels, we have assumed no competition among the species by setting the competition kernel to $C_{ij} = 0$ for $j \neq i$, whereas the within-species competition kernel C_{ii} that models within-species density-dependence is assumed to be equal for all species. In this model, species richness increases gradually towards areas of increasing soil fertility (Figure 4.12A), which is expected, as the occurrence probabilities of all species increase in this direction. Together, the species fill the niche space of available temperatures (Figure 4.11), and thus along the temperature gradient species richness peaks in the middle (Figure 4.12A). The species show marked variation in their prevalences (Figure 4.12C), mainly because some species have a higher baseline establishment probability than others. We note that while the distribution of species-specific prevalences (Figure 4.12C) reflects the amount of heterogeneity among species, the distribution of species richness (Figure 4.12B) reflects the amount of heterogeneity among environmental conditions.

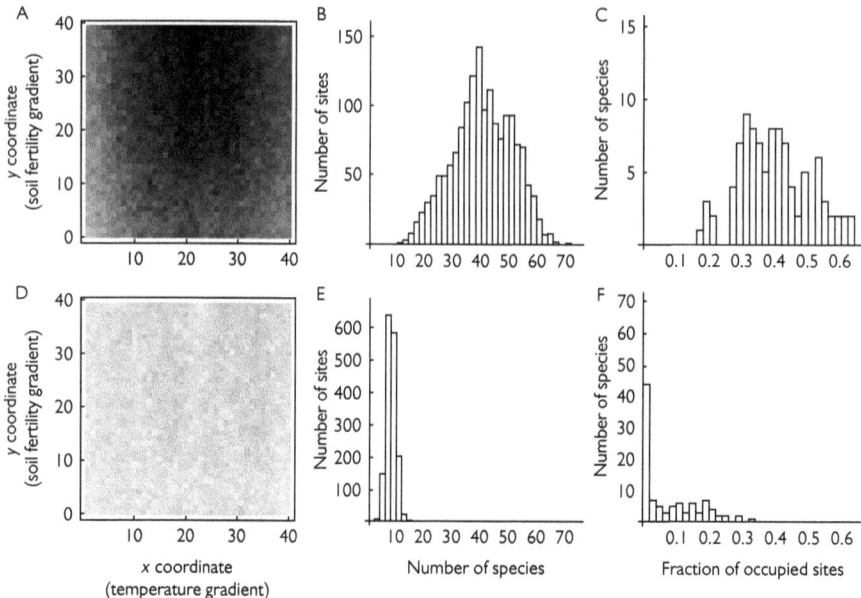

Figure 4.12 Variation in species richness and species-specific prevalence in the plant community model. The upper panels (A–C) corresponds to the case of no interspecific competition, whereas in the lower panels (D–F) some species impose competitive effects to some other species. (A) and (D) show variation in species richness over space, whereas panels (B) and (E) show the same data as histograms of species richness and (C) and (F) as histograms of species-specific prevalences. We have assumed spatial variation in fecundity and establishment as illustrated in Figure 4.11. In the upper panels interspecific ($j \neq i$) interaction kernels are set to $C_{ij} = 0$, whereas in the lower panels we have set $C_{ij} = C$ with probability 0.25 and $C_{ij} = 0$ with probability 0.75 for each (j, i) pair, where the intraspecific competition kernel C is assumed to be equal for all species. All other parameter values equal those of the two-species model of Figure 4.10.

The lower panels in Figure 4.12 illustrate a simulation of otherwise the same community but with interspecific interactions. We have assumed that only a subset of the species compete with each other species, and that competition is not necessarily symmetric. More specifically, we assume that each species i has a competitive effect on each species j with probability 0.25, in which case the interspecific competition kernel C_{ij} equals the intraspecific competition kernel C. In the competitive community, the species reach generally lower prevalences than in the absence of competition (Figure 4.12F vs Figure 4.12C). In particular, many species are extinct now (i.e. their prevalences are 0 in Figure 4.12F), even if the environment includes parts of their fundamental niche. The influence of competition is also seen as lowered species richness (Figure 4.12E vs Figure 4.12B). As the level of competition is highest in the most favourable part of the parameter space, the pattern of species richness increasing with soil fertility is now much less pronounced than in the absence of competition (Figure 4.12D vs Figure 4.12A). In Section 4.5, we will apply a statistical species community model to these data.

4.4 The response of communities to habitat loss and fragmentation

Understanding the responses of entire communities to habitat loss and fragmentation is even more challenging than assessing the response of a single species (Buchmann et al., 2012). This is because on top of its direct effects, habitat loss and fragmentation have indirect effects mediated by interspecific interactions (Götzenberger et al., 2012). In the case of mutualistic interactions, the extinction of one species can make the dependent species go extinct as well (Fortuna and Bascompte, 2006). In the case of competitive interactions, the extinction of one species may leave more resources for the other species, thus benefiting its survival (Bonin et al., 2011; Buchmann et al., 2013). In this section, we use the plant community model to examine how a competitive community may respond to habitat loss.

4.4.1 Endemics-area and species-area relationships generated by the plant community model

We consider again two variants of the plant community model: in one community the species follow their population dynamics independently of each other, and thus only environmental filtering is operating. In the other community also biotic interactions are present, i.e. some species impose competitive effects on other species. We assume that the communities have settled in their stationary states, as they have done in Figure 4.12. We then apply a rapid episode of habitat loss, after which only a fraction p of the original habitat remains (for an example, see Figure 4.13). We assume that individuals left in the unsuitable matrix die immediately, whereas the individuals that remain in the suitable area are not directly influenced by habitat loss. Thus, immediately after habitat loss the distribution of species within the remaining habitat is identical to that before habitat loss. However, due to the individuals lost by conversion of habitat into matrix, generally all species have become rarer, as seen by comparing Figures 4.12 and 4.13. In particular, if a species only occurred in the area that was lost (i.e. was endemic to that area), it went immediately extinct due to habitat loss. This happened to some of the species especially in the community with interspecific interactions, as seen by comparing Figures 4.12F and 4.13F.

While Figure 4.13 depicts the immediate effect of habitat loss, Figure 4.14 depicts its long-term consequences. Here we have continued the simulation after habitat loss so that the community has settled to a new stationary state. As a consequence, some species have further declined in their abundance, and some have even gone extinct (though the difference between Figures 4.13 and 4.14 is only small, the reason for which is explained in the next paragraph). This is because after habitat loss the carrying capacity of the environment is lower, whereas the abundances of the

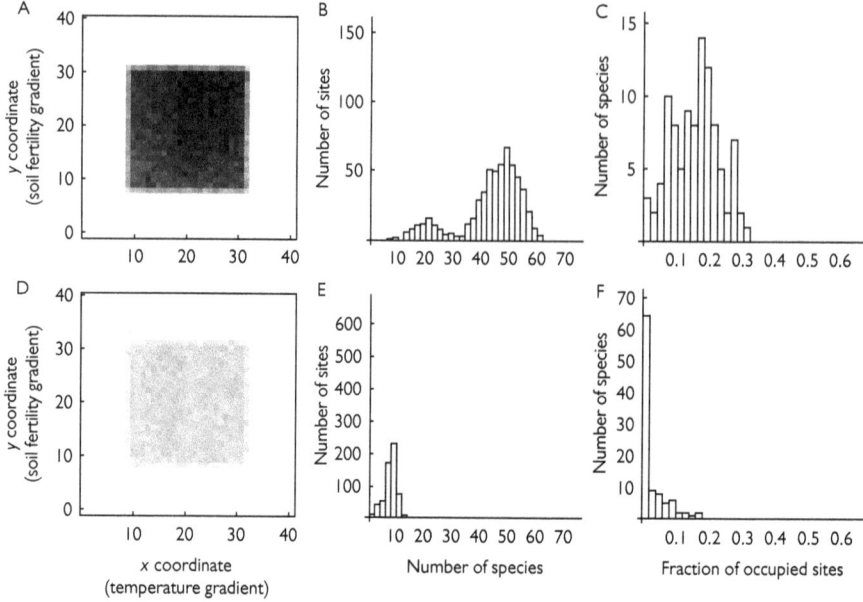

Figure 4.13 Immediate response of the plant community model to habitat loss. We have assumed that the community depicted in Figure 4.12 experiences habitat loss in such a way that 32% of the original habitat area remains in one central block, whereas the rest of the habitat is converted to unsuitable matrix. The panels shown here are otherwise identical to those shown in Figure 4.12, but they depict the situation immediately after habitat loss, so that individuals in the matrix have died. The histograms shown in (B, E) are restricted to sites for which the habitats were not lost, whereas the histograms shown in (C, F) show the fraction of occupied sites among all (suitable and unsuitable) sites. Lower species richness (lighter colours in A, D) at the edge of the remaining habitat is an artefact of discretizing the domain into grid cells, as grid cells at the edge contain both habitat and matrix.

species still reflect the carrying capacity of the environment in the original situation. This illustrates the concept of extinction debt (Tilman et al., 1994; Helm et al., 2006), defined as the number of species that will eventually go extinct after habitat loss but have not yet done so.

The short-term and long-term influences of habitat loss evidently depends on how much of the habitat is lost. For example, the difference between Figures 4.13 and 4.14 is small, because the amount of habitat that still remains is rather large. The effect of the remaining habitat area is illustrated in Figure 4.15, which shows what are called endemics-area relationships and species-area relationships (He and Legendre, 2002; Green and Ostling, 2003). The endemics-area relationships, shown in grey in Figure 4.15, show the numbers of species that remain immediately after habitat loss. Another way of thinking of the endemics-area relationship is that it shows the numbers of species that would be recorded by sampling plots of different sizes

4.4 The response of communities to habitat loss and fragmentation • 147

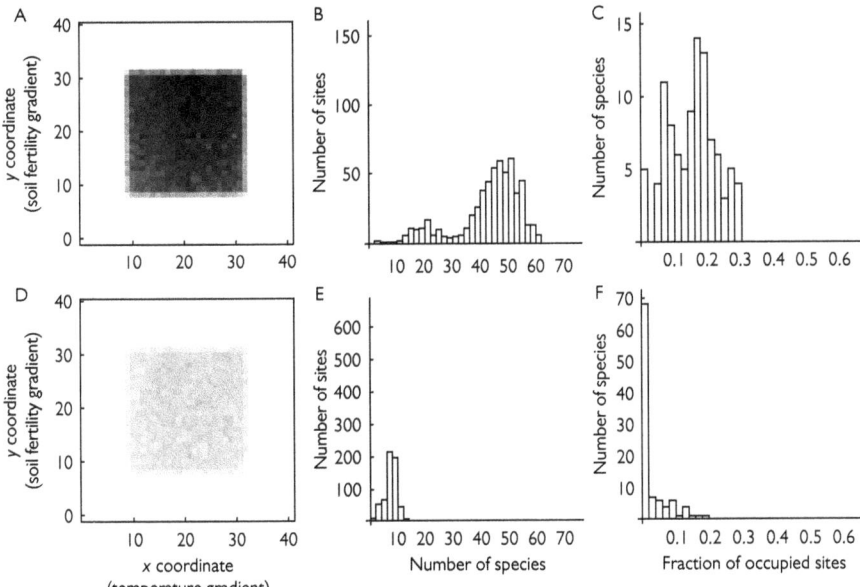

Figure 4.14 Long-term response of the plant community model to habitat loss. The figure is otherwise identical to Figure 4.13, but we have simulated community dynamics for $t = 20$ time units after habitat loss.

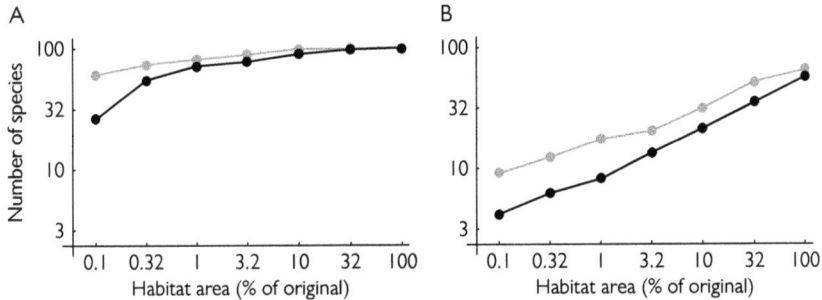

Figure 4.15 Species-area relationships generated by the plant community model. We simulated different degrees of habitat loss, and recorded the number of species that remained immediately after habitat loss (grey dots and lines) and $t = 20$ time units after habitat loss (black dots and lines). (A) shows the results for a community without intraspecific competition (corresponding to upper panels in Figures 4.12, 4.13, and 4.14), whereas in (B) the community is influenced by intraspecific competition (corresponding to lower panels in Figures 4.12, 4.13, and 4.14). Note the logarithmic scales on the axes.

within the undisturbed habitat. Evidently, the larger the area sampled, the larger the number of species found. The black lines show the species-area relationships after the communities have settled to their new stationary states after habitat loss. These curves are necessarily lower than the endemics-area curves, as we have assumed a closed community so that no new species have immigrated to the community, but some of the existing ones may have been lost. The difference between the grey and the black curves represents the magnitude of extinction debt. In the logarithmic scale of Figure 4.15, extinction debt is greatest in the case where only a small fraction of the original habitat remains. This is the case because under such a situation the long-term persistence of the species is most difficult, for the reasons discussed in Section 3.4.

While in ecology it is generally difficult to derive rules of thumb that hold for a large number of ecosystems and taxonomical groups, there is one such rule for the species-area relationship (SAR), namely the power law relationship

$$N = kA^z. \tag{4.9}$$

Here N is the number of species, A is the habitat area, and k and z are parameters. The parameter k measures the number of species found in a habitat of unit area ($A = 1$), and it relates to the overall richness of a species community. The parameter z measures how rapidly the species number increases with the habitat area. If $z = 0$, the number of species is independent of habitat area, whereas with $z = 1$ it increases linearly with habitat area. Empirical studies have typically reported values of z in the range $0.1 < z < 0.5$, with $z = 0.25$ being a typical value.

Transforming the species-area relationship into logarithmic form gives $\log(N) = a + z\log(A)$, where the intercept a is given by $a = \log(k)$. This equation is linear, suggesting that the points plotted in Figure 4.15 should fit into a line. This is indeed the case for the interactive species community (black line in Figure 4.15B), but not for the less realistic community consisting of independent species (black line in Figure 4.15A). In empirical studies, the approximate linearity of the species-area relationship has often been found to be valid over a very broad range of habitat areas, such as the 1,000-fold difference involved in Figure 4.15.

The species-area relationship can be used to predict the consequences of habitat loss. Let us denote by N the number of species and by A the amount of habitat before habitat loss, and by N^* and A^* the corresponding quantities after habitat loss. Let us denote by p the fraction of habitat that remains, so that $A^* = pA$. Then, by the species-area relationship, the number of species that remain after habitat loss is $N^* = k(A^*)^z = k(pA)^z = p^z kA^z = p^z N$. Thus, the prediction of the species-area-relationship is that the fraction p^z of the species will persist, and thus the fraction $1 - p^z$ will be lost. Assuming the value of $z = 0.25$, and that 90% of the habitat is lost, we predict that the fraction $0.1^{0.25} \approx 0.56$ of the species will go extinct. This gives the rough rule of thumb that if 90% of the habitat is lost, half of the species are expected to go extinct.

Not all species are influenced equally by habitat loss (Buchmann et al., 2013). This is partly so because species have dissimilar niches, and thus their distributions are not identical even before habitat loss. Figure 4.16 asks which kinds of species persisted in

4.4 The response of communities to habitat loss and fragmentation • 149

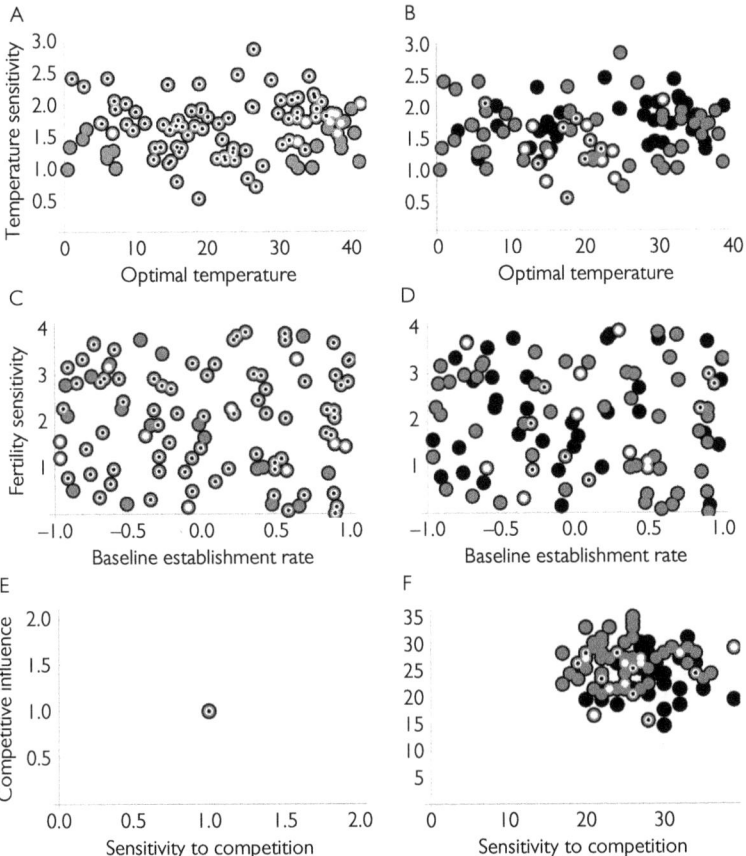

Figure 4.16 The influence of species traits on response to habitat loss. We considered a variant of the scenario simulated in Figures 4.12, 4.13, and 4.14 in which 1% of the original habitat remains after habitat loss. The symbols classify the species into three categories: black dots show those species that went extinct already before habitat loss. Species that went extinct immediately after habitat loss are shown by black-grey (from outside to inside) symbols. Species that went extinct eventually are shown by black-grey-white, whereas species that persisted also after habitat loss are shown by black-grey-white-black. The left-hand panels (A, C, E) show the results for a community without interspecific competition and the right-hand panels (B, D, F) for a community with interspecific competition. In (A, B) the species are located according to their response to the temperature gradient (see Figure 4.11A). In (C, D) the species are located according to their response to the soil fertility gradient (see Figure 4.11B). In (E, F) the species are located according to their sensitivity to competition (how many species have a competitive influence on the focal species) and to their competitive influence (how many species the focal species has a competitive influence on). In the case without interspecific competition, the sensitivities to competition as well as the competitive influences are 1 for all species due to intraspecific competition.

150 • *Community ecology*

the simulated communities before habitat loss, immediately after habitat loss, and in the long term. In the interactive community, some of the species went extinct even before habitat loss (black circles in the right-hand panels of Figure 4.16). As the community dynamics are influenced simultaneously by species niches and by competitive interactions, the patterns from traits to extinction vulnerability are not very clear-cut. However, Figure 4.16 suggests that the following traits characterize the extinct species: high sensitivity to temperature, low baseline establishment rate, and high sensitivity to competition. These traits characterize rare and specialized species, which are generally vulnerable to extinction (e.g. Henle et al., 2004).

Species that were specialized to either a low or a high temperature regime were especially prone to go extinct immediately after habitat loss (Figures 4.16A, B). This is simply because we assumed that the remaining habitat consists of intermediate thermal conditions, and thus species not adapted to such conditions were likely to go extinct. Species that survived in the long term show much variation in their niche, which is to be expected, as it is the niche separation that allows them to coexist.

4.5 Statistical approaches to analysing species communities

Compared to single-species population ecology, community ecology involves a much larger array of questions and data types. Further, due to the greater complexity of the underlying processes, in community ecology it is generally more difficult to derive the links from the underlying processes to the observed patterns than is the case of single-species population ecology. Thus, many statistical methods developed for community ecology are focused on describing patterns rather than in the construction of predictive models. In this section, we start by discussing multispecies time-series analyses based on non-spatial data. We then move to spatial data and exemplify joint species distribution models, ordination analyses, and characterizations of spatial patterns.

4.5.1 Time-series analyses of population size in species communities

We start by illustrating time-series analyses of community data by applying autoregressive models to data generated by the homogeneous-space two-species model of Section 4.2. We analyse three data sets. The first one is generated by the two-species competition model (Figure 4.17), the second by the resource–consumer model (Figure 4.18), and the third one by the predator–prey model (Figure 4.19).

We recall from Section 3.5 that an autoregressive model of order 1, denoted by AR(1), is defined by

$$n_{t+1} - n_t = \alpha + \beta n_t + \varepsilon_t, \qquad (4.10)$$

where $\varepsilon_t \sim N(0, \sigma^2)$ is normally distributed noise. When applying the analyses, we log-transformed the data, so that $n_t = \log N_t$, where N_t stands for the densities of the species, shown by the dots in Figures 4.17–4.19. We started the analysis by

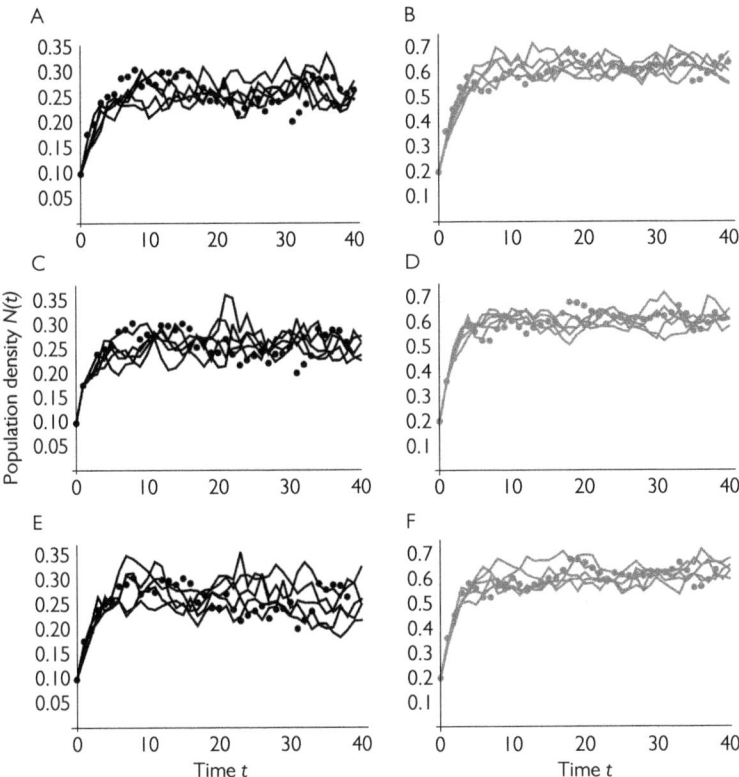

Figure 4.17 Time-series analysis of data generated by a two-species competition model. The data, shown by the dots, were obtained by sampling the stochastic time-series shown in Figure 4.4C to intervals of 1 time unit. The lines show five simulation replicates based on species-specific AR(1) models (A, B), species-specific AR(2) models (C, D), and a bivariate AR(1) model (E, F). The left-hand panels A, C, E (respectively, right-hand panels B, D, F) correspond to species shown by continuous (respectively, dashed) lines in Figure 4.4C.

fitting the AR(1) models separately to each species. The parameter estimates, shown in Table 4.1, are negative for the density-dependence parameter β. This suggests that the underlying process involves density regulation, i.e. that the population growth rates decrease with increasing population densities.

The upmost rows of panels in Figures 4.17–4.19 compare the data to the predictions of the fitted models, obtained by using Eq. (4.10) to generate a simulated time-series of log-transformed population densities, and back-transforming by $N_t = \exp(n_t)$. The models fit well to the data generated by the competition model but not to the data generated by the resource–consumer or predator–prey models. This is because the model of Eq. (4.10) is only capable of predicting monotonous growth (or decline) towards an asymptote, and the dynamics generated by the resource–consumer and predator–prey models are more complex.

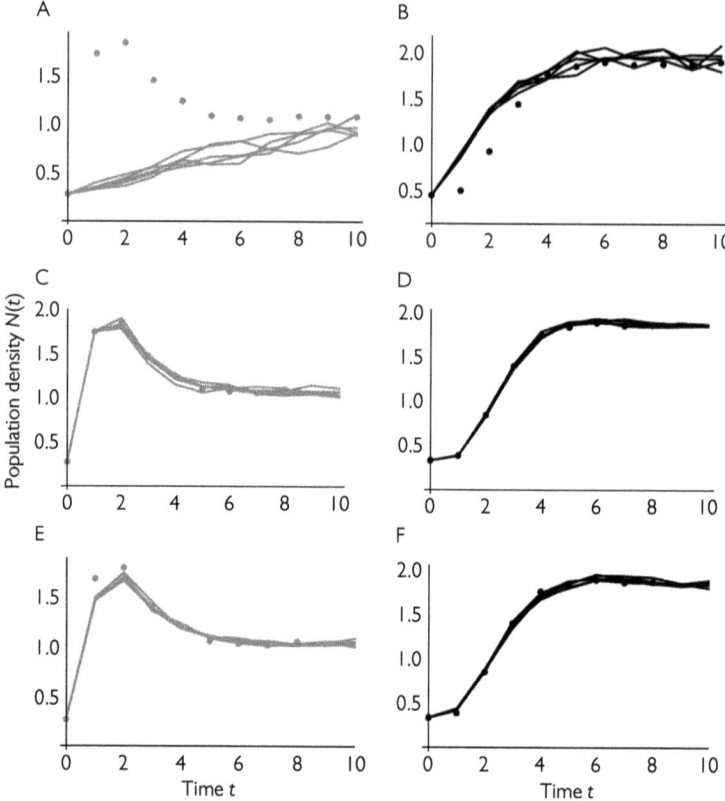

Figure 4.18 Time-series analysis of data generated by a resource–consumer model. The figure is identical to Figure 4.17, but shown for data obtained by sampling the stochastic time-series shown in Figure 4.7G.

To improve the model fits, we increase the orders of the autoregressive models to 2, so that we fit the AR(2) models

$$n_{t+1} - n_t = \alpha + \beta n_t + \gamma n_{t-1} + \varepsilon_t, \qquad (4.11)$$

where the new parameter γ measures the effect of delayed density-dependence. Delayed density regulation can destabilize population dynamics, and in particular it can cause population cycles (e.g. Li et al., 2013 and references therein). The parameter estimates of Eq. (4.11) are given in Table 4.2, and the data are compared simulations of Eq. (4.11) in the middle rows of panels in Figures 4.17–4.19. For the data generated by the competition model, the AR(2) models result in essentially similar fits as the AR(1) models. In contrast, for the data generated by the resource–consumer and predator–prey models, the AR(2) models provide much better fits than the AR(1) model. This is because the AR(2) model is able to capture non-linear dynamics, e.g. the damping oscillations produced by the predator–prey model.

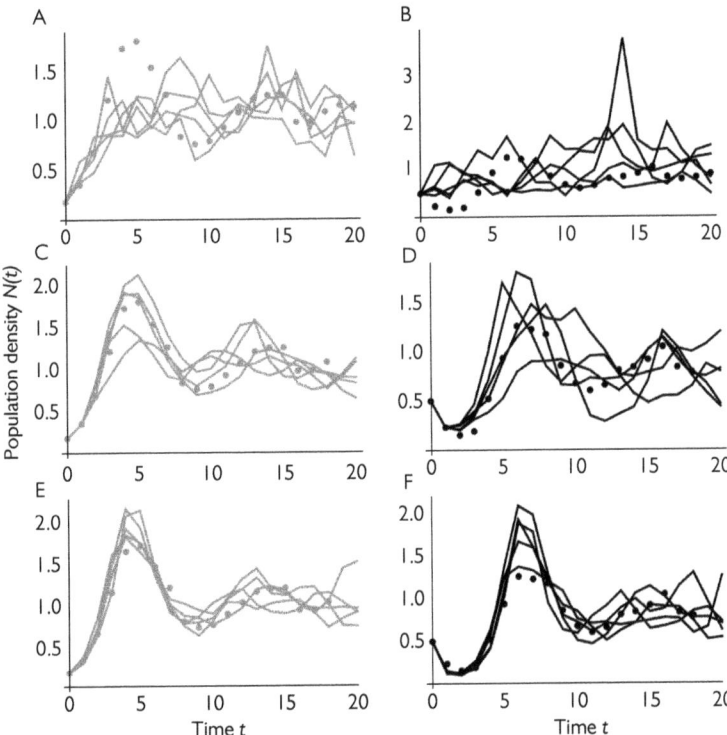

Figure 4.19 Time-series analysis of data generated by a predator–prey model. The figure is identical to Figure 4.17, but shown for data obtained by sampling the stochastic time-series shown in Figure 4.8D.

Table 4.1 Parameter estimates obtained by fitting the single-species AR(1) model (Eq. (4.10)) to data shown by the dots in Figures 4.17–4.19.

Model	Species	α	β	Variance σ^2
Competition	Black	−0.47	−0.35**	0.0063
Competition	Grey	−0.20	−0.40**	0.0018
Resource–consumer	Black (consumer)	0.34	−0.52**	0.0018
Resource–consumer	Grey (resource)	0.0030	−0.23	0.0085
Predator–prey	Predator (black)	−0.019	−0.22	0.090
Predator–prey	Prey (grey)	0.08	−0.49**	0.036

Model fitting was conducted by standard linear model analysis with maximum likelihood estimation. Statistical significances shown for the parameter β are based on the t-statistic, with $0.01 < p < 0.05$ denoted by * and $p < 0.01$ denoted by **.

Table 4.2 Parameter estimates obtained by fitting the single-species AR(2) model (Eq. (4.11)) to data shown by the dots in Figures 4.17–4.19.

Model	Species	α	β	γ	Variance σ^2
Competition	Black	−0.48	−0.38*	0.02	0.0064
Competition	Grey	−0.13	−0.07	−0.20*	0.0015
Resource–consumer	Black (consumer)	0.30	−0.33**	−0.15**	0.00017
Resource–consumer	Grey (resource)	0.036	−0.30**	−0.15**	0.00053
Predator–prey	Predator (black)	−0.089	0.34**	−0.69**	0.028
Predator–prey	Prey (grey)	0.017	0.24	−0.58**	0.0090

Model fitting was conducted by standard linear model analysis with maximum likelihood estimation. Statistical significances shown for the parameters β and γ are based on the t-statistic, with $0.01 < p < 0.05$ denoted by * and $p < 0.01$ denoted by **.

While the species-specific AR(2) models were able to capture the statistical features of the data, these models are not in line with the processes we used to generate the data. This is because the data were generated by a (continuous-time) Markov process, whereas the γ parameter of the AR(2) model measures the effect of delayed density-dependence. In our data, there are no such delays in the model assumptions, but the delays are generated by an interaction between the resources and the consumers, or an interaction between the predators and the prey. If we expect that the species interact, as a more natural alternative we could fit a two-species AR(1) model. Denoting the two time-series by $n_{1,t}$ and $n_{2,t}$, this model reads

$$\begin{cases} n_{1,t+1} - n_{1,t} = \alpha_1 + \beta_1 n_{1,t} + \gamma_1 n_{2,t} + \varepsilon_{1,t}, \\ n_{2,t+1} - n_{2,t} = \alpha_2 + \beta_2 n_{2,t} + \gamma_2 n_{1,t} + \varepsilon_{2,t}, \end{cases} \qquad (4.12)$$

where the β parameters measure the effects of conspecific densities and the γ parameters the effects of heterospecific densities. The residuals have their own variance parameters, $\varepsilon_{1,t} \sim N\left(0, \sigma_1^2\right)$ and $\varepsilon_{2,t} \sim N\left(0, \sigma_2^2\right)$.

The parameter estimates of Eq. (4.12) are given in Table 4.3, and the data are compared to dynamics simulated by this model in the lowest rows of panels in Figures 4.17–4.19. These models provide essentially equally good fits as the single-species AR(2) models. This is to be expected, as the two-species AR(1) model of Eq. (4.12) is mathematically equivalent to the single-species AR(2) model of Eq. (4.11). But now the parameter estimates make more biological sense. In particular, we see that the growth rate of the consumer increases with the density of the resource, whereas the growth rate of the resource decreases with the density of the consumer. Similarly, the growth rate of the predator increases with the density of the prey, whereas the growth rate of the prey decreases with the density of the predators.

4.5 Statistical approaches to analysing species communities • 155

Table 4.3 Parameter estimates obtained by fitting the two-species AR(1) model (Eq. (4.12)) to data shown by the dots in Figures 4.17–4.19.

Model	Species	α	β	γ	Variance σ^2
Competition	Black	−0.50	−0.32**	−0.14	0.0062
Competition	Grey	−0.31	−0.37**	−0.09	0.0017
Resource–consumer	Black (consumer)	0.22	−0.40**	0.33**	0.00025
Resource–consumer	Grey (resource)	0.26	−0.87**	−0.31**	0.00029
Predator–prey	predator (black)	−0.15	−0.47**	0.85**	0.0091
Predator–prey	prey (grey)	−0.05	−0.26**	−0.31**	0.0078

Model fitting was conducted by standard linear model analysis with maximum likelihood estimation. Statistical significances shown for the parameters β and γ are based on the t-statistic, with $0.01 < p < 0.05$ denoted by * and $p < 0.01$ denoted by **.

4.5.2 Joint species distribution models

In Section 3.5, we applied a regression model to analyse snapshot data on a species distribution. We now extend this approach to snapshot data acquired simultaneously for a species community. A common type of such data involves a matrix of environmental covariates **X** and another matrix of species occurrences **Y**. The dimensions of the species matrix **Y** are $n \times m$, where n is the number of sites from which the data have been acquired, and m is the number of species. The elements of this matrix are 1 if a species was found from the site, and 0 if it was not found from the site. The dimensions of the environmental covariate matrix **X** are $n \times k$, where n is the number of sites, and k the number of covariates measured for each site. We include the intercept of the model in this matrix, so that the first column of the matrix **X** consists of 1's, i.e. $x_{i1} = 1$ for all sites i.

We will analyse data generated by the plant community models of Section 4.3. Thus, the relevant environmental variables are temperature and fertility. We assume that the researcher would expect (either *a priori* or as a result of model selection) a monotonous response to fertility and a humped-shaped response to temperature. We thus include as covariates temperature (x_{i2}), temperature squared (x_{i3}), and fertility (x_{i4}).

We recall from Section 3.5 and Appendix B.1 that a probit-regression model for species occurrence can be written as $\Pr(y_i = 1) = \Phi(L_i)$, where Φ is the cumulative distribution function of the standard normal distribution $N(0, 1)$, the linear predictor L_i is defined by $L_i = \sum_{l=1}^{k} x_{il}\beta_l$, and the β_l are the regression coefficients to be estimated. To extend this modelling framework into a multispecies model, we simply add an index j to stand for the species. Thus, the model reads now

$$\Pr(y_{ij} = 1) = \Phi(L_{ij}), \tag{4.13}$$

where

$$L_{ij} = \sum_{l=1}^{k} x_{il}\beta_{lj}. \tag{4.14}$$

156 · Community ecology

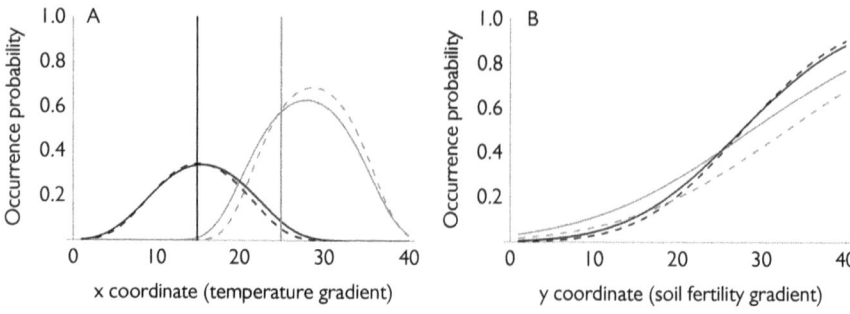

Figure 4.20 Joint species distribution models applied to data generated by the two-species competitive plant population model (data shown in Figure 4.10B). The lines show predicted occurrence probabilities for the two species (colours corresponding to those in Figure 4.10B) as functions of temperature and soil fertility. The dashed lines show the predictions of the two-species model, whereas the continuous lines show the predictions of single-species models in which the presence of the other species has been included as an explanatory variable. The explanatory variables not varied in the panels (fertility in A and temperature in B; and the prevalence of the other species in both panels for the models shown by continuous lines) were set to their mean values. In (A) the vertical lines show temperature optima that maximize the fecundities of the two species (see Figure 4.9).

The environmental covariates x_{il} are the same for all species, but the regression parameters are species-specific, β_{lj} measuring the influence of the covariate l on the occurrence of the species j.

The dashed lines in Figure 4.20 show the predictions of Eq. (4.13) fitted to data generated by the two-species plant population model of Section 4.3. The predictions are in line with the data-generating process in the sense that the species occurrence probabilities peak at intermediate temperatures that are close to the assumed temperature optima of the species, and they increase with increasing soil fertility.

But how does the model of Eq. (4.13) differ from the single-species models of Section 3.5? The models are actually identical; we have just added the index j to have separate parameters for each species. To bring more insight, we consider a joint species distribution model and thus account for the fact that the response variable is of multivariate nature. In other words, we don't consider as the response variable the occurrence $y_{i\cdot}$ of a single species, but $y_{i\cdot}$, the vector of the occurrences of all species in site i. In the case of two species, there are four possibilities: we may have $y_{i\cdot} = (0,0)$, $y_{i\cdot} = (1,0)$, $y_{i\cdot} = (0,1)$, or $y_{i\cdot} = (1,1)$. In the multivariate framework, we can examine the question of co-occurrence: are the two species found more or less often together than would be expected just by chance? This question can be addressed by including in the model of Eq. (4.13) a species-by-species correlation matrix **R**, as detailed in Appendix B.1. This matrix is symmetric ($R_{j_1 j_2} = R_{j_2 j_1}$), and its diagonal elements are one ($R_{jj} = 1$).

In the two-species case, there is only one species-pair, and thus only one parameter R_{12} to be estimated. For the data that we generated by the two-species plant model (Figure 4.10B), the posterior median (95% credibility interval) estimate for R_{12} is −0.08 (−0.21, 0.09) in the model that includes the covariates of temperature and soil fertility, whereas in an alternative model which excludes all covariates it is R_{12} = −0.24 (−0.33, −0.14). The negative correlations mean that the two species are found together less often than expected by random, as one would expect for the case of competing species. The negative correlation is stronger in the model variant that excludes the covariates. This is because the model that accounts for the covariates attributes part of the negative co-occurrence pattern to dissimilarity among the species niches. In general, it is theoretically not possible to infer from snapshot data whether a non-random co-occurrence pattern is due to an ecological interaction or due to the influence of an environmental covariate that is not included in the model (Kraft et al., 2015).

Another possibility for the analysis of community data is to fit species-specific models, but to use the occurrences of other species as predictors (Guisan et al., 2002). We illustrate the results of such analyses by the continuous lines in Figure 4.20. These have been obtained by adding to the predictor matrix **X** a new column, representing the presence–absence of the other species. The regression coefficient that measures the effect of species 1 (shown by black in the figures) on species 2 (shown by grey) is −0.92 (−1.2, −0.55). The negative estimate tells that finding the species 2 from sites occupied by species 1 is less likely than from sites not occupied by species 1. This is consistent with the fact that we assumed that species 1 imposes a competitive effect on species 2. The regression coefficient that measures the effect of species 2 on species 1 is −0.35 (−0.68, 0.04). While the statistical support for this effect being negative is not equally strong, there is a hint that species 1 is also less likely to be found in sites occupied by species 2 than in sites where species 2 is absent. This is to be expected, even though we assumed that species 1 does not experience any competition from species 2. The reason here is that species 2 is more likely to be present if species 1 is not present. Thus the presence of species 2 is an indicator that species 1 is not likely to be present. The take-home message here is that statistical models describe the observed patterns, and thus inferring causal links from their results is challenging. However, here the statistical analyses provided two slight hints about species 1 influencing species 2, not vice versa. The first one is that the negative effect of species 1 on species 2 was stronger than the effect of species 2 on species 1. The second one is that when considering the other species, the niche of species 2 shifted more than that of species 1. Indeed, in Figure 4.20 the grey continuous line is a bit closer to the fundamental niche of species 2 than the dashed line that represents the realized niche.

We next move to analyses of species-rich communities by considering data generated by the heterogeneous-space plant community model with 100 species (Section 4.3). Data acquired for species rich communities are typically characterized by many rare species. Because it is difficult to fit species-specific models to rare species, models are often built for the most dominant species only. This is, however, a limitation, as often the rare species are of particular interest, as they may be e.g. threatened species that are of interest for conservation biology. One solution for the analysis of sparse data is to add a hierarchical layer to the statistical model, e.g. by extending the model of Eq. (4.13) by (Ovaskainen and Soininen, 2011)

$$\beta_{\cdot j} \sim N(\mu, \Sigma). \tag{4.15}$$

Here $\beta_{\cdot j}$ is the vector of all regression coefficients for species j, and thus it describes how that species responds to environmental variation. By Eq. (4.15), we assume that $\beta_{\cdot j}$ is a sample from a community-level distribution, which has a mean vector μ and a variance-covariance matrix Σ. As in our example we have $k = 4$ covariates (intercept, temperature, temperature squared, and fertility), the dimension of μ is 4×1 and the dimension of Σ is 4×4. As the matrix Σ is symmetric, it has 10 free parameters (4 diagonal parameters and 6 off-diagonal parameters).

Figure 4.21 shows species-specific occurrence profiles as a function of the covariates, based on fitting the model of Eqs. (4.13)–(4.15) to data generated by the plant community model. The results of the statistical model reflect the assumptions made

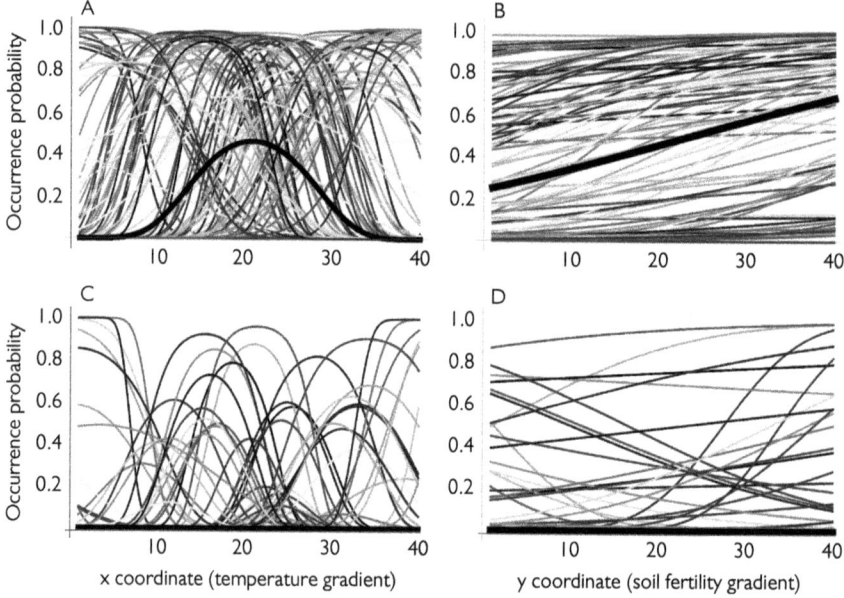

Figure 4.21 Joint species distribution models applied to data generated by the plant community model. The lines show species-specific occurrence probabilities as functions of the temperature and fertility covariates. The covariates not varied in the panels have been set to their mean values. The thin lines show the species-specific predictions based on the regression parameters (β), and the thick lines show the results based on the community-level mean parameter μ. (A, B) show the results for the data generated by the model without interspecific competition (data illustrated in the upper row of panels in Figure 4.12), whereas (C, D) show the results for the data generated by the model with interspecific competition (data illustrated in the lower row of panels in Figure 4.12).

by the data-generating process: the species reach the highest occurrence probability at some intermediate temperature, and vary in their temperature optimum and their sensitivity to temperature. Similarly, the occurrence probabilities of the species generally increase with increasing soil fertility. There is great variation in species abundance, some being common and some others being rare. In particular, in the case of the competitive community (Figures 4.21C, D), many of the species went extinct, and thus their predicted occurrence probabilities are very low. Note that Figure 4.21 involves the predictions for all of the 100 species, also those that went extinct, and thus many some curves are very close to 0.

There are two reasons why it is beneficial to glue the species-specific models (Eq. (4.13)) together with community-level model (Eq. (4.15)). The first reason is that the community-level model aids the parameterization of the species level models, as while estimating the parameters for a given species, the model can borrow information from the other species. As shown with both simulated and real data in Ovaskainen and Soininen (2011), this feature improves the estimates of the species-specific parameters (i.e. the β parameters) especially for rare species. The second reason why the inclusion of the community-level component of Eq. (4.15) is beneficial is that it provides a parameter sparse and compact description of the whole community. In our example, there are 100 species, and thus 400 species-specific β parameters. While these parameters jointly describe how the occurrences of all species depend on environmental conditions, it is very difficult to see the big picture from such a large number of parameters. In contrast, there are only 14 parameters at the community level. The 4 parameters included in the vector μ describe how the occurrence probability of a typical species depends on the environmental conditions, as illustrated by the thick lines in Figure 4.21. Note that in the lower panels of Figure 4.21 the thick lines are almost at 0, illustrating that in the competitive community a typical species is very rare. The 4 diagonal parameters of the matrix Σ measure how much variation among species there are in their responses to the individual covariates, whereas the 6 off-diagonal elements measure possible covariation between responses to the individual covariates.

Two community-level properties of major interest are species richness (i.e. the number of species) and species composition. Figure 4.22 shows how these can be inferred from the estimates of μ and Σ. The upper panels show the species richness S predicted by the model, measured here as the proportion of species out of all 100 species. For an environment described by the vector x of environmental covariates, S can be computed as

$$S = \Phi\left(\frac{x^T \mu}{\sqrt{x^T \Sigma x + 1}}\right). \tag{4.16}$$

As expected, species richness peaks at intermediate temperature and increases with soil fertility, the pattern being stronger in the community with independent species than in the community of competitive species.

The lower panels of Figure 4.22 ask how similar two communities are in their species composition. Let us denote by the vectors x_1 and x_2 the environmental

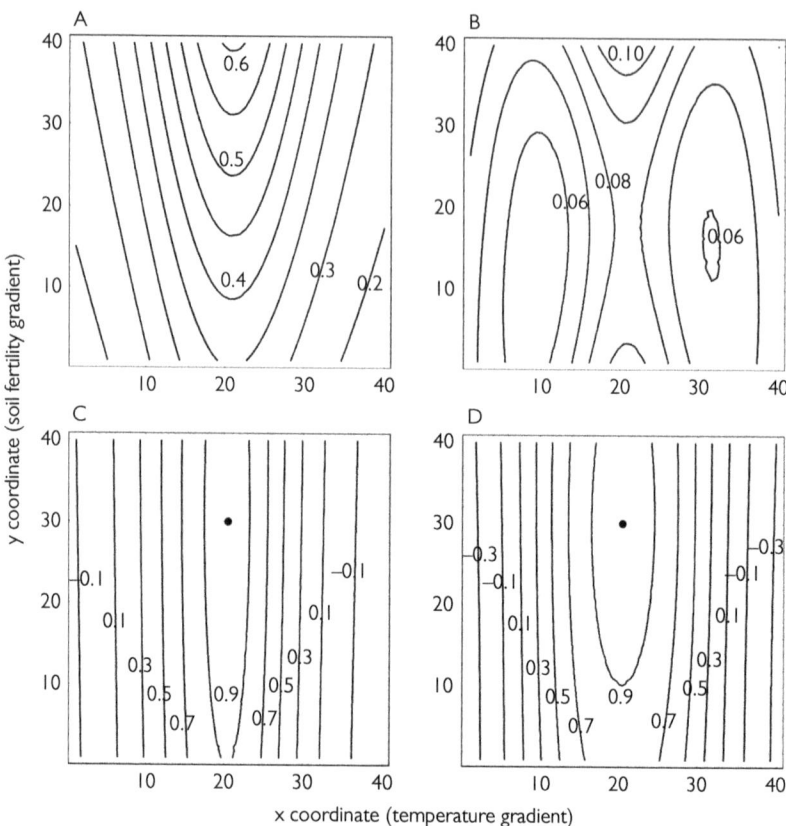

Figure 4.22 Predictions of species richness and community similarity generated by joint species distribution models. (A, B) show predictions of species richness, defined here as the fraction of the species in the community predicted to be present in a given locality. (C, D) show predictions of community similarity, defined as the correlation between the focal community and a reference community, the location of which is shown by the grey dot. The results are shown for the plant community model without (A, C) and with (B, D) interspecific competition.

conditions at localities occupied by two communities. We measure the similarity among these communities by the correlation in the linear predictors, given by $\rho_{12} = \text{Cov}_{12}/\sqrt{\text{Var}_{11}\text{Var}_{22}}$, where the covariance between the two communities is given by (Ovaskainen and Soininen 2011)

$$\text{Cov}_{12} = \text{Cov}(L(x_1), L(x_2)) = x_1^T \Sigma x_2. \quad (4.17)$$

In this equation, $L(x_1)$ and $L(x_2)$ refer to the linear predictors for the two communities. In Figure 4.22 we have measured the similarity of communities to an arbitrarily selected reference community, which is represented in the figure by the grey dots.

Community similarity decreases fast along the temperature gradient, as some species are specialized to low and others to high temperatures. In contrast, community similarity decreases only slow along the soil fertility gradient, as all species generally prefer high soil fertility. Thus, compared to the reference community, a community found from an environment with the same temperature but with lower soil fertility is likely to have fewer species, the species identities being a random subset of the more species rich community.

4.5.3 Ordination methods

Various kinds of ordination methods are applicable to the same kind of data as the joint species distribution model analyses illustrated earlier. Unconstrained ordinations, such as correspondence analysis (CA) and principal component analysis (PCA), are based on the species matrix y only. Constrained ordinations, such as constrained (or canonical) CA (CCA) and redundancy analysis (RDA), also account for the environmental covariates present in the X matrix.

Ordination methods are covered in great detail e.g. in the books by Borcard et al. (2011) and Legendre and Legendre (2012). Instead of going through the different kinds of ordination methods and the choices of the distance measures involved in those, we exemplify ordinations by generating RDA plots for the same data that we used for the joint species distribution model. In the RDA plots of Figure 4.23, the circles denote the sites, the crosses the species, and the vectors the environmental covariates. The RDA plots have correctly identified that most of the variation in community structure, represented by the RDA axis 1, relates to variation in temperature. The length of the soil fertility vector is much smaller, consistently with the fact that we found soil fertility to influence community dissimilarity only little with the joint

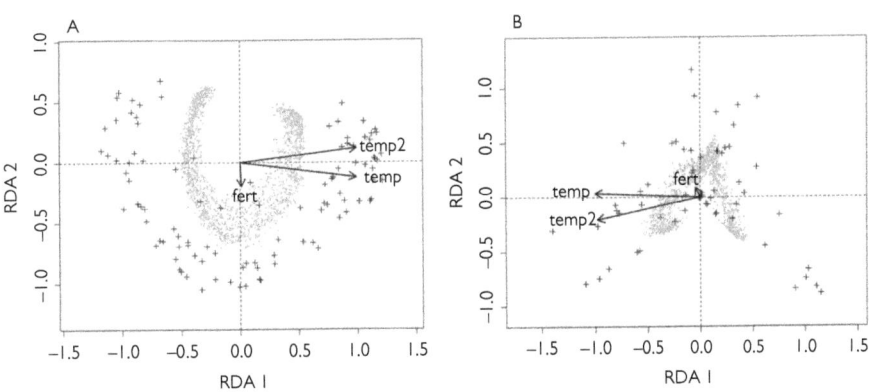

Figure 4.23 RDA plots illustrating community data generated by the plant community models. The grey dots correspond to the sampling sites, the black crosses to the species, and the vectors to the covariates. The results are shown for the plant community model without (A) and with (B) interspecific competition.

species distribution model analyses. Often the species form separate clusters in ordination plots, but here they form a continuum, reflecting the fact that they all share the same life-history (i.e., the same model structure) and their parameter values form a continuum.

4.5.4 Point-pattern analyses of distribution of individuals

We finish this section by returning to the issue of co-occurrence patterns. Instead of presence–absence data on a grid considered earlier, we assume that we have collected accurate spatial data on the locations of the individuals. The data that we utilize are final distributions of individuals simulated by the two-species models, assuming competitive interactions, consumer–resource interactions, or predator–prey interactions (Figures 4.6–4.8). As the models were run in homogeneous space, the point patterns are influenced only by population dynamical processes, making this example more simplistic than is usually the analysis of real data.

Statistical point process analyses involve a large array of methods (see e.g. Illian et al., 2007; Bivand et al., 2013). Our aim is to examine whether the point patterns show evidence of spatial aggregation or segregation, either within or among species. Among the many metrics and methods available for addressing this question, such as variograms and K-functions, we apply here the second-order spatial moment, by which we already characterized spatial aggregation in the single-species model (Figure 3.3C).

The second-order spatial moment $k_{ij}^{(2)}(d)$ measures the density of pairs of individuals that are at distance d from each other, one of which belongs to the species i and the other to the species j. In a random pattern, this equals the product of the population densities of the two species, which are given by the first-order moments $k_i^{(1)} k_j^{(1)}$. To characterize the level of spatial correlation $\rho_{ij}(d)$, we subtract the expectation and normalize, thus defining

$$\rho_{ij}(d) = \frac{k_{ij}^{(2)}(d) - k_i^{(1)} k_j^{(1)}}{k_i^{(1)} k_j^{(1)}}. \tag{4.18}$$

Defined in this way, spatial autocorrelation measures the excess ($\rho_{ij}(d) > 0$) or shortage ($\rho_{ij}(d) < 0$) of individuals of species i and j at a distance d from each other, compared to the random expectation.

In the two-species competition model, the level of intraspecific aggregation increases whereas the level of interspecific segregation increases when increasing level of competition (Figures 4.24A–D). This is to be expected, as interspecific competition separates the species to their own parts of the space, as seen by visual inspection of Figure 4.6. Note that in Figure 4.6D the species depicted by black colour is present only in one remnant population. This is reflected in Figure 4.24D by the black species showing a very high level of intraspecific aggregation.

In the predator–prey models, the very clear spatial patterns visible in Figure 4.8 are translated to high levels of intraspecific aggregation (Figures 4.24E, F).

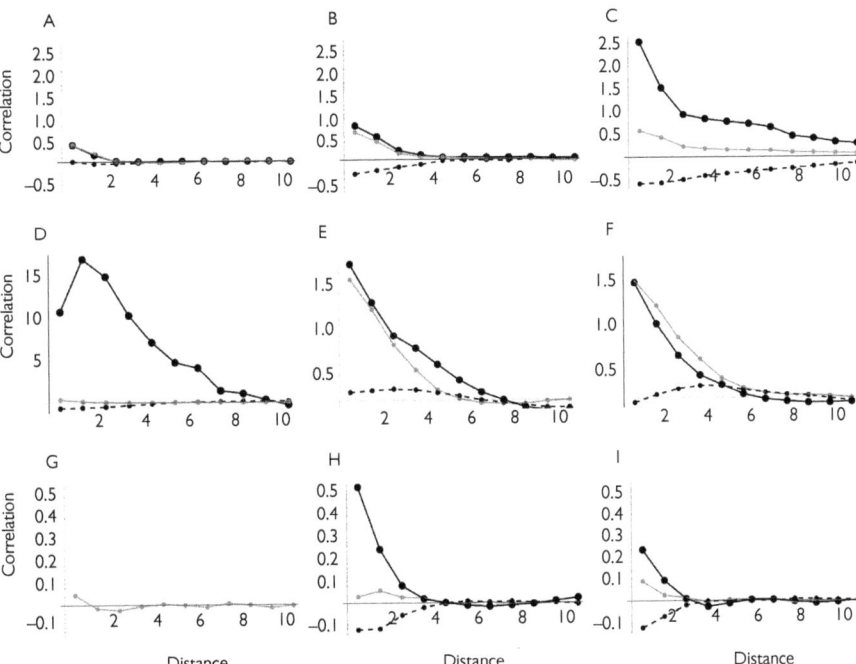

Figure 4.24 Within- and between-species patterns of spatial aggregation measured by spatial correlation functions (Eq. (4.18)). The panels show correlation functions for distributions of individuals generated by the homogeneous-space two-species models of Section 4.2. (A–D) correspond to two-species competition models, shown respectively in Figures 4.6A–D. (E, F) correspond to predator–prey models, shown respectively in Figures 4.8C, F. (G–I) correspond to resource–consumer models, shown respectively in Figures 4.7C, F, I. In all panels, black dots (respectively, grey dots) connected by solid lines show the correlation functions for species depicted by black (respectively, grey) in the figures of Section 4.2. Black lines connected by dashed lines show between species correlation functions.

The interspecific correlation is influenced by two counteracting forces: the predators increase in areas where the prey is abundant, but the prey remains abundant only in areas where the predator is absent. With the parameters assumed in the simulation, the balance of these two forces results in the slightly positive interspecific correlation (Figures 4.24E, F). In the case of the resource–consumer model (Figures 4.24G–I), the distribution of resources is less aggregated than the distribution of the prey. This is because new resources appear in random locations rather than in locations close to existing resources. Thus, their non-random pattern is generated only by consumers, which decrease resource abundance more in some parts of the space than in others. Due to the same reason, there is negative spatial correlation between the resources and the consumers (Figures 4.24H, I).

In the consumer–resource model, the consumer-free equilibrium results in a fully random spatial pattern of resources. Thus, the deviation of the correlation function shown in Figure 4.24G from zero represents random noise. To test for the statistical significance of any of the patterns shown in Figure 4.24, the empirically observed correlation functions should be compared to the corresponding null-expectations, which can be done using e.g. the R-package spatstat (Baddeley and Turner, 2005).

4.6 Perspectives

In this chapter, we have introduced mathematical and statistical approaches in the context of community ecology. Real communities consist of a large number of interacting species and they are influenced by many kinds of spatially and temporally varying environmental conditions. This is the reason why theoretical community ecology involves a large number of choices related to model structures and parameterization (Vellend, 2010). To avoid ending up in a complete mess, it is necessary to start from the simplest models, and then examine which aspects of model behaviour change and which kinds of new phenomena appear when additional layers of complexity are introduced. To follow this recipe, we started this chapter with two-species models in homogeneous space, and then expanded the models in various ways, to finally consider a large community of interacting species in heterogeneous space.

Like with the previous chapters, we have made a large number of choices and assumptions when building our models. For example, we considered simple two-species models, and a model of a large competitive community where all species belong to the same trophic level. Therefore, we largely ignored the influence of the food web structure itself, the question of which has been much studied in the theoretical community ecology literature (e.g. Cohen and Newman, 1985; Williams and Martinez, 2000; Cattin et al., 2004; Allesina et al., 2008). As a second example, we assumed in all models stationary individuals rather than modelling interactions that may result from encounters generated by movement processes, such as predators searching for prey (Avgar et al., 2011). As a third example, we continued to work within the frameworks of Markov processes and differential equations, and thus did not consider time lags generated e.g. by handling times or saturation of the consumers (Li et al., 2013).

Rather than continuing to list all the caveats and limitations, let us next place the models of this chapter into a broader context by discussing how they relate to the metacommunity paradigms introduced in Section 4.1.

4.6.1 Back to the metacommunity paradigms

Even though the concepts of metapopulations and metacommunities are theoretically appealing thanks to their simplicity, in real environments it may be difficult to partition the space into a well-defined set of discrete local populations or local communities. For example, the spatially explicit plant community model that we considered in this chapter is not a model of a single well-mixing local community, but it is neither a metacommunity model of a collection of local communities. In Section 4.1

we introduced four metacommunity paradigms: the species-sorting, the mass-effect, the patch-dynamics, and the neutral paradigms (Leibold et al., 2004; Logue et al., 2011). We have not built our models strictly following the assumptions of any of these paradigms. However, they include elements from all of these, and thus illustrate that the different paradigms are not isolated from each other.

First, the species-sorting perspective (Shmida and Wilson, 1985; Leibold, 1998; Cottenie et al., 2003) was evidently present in our heterogeneous-space models. This is because we assumed variation in the fundamental niches of the species, and this variation made them thrive in different parts of the space. If considering a small subset of the simulation domain, the dynamics are dominated by niche variation and local interactions, but over larger spatial scales, the model allows for occasional colonization events. This brings in the mass-effect paradigm (Mouquet and Loreau 2002, 2003): a species that is abundant in its core area will also send propagules to the neighbouring areas, thus potentially extending its distribution outside its fundamental niche, influencing also the dynamics of other species.

The patch-dynamics perspective allows for regional coexistence of species by a trade-off between competitive ability and colonization (Levins and Culver, 1971; Hastings, 1980; Tilman, 1994). Even though we did not consider differences in dispersal ability, this perspective was present in our plant models in two ways. First, we assumed variation in competitive ability, as only a subset of species imposed competitive effects on other species. Second, we assumed variation in colonization ability, as the parameters controlling for fecundity and establishment varied among the species. Thus, some species may have persisted in our simulations because of their high colonization ability, whereas others may have persisted due to their high competitive ability.

The neutral perspective emphasizes that species communities are influenced by inherent stochasticity in population processes (Hubbell, 2001). Although we did not explore it here, our plant models involve both the fully neutral- and fully niche-based models as two extremes. A fully neutral model is obtained by assuming that the species have identical parameters, in which case the identities of the species that would persist in the system would be based solely on randomness. The fully niche-based model could be obtained by taking the deterministic (either spatial or non-spatial) limit of the stochastic model presented here, using the techniques introduced in Chapter 3. In that case, the community would not involve ecological drift, and its dynamics would thus be fully niche-based.

4.6.2 Some insights derived from community models

As we showed in Section 4.2, even very simple two-species models can lead to rather complex dynamics, both in space and in time. Further, small changes in model structure can lead to qualitatively different dynamics, making it difficult to obtain general and robust insights. This was exemplified with the resource–consumer and the predator–prey models, which differ only in their assumption on whether new preys (or resources) were introduced to the system at a constant rate or through a population dynamical process. As a consequence of this small but important difference, the resource–consumer model showed stable dynamics, whereas the predator–prey showed oscillatory behaviour.

When moving to models in heterogeneous space, we included processes related both to ecological interactions and to environmental filtering, the processes of which jointly determine the composition and structure of a community (Leibold, 1995; Kraft et al., 2015). As illustrated by our models, niche differentiation combined with spatial environmental heterogeneity is one straightforward mechanism that allows species to coexist (Leibold, 1995; Chesson, 2000), through either classical resource partitioning (MacArthur and Levins, 1967) or spatial niche partitioning (May and Hassell, 1981). As reviewed by Amarasekare (2003), there are many other mechanisms that enable the coexistence of species, too. As one example, species can exploit the same resource but at different times, a process called temporal niche partitioning (Chesson, 2000). Indeed, many abiotic factors fluctuate over time, allowing the coexistence of species due to e.g. diurnal or seasonal variation (Hsu, 1980).

We simulated the responses of ecological communities to habitat loss in order to generate species-area relationships, which quantify how the number of species expected to persist in an area depends on the area of the habitats available (He and Legendre, 2002). The reasons for the general validity of the species-area relationship (Eq. (4.9)) is still debated (He and Hubbell, 2011), but it is generally agreed that two factors play a major role. The first one is related to environmental filtering: a larger area contains more habitat heterogeneity, and thus it contains niches for a larger number of species (Holt et al., 1999). The second one is the link between population persistence and habitat amount: in a larger area, a species will have a larger population size, which decreases its risk of extinction for the reasons discussed in the single-species context of Chapter 3. In our simulated communities, both of these processes were operating. At still larger spatial and temporal scales, also evolutionary dynamics become relevant in generating species-area relationships (Lawson and Jensen, 2006), a process that we did not consider here.

As discussed in the single-species context of Section 3.4, fragmentation can have an additional effect on population dynamics on top of the effects of habitat loss. Such effects are operating also at the community level (Bonin et al., 2011), even though we did not study them here. Let us assume that a conservation biologist could not influence the amount of habitat loss, but there would be the option of keeping the remaining habitat as one block (as was assumed in our simulations) or distributing it as a set of smaller fragments. The optimal arrangement of habitats is not an easy question, as there is a trade-off (Lindenmayer and Fischer, 2007). If protecting all in one block, the dynamics of the species within the protected area are likely to remain relatively uninfluenced, but some of the species may have been endemic to the lost areas. If distributing the protected sites evenly throughout the region, a larger number of species will be covered, but in the fragmented network of patches their persistence may be compromised. This question was much debated in the context of the SLOSS (single large or several small) discussion (Wilcox and Murphy, 1985, and references therein), it has been addressed by methods developed for the design of conservation networks (Hokkanen et al., 2009), and it is in the focus of theory for species fragmented-area relationships (SFAR; Hanski et al., 2013).

4.6.3 The many approaches to modelling community data

Community ecology involves a large array of questions, and this is reflected by the diversity of data types and consequently also the statistical methods that have been developed. For all analysis introduced in Section 4.5, a large number of extensions are available. For time-series analysis of multispecies data, the same alternatives that were discussed in the single-species context (Section 3.6) apply as well. For joint species distribution models, we could have considered abundance data, related the responses of the species to the environmental covariates to their measured traits (Brown et al., 2014), let the model classify the species to discrete groups instead of assuming a continuous community-level distribution (Hui et al., 2013), accounted for phylogenetic relationships (Ives and Helmus, 2010), or extended to time-series data (Sebastián-González et al., 2010). In case of ordinations, the list of different options is very long (see e.g. the books Borcard et al., 2011; Legendre and Legendre, 2012), as is the case with point-pattern analyses (Illian et al., 2007; Bivand et al., 2013).

The emergence of the neutral theory of biodiversity has provoked much interplay between theory and data. Neutral theory highlights the importance of dispersal limitation, speciation, and ecological drift, and provides a quantitative null model for assessing the roles of niche-based processes, including adaptation and natural selection (Rosindell et al., 2011). Many fundamental patterns, such as species-area and species-abundance relationships, can be related to parameterizations of neutral models (Rosindell et al., 2011), implying that these patterns can be explained without explicitly invoking species' differences (Wennekes et al., 2012). Models that include both neutral and niche-space processes have been fitted to data to estimate the relative influences of neutral and non-neutral processes in shaping community assembly (e.g. Mutshinda et al., 2009; Pollock et al., 2014; van der Plas et al., 2015). The results have been found to depend on the spatial scale considered (Chase, 2014; Garzon-Lopez et al., 2014). With increasing spatial scale, the importance of habitat association generally increases, whereas the importance of stochastic processes generally decreases. Thus, communities appear niche-structured at large scales and neutrally structured at small scales (Chase, 2014).

Rather than attempting to master all the details of all possible statistical methods, the main challenge for a quantitative community ecologist is to define the key questions to address, and to acquire the data and apply the statistical methods adequate for addressing those questions. As partly illustrated by the analyses of this chapter, different statistical methods can result in similar insights if the data contains a strong signal of the studied phenomenon.

5

Genetics and evolutionary ecology

5.1 Inheritance mechanisms and evolutionary processes

In the previous chapters, we have dealt with the ecology of individuals, populations, and communities. But as famously remarked by Dobzhansky (1973), nothing in biology makes sense except in the light of evolution. Thus, to understand why animals, plants, and other organisms move as they do (Chapter 2), and why they show such remarkable variation in their life-history traits, ecological niches, and interactions with other organisms (Chapters 3 and 4), one needs to take an evolutionary point of view. This is the focus of the present chapter.

For an overview on how the history of genetics and evolutionary biology unfolded, we refer to Provine (1971) and Edwards (2011). Very briefly, the foundations for modern genetics and evolutionary biology were set by the works of Gregor Mendel (Henig, 2001; Ellis et al., 2011), Hugo de Vries (1914), Sewall Wright (1931, 1932, 1937, 1942, 1948), John Haldane (1932a, b), and Ronald Fisher (1930, 1932, 1936). These pioneering works built theory upon 'heredity' of traits, i.e. the propensity of traits of a parent to be passed on to its offspring. Now we know that genes are the agents that carry traits through generations and that they are made of DNA (Watson and Crick, 1953).

Studies in evolutionary ecology examine how natural selection and other genetic processes generate and maintain genetic variation within species and among species (Figure 5.1), enabling them to adapt to changing environments. In this introductory section, we provide a brief overview on the mechanisms of heritability and discuss some fundamental principles of evolutionary and eco-evolutionary dynamics.

5.1.1 Genetic building blocks and heritability

Genetic code can be represented with the help of the letters G, A, T, and C, which stand respectively for guanine, adenine, thymine, and cytosine. These are nucleobases that combine with deoxyribose to form nucleotides, which are joined together into chains called polynucleotides. Two of such chains form a double helix structure called deoxyribonucleic acid, i.e. DNA. DNA contains genes that are the units of genetic information. Genes contain the instructions for making ribonucleic acid (RNA), which in turn makes proteins, i.e. the molecules that an organism needs for all of its functions throughout its lifetime. Other parts of DNA contain regulatory elements

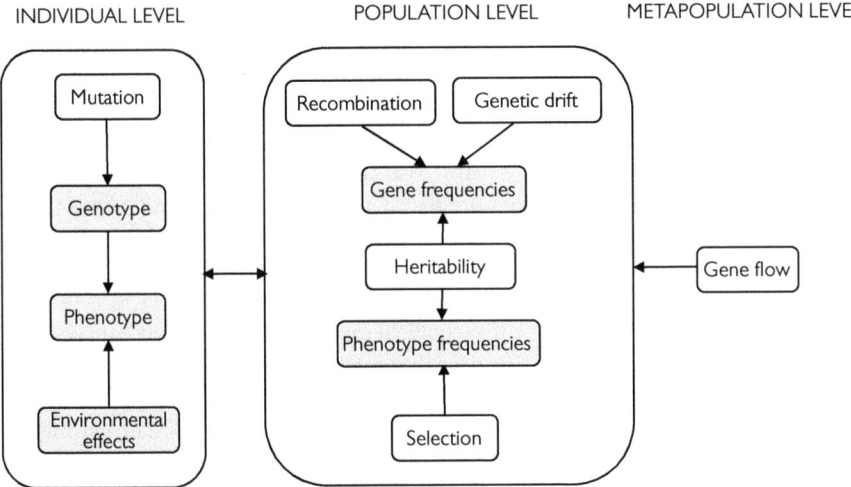

Figure 5.1 An individual's phenotype is jointly determined by its genotype and the environment to which the individual is exposed during its development. The degree by which the phenotype is determined by the genotype is called heritability. Evolution is defined as the change in gene frequency at the population level, and it is influenced by the processes of recombination, mutation, gene flow, and selection. Recombination and mutation act directly on the genotypes, whereas natural selection acts on phenotypes, and thus the effect of selection is carried over to the next generation only to the extent by which the trait under selection is heritable. Gene flow adds or removes genetic material from a population through immigration and emigration, respectively.

that control the process of making these molecules. Genes, regulatory elements, and non-coding parts of DNA are packaged together into larger structures called chromosomes. Some cells contain two sets or copies of each chromosome (called diploid cells), whereas other cells have only one copy of each chromosome (haploid cells, e.g. in the case of egg and sperm cells in sexually reproducing organisms), and some have more than two copies (called polyploid cells, common in plants). Eukaryotes (organisms whose cells contain a nucleus, e.g. animals and plants) have several chromosomes, whereas prokaryotes (single-celled organisms that lack a membrane-bound nucleus or other organelles, e.g. bacteria and archaea) have only one.

The collection of all of an organism's genetic material is called its genome. The location of a gene within a chromosome is called locus, and the variants or alternative forms of a gene on a specific locus are called alleles. Individuals belonging to the same species generally have the same genes, but they differ in their genotypes, i.e. the alleles that they have for each gene. The phenotype of an organism refers to all of its traits, such as its morphology and behaviour, which are determined jointly by the genotype and the environmental conditions that an individual experiences during its development. As a first approximation, the phenotype can be considered as the result of the genetic effects plus the environmental effects.

We will focus mainly on sexually reproducing diploid organisms. If a given locus has e.g. two allelic variants, denoted by A_1 and A_2, there are three possible allelic combinations. The locus is said to be homozygous if there are two copies of the same allele, i.e. if the genotype is A_1A_1 or A_2A_2. The locus is said to be heterozygous if the alleles are different, so that the genotype is A_1A_2. In a population in which the frequencies of the alleles A_1 and A_2 are 90% and 10%, respectively, the expected frequencies of these three genotypes are 81% (A_1A_1), 1% (A_2A_2), and 18% (A_1A_2). Here we have assumed that the population is in the Hardy–Weinberg equilibrium (Weinberg, 1908; Hardy, 1908), i.e. that the alleles are randomly distributed over the individuals.

Allelic effects are called additive if each allele makes its independent contribution to a phenotypic trait so that the effects of different alleles can be summed together. In this case, the expected value of the phenotypic trait of the heterozygote A_1A_2 is the average of those of the homozygotes A_1A_1 and A_2A_2. At the other extreme is the case where one of the alleles is dominant (often denoted by capital letters such as A) over the other alleles, which are called recessive (often denoted by lowercase letters such as a). Mendel's law of dominance states that recessive alleles will always be masked by dominant alleles, so that both genotypes AA and Aa display the effect of the dominant allele, whereas only the genotype aa displays the effect of the recessive allele. Although some phenotypic traits depend only on a single gene, many traits are polygenetic, i.e. they are influenced by more than one gene. Epistasis means that the allelic effects of one gene are modified by the allelic state of another gene, which is then called a modifier gene. If there is no epistasis, the contributions that the different genes make to a phenotype can be simply summed together.

In asexually reproducing species, an individual inherits its genes clonally from a single progenitor, whereas in sexually reproducing species, an individual inherits its genes from its parents through recombination. In the latter case, pairs of genes are separated (or segregated) into reproductive cells (gametes), so that the individual inherits for each gene one allele from one parent and the second allele from the other parent. The law of independent assortment dictates that separate genes are passed on independently from parents to offspring. However, alleles located near each other within the same chromosome are often linked, i.e. inherited together. In this case, the combinations of alleles that are inherited as one unit are called haplotypes. For a more detailed treatment about the fundamentals of genetic architectures and inheritance mechanisms, we refer the interested reader to one of the many textbooks on this topic, e.g. the one written by Daniel Hartl (2014).

5.1.2 Selection, drift, mutation, and gene flow

The concept of evolution includes all processes that result in a change in the genetic composition of a population or a species over time. Genetic composition can be measured by gene frequency, also called allele frequency. For example, if the frequencies of the alleles A_1 and A_2 were 80% and 20% in the previous generation, and they are 90% and 10% in the present generation, evolution has taken place.

Such an evolutionary change may have occurred because of selection, meaning that the allele frequency of A_1 increased because it was positively associated with fitness. In this case individuals with allele A_1 had a higher chance of producing offspring than individuals with allele A_2, and through inheritance the frequency of A_1 thus increased in the next generation.

As survival and reproduction depend on the phenotype, selection acts on phenotypes and not directly on genotypes. This is why the influence of selection on the next generation depends on the heritability of the trait: only the heritable part of the effect of selection is carried over to the next generation. In the case of animal and plant breeding, the choice of which individuals reproduce are made by the breeder, i.e. by artificial selection, whereas in the case of natural selection, the fitness of an individual depends on the interplay between its traits and its environment (Darwin, 1859).

Evolution does not operate through natural selection only, but it also takes place through a neutral process called genetic drift. In a sexually reproducing organism, the reproductive cells contain only one copy of each gene, so that an offspring obtains one copy from each parent. For example, offspring from a father with genotype A_1A_1 and a mother with genotype A_2A_2 will necessarily have genotype A_1A_2. But if the mother instead has genotype A_1A_2, the offspring will have genotype A_1A_1 with probability 0.5 and genotype A_1A_2 with probability 0.5. Recombination thus shuffles genetic material so that offspring can have different combinations of genes than their parents and their siblings. Recombination is neutral in the sense that it does not change allele frequencies in any specific direction over time, and it thus generates a random 'drift'. Neutral genetic changes accumulate over time and thus make populations eventually distinct from each other. Genetic differentiation provides tools for research, as it makes it possible to identify individuals that have common ancestors and to place species in the phylogenetic tree of life.

In addition to selection and recombination, mutation and gene flow are fundamental components of evolution. Mutation changes DNA sequences by generating errors in the replication process. A mutation becomes heritable if it is passed down to a gamete, i.e. a sperm or egg cell in sexually reproducing organisms. Mutations can introduce new alleles into a population and thereby increase its genetic variation. Gene flow changes a population's genetic composition through alleles carried by migrants originating from other populations (Slatkin, 1985; Bohonak, 2012). Thus, the rate of movements between local populations (Chapter 2) determines not only the degree to which they are demographically independent (Chapter 3), but also the rate of gene flow and thus the extent to which the local populations are genetically independent (Slatkin, 1985; Hanski, 1999). Both mutation and gene flow can lead to either neutral or adaptive evolution, depending on whether the genes they modify are under a selection pressure.

Taken together, mutation and gene flow have the potential of introducing novel genetic variation into a population, whereas recombination and the associated genetic drift shuffle the existing variation. Natural selection filters the genetic variation by amplifying the frequencies of beneficial genes (positive selection) and reducing the frequencies of harmful genes (negative selection).

5.1.3 Connections between ecological and evolutionary dynamics

Ecological and evolutionary dynamics are intimately linked because the ecological context determines the fitness of an individual (Pelletier et al., 2009), and because the persistence of organisms in changing environments ultimately depends on their ability to adapt through evolutionary changes (Post and Palkovacs, 2009). The reciprocal coupling of ecological and evolutionary dynamics is called eco-evolutionary dynamics.

In Chapter 4, we discussed how species composition in a local community is determined by the balance between environmental filtering, species interactions, and dispersal. In brief, dispersal introduces species from the regional species pool into a local community, after which environmental filtering and species interactions determine which species are able to persist there. In the same way, the interplay between evolutionary forces and environmental variation influences the outcome of evolutionary dynamics. Sufficient levels of gene flow, recombination, and mutation are needed to generate genetic variation, on which selection acts to amplify the genotypes that are best adapted to the prevailing environmental conditions. For ecological dynamics, a high level of dispersal brings a source-sink effect, making it possible that species are found also outside their fundamental niche. For evolutionary dynamics, a high level of gene flow may sustain genes that are maladapted in the focal population but adapted in the source population.

As another example of connections between ecological and evolutionary dynamics, in Chapter 3 we discussed the influence of the temporal scale of environmental variation on population dynamics. In slowly varying environments, populations closely track variation in the carrying capacity of the environment, whereas the effects of environmental variation are essentially averaged out in fast varying environments. In an analogous way, the potential and need of a population to adapt to environmental changes depends on the rate at which new genetic variation is created. Under the ongoing global change, natural habitats are being converted into anthropogenic habitats, whilst changing climatic conditions rearrange the extent and spatial distribution of entire biomes (e.g. Loarie et al., 2009). While this change leads to population declines in some species, which eventually might go extinct, other species are able to adapt to the new conditions—a phenomenon called the evolutionary rescue effect (Bell and Gonzalez, 2009, 2011; Bell, 2012; Gonzalez et al., 2013; Carlson et al., 2014).

The joint influence of demographic and genetic processes on population viability is perhaps best illustrated by the dynamics of small and isolated populations. In Chapter 3, we discussed why such populations have a high probability of extinction for ecological reasons (Hanski, 1999; Lande et al., 2003; Ovaskainen and Meerson, 2010). On top of these, also genetic mechanisms can negatively influence the viability of small populations. First, the rate of loss of genetic variation due to drift is inversely related to population size (Kimura and Ohta, 1969; Falconer and Mackay, 1996; Amos and Balmford, 2001; Hartl, 2014), thereby decreasing the future potential for adaptation especially for small populations (Willi et al., 2006). Second, small population size increases the chance of inbreeding, i.e. the mating between relatives.

This reduces the level of heterozygosity and can give rise to inbreeding depression, i.e. reduced fitness of the individuals (Mattila et al., 2012; Hartl, 2014), which can have negative consequences for population growth (Amos and Balmford, 2001; Keller and Waller, 2002). The interplay between ecological and genetic mechanisms in making small populations vulnerable to extinction has been called the extinction vortex (Courchamp et al., 1999; Keller and Waller, 2002; Fagan and Holmes, 2006; Blomqvist et al., 2010).

5.1.4 The outline of this chapter

In the rest of this chapter, we present a selection of mathematical and statistical approaches that can be used to address questions in evolutionary ecology. We will focus especially on patterns generated by neutral versus adaptive evolution. Much of the work presented here has been motivated by the work of Juha Merilä and his colleagues on the evolutionary ecology of nine-spined sticklebacks, illustrated in Box 5.1

Box 5.1 An empirical example of research in evolutionary ecology

Sticklebacks are one of the best studied model species for evolutionary biology, in particular the three-spined stickleback (e.g. Bell and Foster, 1994), but also the nine-spined stickleback (e.g. Merilä, 2013). Juha Merilä and colleagues have conducted a large number of studies on the latter species in Fennoscandia, where it is present in both marine and freshwater environments. The present populations diverged from a common ancestral population less than 8,000 years ago during the postglacial period. The most contrasting case is the comparison between the large marine populations and the small populations found in isolated ponds. The small pond populations experience a much higher level of genetic drift than the large marine populations, and consequently they sustain a much lower level of genetic variation (Shikano et al., 2010). The populations show a great amount of phenotypic divergence, much of which correlates with habitat type. For example, fish found from small ponds are larger (Figure 5.2; Herczeg et al., 2009a), they have less body armour (Välimäki et al., 2012), they are bolder (Herczeg et al., 2009b), and they differ in their brain size and structure (Gonda et al., 2009) from fish originating from marine populations. This is largely due to the higher predator abundance in the marine environments, which imposes a strong selective pressure towards traits promoting survival under high predation risk (e.g. Merilä, 2013).

To show that the differences among the marine and freshwater populations are determined genetically rather than environmentally, Merilä and colleagues have conducted common garden experiments. In these experiments, fish originating from the wild formed the parental generation, and their offspring were grown under identical conditions (e.g. Herczeg et al., 2009a). As the differences among the populations were present also in the offspring generation, they must have a genetic basis. The fact that freshwater populations differ generally from marine populations makes the nine-spined stickleback a prime example of convergent evolution: evolution has taken the same route in many freshwater populations even if they represent independent postglacial colonizations (e.g. Merilä, 2013).

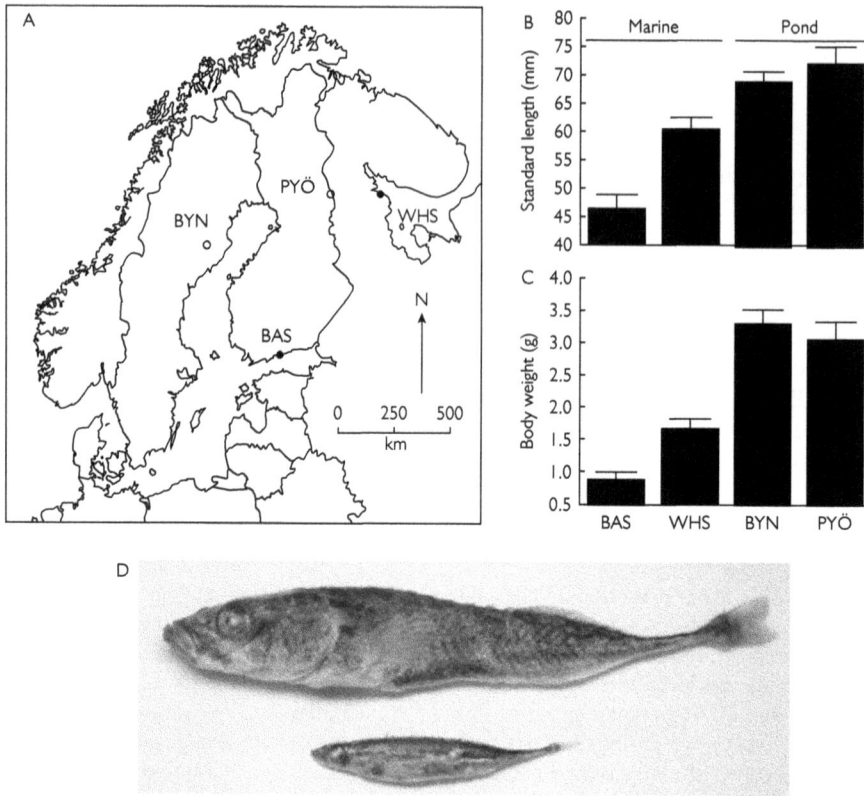

Figure 5.2 An empirical case study on the evolutionary ecology of nine-spined sticklebacks (*Pungitius pungitius*). The map of (A) (from Karhunen et al., 2014) shows the location of four study populations, out of which two are in marine (BAS = Baltic Sea and WHS = White Sea) and two in freshwater (PYÖ = Pyöreälampi and BYN = Bynästjärnen) environments. (B, C) show the results of a common garden study, in which fish originating from the four populations formed the parent generation, and their offspring were grown in the laboratory under identical conditions. The bars show the mean body length (B) and mean body weight (C), and the error bars correspond to one standard error (Herczeg et al., 2009a). (D) illustrates the great amount of variation in the sizes and shapes among individuals. The large individual originates from a freshwater population, whereas the small individual originates from a marine population. Photo by Juha Merilä. (A), (B), and (C) reproduced with permission from John Wiley and Sons.

and Figure 5.2. In Section 5.2, we introduce the basics of quantitative genetics in the context of neutral dynamics for a diploid, sexually reproducing species. In particular, we discuss why and how the degree of phenotypic similarity between individuals and populations depends on their relatedness. In Section 5.3, we bring the process of natural selection into the picture, and thus discuss how evolution is shaped by the interplay between selection and genetic drift. In Section 5.4, we look into the evolutionary consequences of habitat loss and fragmentation, specifically focusing on the evolution of dispersal. Finally, in Section 5.5 we discuss how evolutionary questions can be addressed with statistical approaches: how to use genetic data to quantify the relatedness between individuals and populations; how to find the loci that influence quantitative phenotypic traits; and how to disentangle the contributions of genetic drift and natural selection as causes of evolutionary change.

5.2 The evolution of quantitative traits under neutrality

In this section, we focus on population genetics under the assumption of neutrality. Neutrality means that the genes under consideration do not influence the fitness (survival or reproductive success) of the individuals. In this case, there is no feedback from evolution to ecology, and hence population genetic processes can be considered as an independent additional layer on top of demographic population dynamics.

As discussed in Section 5.1, evolution has generated multiple types of reproductive and inheritance mechanisms (such as clonal and sexual), and multiple types of genetic architectures (such as haploid, diploid, and polyploid). We focus here on the case of a sexual species with two mating types (to be called sire and dam to follow the usual jargon in population genetics; van der Werf et al., 2009), and diploid genetic architecture. Therefore, each individual possesses two alleles for each gene. Following Mendel's first law of segregation (Henig, 2001), the offspring receives one allelic copy from the dam, selected randomly from her two copies, and similarly one allelic copy from the sire. Mendel's second law (Henig, 2001) assumes independent assortment: different genes are passed on to the offspring independently from each other. This is not generally true, as genes located close to each other on the same chromosome often show linkage, i.e. they can be inherited in one block (Grant, 1966). However, to simplify the mathematical treatment, we will assume Mendel's second law in the following sections.

5.2.1 An additive model for the map from genotype to phenotype

We focus here on the so-called infinitesimal model for a polygenetic trait, which assumes that the trait is influenced by many genes, each having a small influence (Falconer and Mackay, 1996; Lynch and Walsh, 1998). A much studied example of a polygenetic quantitative trait is body size; other examples include e.g. the colour (Bradshaw et al., 1998; Tripathi et al., 2009) and shape (Zimmerman et al., 2000; Klingenberg, 2004) of body parts, as well as competitive ability (Collins, 2003). Considering for a moment the case of a single trait, the additive genetic value of individual i can be modelled as

$$a_i = \sum_{j=1}^{n} \sum_{k=1}^{2} \sum_{u=1}^{m_j} x_{ijku} v_{ju}. \tag{5.1}$$

Here the index j runs over the (large) number n of loci that influence the trait in question, and the index k runs over the two allelic copies that the individual has.

The index u runs over the m_j allelic variants present in locus j, and x_{ijku} is an indicator variable with value $x_{ijku} = 1$ if the allelic type of the kth copy is u and otherwise $x_{ijku} = 0$. The term v_{ju} is the additive value of the allele u in locus j. Since much of the theory for population genetics has been developed in the context of animal and plant breeding rather than evolutionary ecology of natural populations, the additive genetic value a_i is commonly called breeding value (Lynch and Walsh, 1998).

In words, Eq. (5.1) says that the breeding value of an individual is the sum of the effects of those allelic variants that the individual holds in the loci that influence the trait in question. As the use of the indicator variables x_{ijku} may make Eq. (5.1) look somewhat cumbersome, let us go through it with an example. Assume that body size is influenced by two genes, denoted by A and B, where gene A has two allelic variants (denoted by A_1 and A_2), whereas gene B has three allelic variants (denoted by B_1, B_2, and B_3). Let us further assume that the effects of these alleles on body size are $v_{A1} = -1$, $v_{A2} = 1$, and $v_{B1} = -2$, $v_{B2} = 1$, $v_{B3} = 1$. Consider two individuals, of which individual 1 has alleles (A_1, A_1) in locus A and alleles (B_1, B_1) in locus B, whereas individual 2 has alleles (A_1, A_2) and (B_2, B_3). The reader may sum up the effects of these alleles to verify that the breeding values of the individuals are $a_1 = -6$ and $a_2 = +2$. Thus, based on their genes, individual 1 will have a smaller body size than individual 2. To exemplify the indicator function notation x_{ijku}, consider the locus B of individual 2. As the allelic state of the first copy is B_2, and as there are three possible variants, we have $x_{2B11} = 0$, $x_{2B12} = 1$, and $x_{2B13} = 0$. Similarly, as the allelic state of the second copy is B_3, we have $x_{2B21} = 0$, $x_{2B22} = 0$, and $x_{2B23} = 1$.

The additive model is a great simplification of reality, as it ignores dominance and epistasis. The assumption of no dominance is seen in Eq. (5.1) by the fact that we just added together the effects of the two allelic copies, indexed here by $k = 1$ and $k = 2$. In the presence of dominance, the alleles may interact, so that their joint effect can be greater or smaller than what would be expected by the effects of the individual genes (Hartl, 2014). The lack of epistasis is seen by the fact that in Eq. (5.1) we just add together the effects of the individual loci. In the presence of epistasis, the effect of one gene is modified by the allelic state of another gene (Hartl, 2014). Thus, the effects of the allelic variants B_1, B_2, and B_3 would depend on the combination of alleles in locus A.

As discussed in Section 5.1, a phenotype is jointly influenced by the genotype and the environment. Figure 5.3 illustrates such a decomposition, and also makes the point that even genetically identical individuals that have developed in identical environments are not expected to be identical in their phenotypes. This is due to developmental instability, i.e. random events in development influencing the phenotype.

Assuming that the genetic and environmental effects are independent of each other (illustrated in Figure 5.3C), the phenotypic value of individual i can be modelled as

$$p_i = \mu_i + a_i + e_i. \tag{5.2}$$

Here μ_i is the expected value of the trait (which may include e.g. the effect of the sex of the individual), a_i is the breeding value, and e_i is the environmental effect. Due to

5.2 The evolution of quantitative traits under neutrality • 177

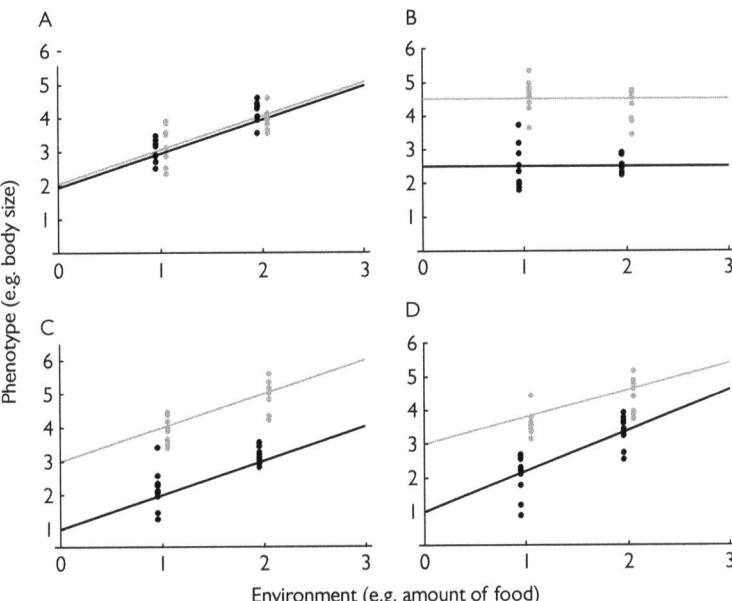

Figure 5.3 Illustrations of genotype–phenotype maps under different assumptions. The black and grey colours correspond to two genotypes. The lines show the expected phenotype as a function of the environment, whereas the dots show the realized phenotypes of ten individuals for each genotype under environments $E = 1$ and $E = 2$. In (A) genetic effects are negligible, whereas in (B) environmental effects are negligible. In (C) the effects of the genotype and the environment are additive, whereas (D) involves an interaction between the genetic and environmental effects. The individual phenotypes were sampled by adding normally distributed noise with zero mean and standard deviation 0.5 on top of the expected values. The dots have been slightly shifted in the horizontal direction to enable easier separation of black and grey dots.

its additive nature, Eq. (5.2) excludes genotype-environment ($G \times E$) interactions, which are illustrated in Figure 5.3D.

In full generality, the genetic value of a quantitative trait can be any non-linear function of the genotype. Likewise, the map from genotype to phenotype can be any stochastic function of the genotype and the environment. However, to maintain mathematical tractability, we will assume the additive model given by Eqs. (5.1) and (5.2).

5.2.2 Coancestry and the additive genetic relationship matrix

Before we can consider how neutral traits evolve over time, and to assess their distribution in a set of populations, we need the concept of coancestry, i.e. the relatedness between individuals. Let us thus consider a pedigree, such as that illustrated in Figure 5.4A, which shows the sire and dam for each individual. This population

178 • Genetics and evolutionary ecology

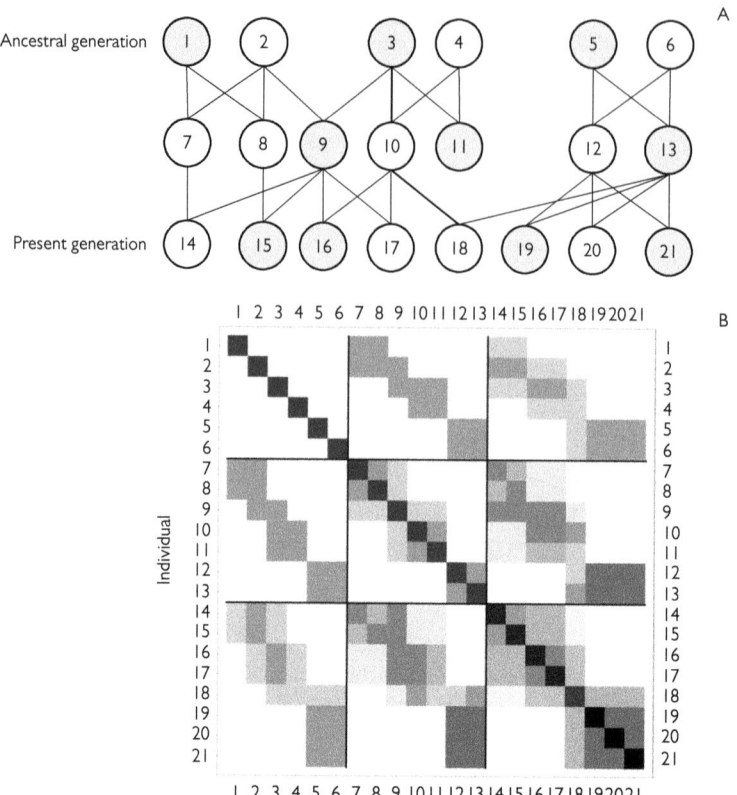

Figure 5.4 Illustration of a pedigree and the corresponding genetic relationship matrix. In the pedigree shown in (A), the grey dots represent males (sires) and white dots females (dams), and the numbers refer to the individual identities. (B) shows the genetic relationship matrix **A** derived from the pedigree of (A) using Eq. (5.3) in a recursive manner. The colours are scaled so that white corresponds to 0 and black to the highest value among the elements of **A**, which is $A_{19,19} = A_{20,20} = A_{21,21} = 1.25$ (see text).

consists of three non-overlapping generations, out of which we have named the first one as the ancestral population and the last one as the present generation. A pedigree may be observed under natural conditions, or it may arise through a breeding experiment conducted in a laboratory, or it may be generated by a population dynamical model. In evolutionary ecology of natural populations, the pedigree is often unknown to the researcher, and it may consist of millions of individuals over hundreds of generations, as would be the case for the pedigree connecting nine-spined sticklebacks found in ponds and marine environments (Box 5.1). In this section, our aim is to develop theory, and thus we may assume that the pedigree is known. When relating the theory to data in Section 5.5, we will assume that the pedigree is unknown, and therefore we will need to estimate the pedigree, or some summary statistics of it, from data.

We denote the coancestry coefficient between individuals i_1 and i_2 by $\theta_{i_1 i_2}$. Coancestry can be defined as the probability that randomly chosen alleles in a neutral locus are identical by descent (Whittemore and Halpern, 1994), i.e. that they are derived from the same allelic copy in the ancestral population. For example, the coancestry between a dam and her offspring is $\frac{1}{4}$, because the probability by which an allele picked randomly from the offspring originates from the dam is $\frac{1}{2}$, and the probability that the allele picked randomly from the dam is the copy that was transferred to the offspring is also $\frac{1}{2}$ (Roff, 1997). Note that the coancestry between the individual and itself is $\frac{1}{2}$, because this is the probability by which the same allelic copy is picked twice.

The genetic relationships among individuals can be described by a symmetric square matrix **A**, the rows and columns of which correspond to the individuals. The element corresponding to individuals i_1 and i_2 is defined as twice the coancestry coefficient, $A_{i_1 i_2} = 2\theta_{i_1 i_2}$. It is easy to see that the elements of the genetic relationship matrix can be calculated recursively using the formula

$$\begin{cases} A_{ii} = 1 + \dfrac{A_{d(i)s(i)}}{2}, \\ A_{i_1 i_2} = \dfrac{A_{i_1 s(i_2)} + A_{i_1 d(i_2)}}{2}, \end{cases} \quad (5.3)$$

where $d(i)$ and $s(i)$ refer to the dam and the sire of the individual i, respectively (Lynch and Walsh, 1998).

Applying Eq. (5.3) to the pedigree of Figure 5.4A generates the genetic relationship matrix **A** illustrated in Figure 5.4B. This figure illustrates how the coancestry coefficients reflect the degree of relatedness generated through all earlier generations. For example, continued inbreeding within a small population results eventually in the fixation of some allelic variants, thus making both the within-individual and among-individual coancestry coefficients approach 1. In the example of Figure 5.4, the individuals in the first generation are assumed to be non-inbred, so that $A_{ii} = 1$, and unrelated, so that $A_{ij} = 0$ for $i, j = 1, \ldots, 6$ and $j \neq i$. As one example, the individuals 7 and 8 are full siblings of unrelated parents, and thus $A_{7,8} = 1/2$. As another example, the parents of the individual 14 are related, resulting in inbreeding, and thus $A_{14,14} = 1.125$ is greater than 1.

5.2.3 Why related individuals resemble each other?

Let us make the usual assumption that the individuals in the ancestral population are unrelated, and let us denote the amount of additive genetic variation in the ancestral population by V_A. Assuming further that no mutations have taken place since the ancestral generation, a fundamental result of quantitative genetics (see Lynch and Walsh, 1998) is that the genetic covariance between two individuals i_1 and i_2 is

$$\text{Cov}\left(a_{i_1}, a_{i_2}\right) = 2\theta_{i_1 i_2} V_A = A_{i_1 i_2} V_A. \quad (5.4)$$

We show in Appendix A.3 how Eq. (5.4) can be derived from the additive model of breeding values (Eq. (5.1)) by considering the allelic states x_{ijku} as random variables.

The derivation shows that the amount of additive genetic variation in the ancestral population can be written as

$$V_A = 2 \sum_{j=1}^{n} \sum_{u_1,u_2=1}^{m_j} \left(\delta_{u_1 u_2} p_{j u_1} - p_{j u_1} p_{j u_2} \right) v_{j u_1} v_{j u_2}, \quad (5.5)$$

where p_{ju} is the frequency of the allele u in locus j, and $\delta_{u_1 u_2}$ is Kronecker's delta with value 1 if $u_1 = u_2$ and with value 0 if $u_1 \neq u_2$.

To see that it makes sense to measure the amount of additive genetic variation by Eq. (5.5), we first note that this definition of V_A depends on the allele frequencies (p_{ju}) and the additive values of the alleles (v_{ju}), which indeed are the necessary and sufficient ingredients for measuring the amount of genetic variation in a population. Further, consider a representative individual of the ancestral population, i.e. an individual that is constructed by sampling the alleles randomly based on their frequencies p_{ju}. Then it holds that $\text{Var}(a_i) = V_A$, with V_A defined by Eq. (5.5).

To further illustrate Eq. (5.5), assume that there are two allelic variants in each locus, which are equally frequent in the ancestral generation, so that $p_{j u_k} = 0.5$ for all $j = 1, \ldots, n$ and $k = 1, 2$. Let us further assume that the allelic effects are distributed normally as $v_{j u_k} \sim N(0, \sigma^2)$, where the variance σ^2 is the same for all loci. Then the reader may verify that the expected amount of additive variance in the ancestral generation behaves as

$$E[V_A] = n\sigma^2. \quad (5.6)$$

Thus, the amount of additive genetic variation increases with increasing number of loci and with increasing variance among the allelic effects per locus, as expected.

Equation (5.4) shows that the amount of covariance between the breeding values of two individuals is proportional to the coancestry between those individuals. In other words, related individuals resemble each other regarding their quantitative traits. This is because related individuals share a larger fraction of their alleles than unrelated individuals.

5.2.4 The animal model

Equation (5.4) is very general in the sense that its derivation does not assume any particular distribution of the trait values. However, to build a framework that allows for both mathematical and statistical developments, further assumptions are needed. We will make the usual assumption that the trait under consideration is normally distributed. In this case, Eq. (5.4) is equivalent with

$$\boldsymbol{a} \sim N(0, V_A \boldsymbol{A}), \quad (5.7)$$

where $\boldsymbol{a}$ is a vector containing the breeding values of all individuals. The breeding values are thus distributed according to the multivariate normal distribution (for basic concepts of multivariate distributions, see Appendices A.3 and B.1), the mean of which is a zero vector, and the variance-covariance matrix of which is $V_A \boldsymbol{A}$.

Assuming that the environmental effects are independent among the individuals and normally distributed, we may write

$$e \sim N(0, V_E \mathbf{I}), \tag{5.8}$$

where V_E is the amount of environmental variation. The matrix $\mathbf{I}$ is the identity matrix, i.e. a matrix with 1s at the diagonal and 0s at off-diagonals, reflecting the assumption of independence among individuals (see Appendix B.1). Combining Eqs. (5.7) and (5.8), we may write Eq. (5.2) in matrix notation as

$$\boldsymbol{p} \sim N(\boldsymbol{\mu}, V_A \mathbf{A} + V_E \mathbf{I}), \tag{5.9}$$

where $\boldsymbol{\mu}$ is the mean trait value. Equation (5.9) is the so-called 'animal model', which has a central role in animal breeding sciences as well as in evolutionary ecology (Henderson, 1950, 1976; Meyer, 1985; Shaw, 1987; Lynch and Walsh, 1998; Kruuk, 2004). While in animal and plant breeding the major interest is in the prediction of the breeding values a_i of the individuals, in evolutionary biology the interest is mostly at the population level, i.e. the additive genetic and environmental variances V_A and V_E. These two are needed e.g. to determine how heritable a trait is. Heritability, denoted by h^2, is defined as the proportion of phenotypic variation that can be explained by genetic effects, $h^2 = V_A/(V_A + V_E)$ (Roff, 1997). In the presence of non-additive genetic variation, broad-sense heritability H^2 includes all genetic effects combined, whereas narrow-sense heritability h^2 accounts only for additive effects (Hartl, 2014).

Thus far, we have considered the case of a single trait. Often the interest is in multiple traits, which may show both genetic and environmental correlations among each other. In such a case, the variance components V_A and V_E are replaced by variance-covariance matrices, which we denote by $\mathbf{G}$ and $\mathbf{E}$. The diagonal elements of these matrices measure the amounts of genetic and environmental variances of the individual traits, whereas the off-diagonal elements measure the amounts of co-variation between pairs of traits. Including in the vector $\boldsymbol{p}$ the phenotypic values for all traits and all individuals, we may generalize Eq. (5.9) to the multivariate version

$$\boldsymbol{p} \sim N(\boldsymbol{\mu}, \mathbf{G} \otimes \mathbf{A} + \mathbf{E} \otimes \mathbf{I}), \tag{5.10}$$

where $\otimes$ denotes the Kronecker product (see Appendix B.1).

5.2.5 Why related populations resemble each other?

In evolutionary genetics, it is often of major interest to decompose genetic variation to different hierarchical levels, such as among populations and among individuals within a population. We will next derive such decomposition by assuming that the individuals belong to a discrete set of populations originating from a common ancestral population, as illustrated in Figure 5.5. As the ancestors of all individuals can be traced back to the same ancestral population, their degrees of relatedness are determined by the underlying individual-level pedigree. As noted earlier, constructing such a pedigree in practice may be difficult or impossible for most natural populations, but this does not prevent us from using the concept of the underlying pedigree to develop the theory.

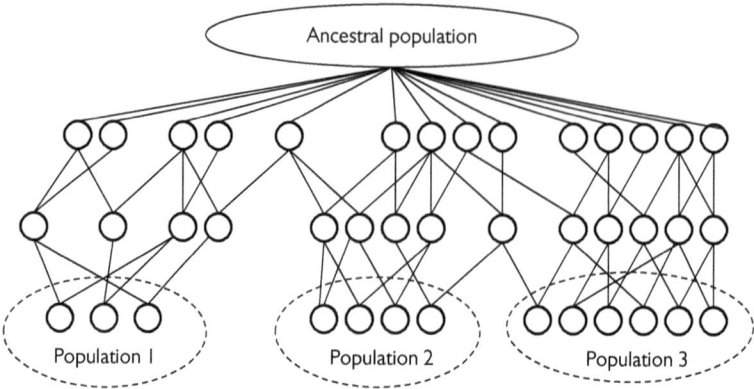

Figure 5.5 A set of local populations that have diverged from a common ancestral population. The circles represent individuals and the lines connect the offspring to their parents.

We denote the relatedness between any two individuals i_1 and i_2 in the present generation by $\theta_{i_1 i_2}$, whether the individuals belong to the same population or to different populations. We measure the population-level coancestry coefficient between the two populations by the average coancestry among all pairs of individuals,

$$\theta_{XY}^{\mathcal{P}} = \frac{1}{n_X n_Y} \sum_{i_X \in X, i_Y \in Y} \theta_{i_X i_Y}, \qquad (5.11)$$

where n_X and n_Y denote the numbers of individuals in the populations X and Y, and the superscript $\mathcal{P}$ refers to the population level. This formula also applies for $\theta_{XX}^{\mathcal{P}}$, which measures the average relatedness within a population.

We first proceed with the case of a single trait, and denote the mean breeding value in population X by

$$a_X^{\mathcal{P}} = \frac{1}{n_X} \sum_{i \in X} a_i. \qquad (5.12)$$

Our interest is in quantifying how much different populations resemble each other, i.e. how large is the covariance $\mathrm{Cov}\left[a_X^{\mathcal{P}}, a_Y^{\mathcal{P}}\right]$ under the assumption of neutrality. To derive this, we utilize basic properties of random variables (Appendix A.3) to yield

$$\mathrm{Cov}\left[a_X^{\mathcal{P}}, a_Y^{\mathcal{P}}\right] = \mathrm{Cov}\left[\frac{1}{n_X}\sum_{i \in X} a_i, \frac{1}{n_Y}\sum_{j \in Y} a_j\right] = \frac{1}{n_X n_Y}\sum_{i \in X, j \in Y} \mathrm{Cov}\left[a_i, a_j\right]. \qquad (5.13)$$

Employing Eqs. (5.11) and (5.4) then yields

$$\mathrm{Cov}\left[a_X^{\mathcal{P}}, a_Y^{\mathcal{P}}\right] = 2\theta_{XY}^{\mathcal{P}} V_A. \qquad (5.14)$$

This equation is the population-level analogue of Eq. (5.4), and it describes that the covariance in the mean breeding values between two populations is proportional to the population-level relatedness coefficient. Thus, not only related individuals but also related populations resemble each other.

Denoting the vector of population-level breeding values by $a^\mathcal{P}$ and assuming normally distributed traits, we can write

$$a^\mathcal{P} \sim N(0, V_A \mathbf{A}^\mathcal{P}), \qquad (5.15)$$

where $\mathbf{A}^\mathcal{P} = 2\mathbf{\theta}^\mathcal{P}$ and the matrix $\mathbf{\theta}^\mathcal{P}$ consists of the elements $\theta^\mathcal{P}_{XY}$. This equation is the population-level analogue of Eq. (5.7).

In the multivariate case, we denote by the vector $a^\mathcal{P}$ the population mean breeding values for all traits and all populations, in which case Eq. (5.15) generalizes to

$$a^\mathcal{P} \sim N(0, \mathbf{G} \otimes \mathbf{A}^\mathcal{P}). \qquad (5.16)$$

To illustrate the concepts and relations we've just discussed, consider a single trait influenced by $n = 100$ loci. Assume that each locus has two allelic variants that are equally frequent in the ancestral population, and that their allelic effects are distributed as $v_{ju} \sim N(0, \sigma^2)$, with $\sigma^2 = 1/n$. We recall from Eq. (5.6) that in this case the expected amount of additive genetic variation in the ancestral population is $E[V_A] = 1$.

Figure 5.6 shows simulations of populations consisting either of 10 or 100 individuals. The population mean breeding value performs a random walk over time (Figures 5.6A, B). This is due to genetic drift, i.e. chance events associated with which individuals reproduce and which of their allelic copies are passed on to their offspring. The influence of genetic drift is greater in a small population than in a large population, as in the latter case positive and negative chance events tend to average out. The amount of additive genetic variance V_A does not stay constant, but it also varies over time (Figures 5.6C, D). It decreases in the long term because eventually some of the allelic variants become fixed, and because we have not incorporated mutations that would produce new variation to replace the variation lost. Let us denote the amount of genetic variation in the present population X by $V_A(X)$, to distinguish it from the amount of genetic variation in the ancestral population denoted by V_A. Without mutation, the expected amount of the remaining genetic variation behaves as $E[V_A(X)] = (1 - \theta^\mathcal{P}_{XX})V_A$ (Ovaskainen et al., 2011). Thus, the amount of genetic variation decreases as the within-population relatedness increases. On top of the long-term decreasing trend, V_A shows stochastic fluctuations (Figure 5.6), illustrating that it should be treated as a random variable.

In Figures 5.6E and F, we have assumed that another habitat patch becomes available after 20 generations, and consequently the population splits into two populations. We assume that the two populations become isolated from each other for the subsequent generations, i.e. there is no gene flow between them. Over time, within-population relatedness will increase in comparison to between-population relatedness, and thus genetic drift makes the two populations drift away from each other (Eq. (5.14)).

Equations (5.15) and (5.16) can be considered as statistical statements about the distribution of population mean values under neutrality. As we will discuss in

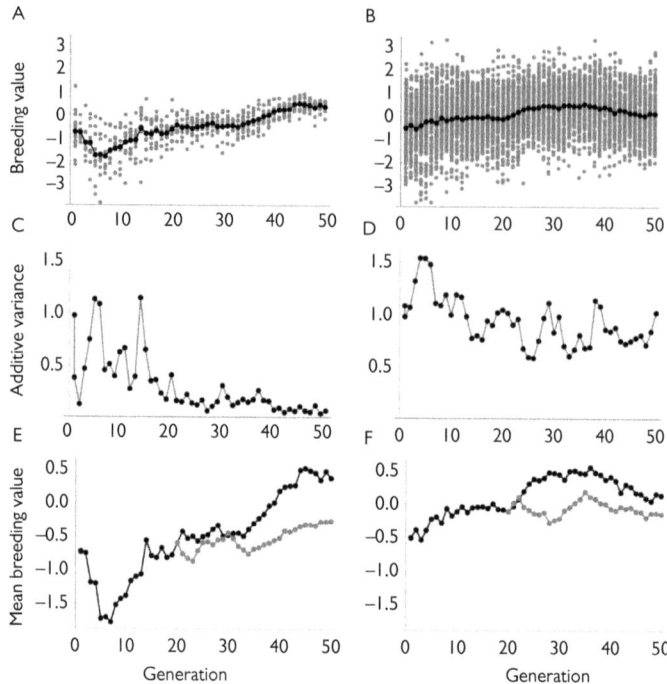

Figure 5.6 Simulations of neutral population genetics. We have assumed that the population size remains constant over the generations, being 10 individuals in (A, C, E) and 100 individuals in (B, D, F). For simplicity, we have assumed a hermaphrodite species, so that each individual can act both as sire and as dam. The sire and the dam for each individual were selected randomly among the individuals of the previous generation. The genetic architecture consists of $n = 100$ loci, each of which has two allelic variants which are equally frequent in the ancestral population, and the allelic effects of which are distributed as $N(0, 1/n)$. (A, B) show how individual breeding values (grey dots) and the mean population breeding value (black dots) evolve over time. (C, D) show the amount of additive genetic variation present in each generation. (E, F) show the time evolution of population mean breeding values assuming that at generation 20 the population diverges into two populations, both of the size of 10 (E) or 100 (F), which evolve independently of each other for the generations 21–50.

Section 5.5, they provide a useful null model for statistical analyses aimed at asking whether and how the observed pattern of population divergence deviates from neutrality, i.e. whether the data show a signal of selection or whether the data are compatible with the null model of genetic drift.

5.3 The evolution of quantitative traits under selection

Evolution is defined as a change in gene frequency over time (Nei, 1975). In Section 5.2, we considered the evolution of neutral traits, and thus assumed that gene frequencies change over time solely due to genetic drift. In addition to drift,

5.3 The evolution of quantitative traits under selection

evolution is shaped by natural selection, as well as the processes of gene flow, recombination, and mutation. In this section, we extend the neutral perspective of the previous section by incorporating these other forces of evolutionary dynamics.

5.3.1 Evolution by drift, selection, mutation, recombination, and gene flow

To illustrate the different ingredients that shape evolutionary processes, we continue with the infinitesimal model in which the breeding values of the individuals are determined additively by their allelic states. We recall from Eq. (5.1) that the breeding values are modelled as a sum of contributions over the n loci that influence the trait in question, that each individual has two allelic copies for each locus j, and that the allelic effects and frequencies of the different allelic variants are denoted by v_{ju} and p_{ju}, respectively.

Figure 5.7 illustrates the different evolutionary processes by showing how they modify the genetic and phenotypic composition of a population over a single generation. The transition from population state A to population state B illustrates gene flow. Here the immigrants have higher breeding values than the residents, and thus they shift the population mean to a higher value and increase the genetic diversity of the population. The transition from population state B to population state C illustrates recombination: the parents from population state B have mated to produce the offspring generation C. The expected breeding value of an offspring is the mean

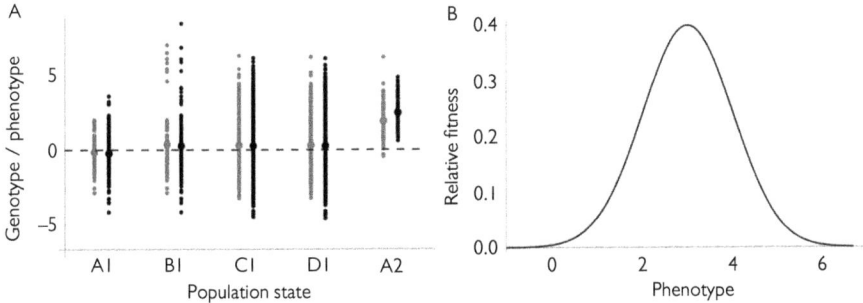

Figure 5.7 Evolution shaped by drift, selection, gene flow, recombination, and mutation. (A) shows how breeding values (shown in grey; small dots correspond to individuals, large dots to population means) and phenotypic values (shown in black) evolve over one generation. In state A of generation 1, the population consists of 100 individuals. The breeding values of the individuals are based on $n = 100$ diploid loci. Each locus has two allelic variants with equal frequencies, and the allelic effects distributed as $N(0, \sigma^2)$, with $\sigma^2 = 1/n$. The phenotypic values were obtained by adding to the breeding values $N(0, 1)$ distributed environmental effects. In state B, 10 immigrants with higher breeding values have arrived. State C consists of 1,000 propagules generated under the assumption of random mating. In state D, mutation has changed the state of each allele with probability 0.01. State A of generation 2 consists of 100 individuals that were selected among the 1,000 propagules based on the fitness function shown in (B).

breeding value of its two parents. As some of the offspring have one resident and one immigrant parent, recombination has produced intermediate genotypes and phenotypes that were not present in the parental generation. The transition from population state C to population state D illustrates mutation, which has randomly modified some of the breeding values in the offspring generation. While illustrated here separately, both recombination and mutation take place during reproduction. The next process that we consider is natural selection. The fitness function shown in Figure 5.7B depicts the relative probability that an individual in the offspring generation survives from juvenile to adult. We thus assume that the optimal phenotype has value $p_i = 3$, and that the probability of survival decreases with an increasing distance from the optimum. The individuals that survive form the population state A of generation 2. Selection has shifted the distributions of both phenotypes and genotypes towards higher values.

In summary, Figure 5.7 shows how the distribution of breeding values changes from one generation to the next generation as a result of gene flow, recombination, mutation, natural selection, genetic drift, and heritability. While we did not discuss genetic drift in the previous paragraph, it is present in the transition from state B to state C due to randomness associated to mating and recombination, as well as in the transition from state D to state A of the next generation due to randomness in exactly which individuals survive from juveniles to adults. Likewise, while we did not explicitly model heritability, it was present as we applied selection to phenotypes, not to genotypes. As phenotypes have random variation on top of that of genotypic variation, the effect of selection is carried out only partially to the next generation. This is why the breeding values of the selected individuals did not increase as much as their phenotypic values.

Note that we have assumed a somewhat arbitrary chronological sequence in which the events took place. For example, we assumed that dispersal takes place in the adult generation before mating, but it could equally well take place in the adult generation after mating, or in the offspring generation. Likewise, we assumed that selection operates on juvenile survival, but it could equally well operate in determining the reproductive success of the adults. Further, we assumed a species with sexual reproduction and diploid genetic structure, whereas e.g. in clonally reproducing species the process of recombination would be lacking.

5.3.2 Selection differential and the breeder's equation

Let us now focus more analytically on the selection part of the evolutionary process. To simplify, we exclude the effect of recombination by assuming clonal reproduction. To exclude the effect of genetic drift, we assume an infinitely large population size.

We denote by $q_P(p)$ and by $q_O(p)$ the frequency distributions (i.e. the probability density functions) of phenotypes in the parent and offspring generations, respectively. Assuming a single quantitative trait, the mean phenotype in the parent generation can be computed as $z_P = \int_{p=-\infty}^{\infty} q_P(p)p\,dp$, i.e. averaging the trait values over the distribution of individuals. Like we did in Eq. (5.8), we assume also here that the mean environmental effect is 0, and thus z_P is also the expected breeding value of the parent generation. What we attempt to do next is to evaluate the influence of selection by deriving the mean breeding value in the offspring generation.

In the context of Figure 5.7B, the fitness function $f(p)$ relates specifically to juvenile survival, but here we let it describe more generally the relative number of offspring that an individual with phenotype p will contribute to the next generation. By this definition of the fitness function, a parent with phenotype p will have $Cf(p)$ offspring, where C is a proportionality constant. But what can we say about the genotype of a parent individual with phenotype p? To derive this, we need the following property of the bivariate normal distribution: assume that a vector with the two elements x_1 and x_2 is bivariate normally distributed as

$$\begin{pmatrix} x_1 \\ x_2 \end{pmatrix} \sim N\left(\begin{pmatrix} \mu_1 \\ \mu_2 \end{pmatrix}, \begin{pmatrix} \sigma_1^2 & \rho\sigma_1\sigma_2 \\ \rho\sigma_1\sigma_2 & \sigma_2^2 \end{pmatrix}\right). \tag{5.17}$$

Then, conditionally on the value of x_2, x_1 is distributed as

$$x_1|x_2 \sim N\left(\mu_1 + \frac{\sigma_1}{\sigma_2}\rho(x_2 - \mu_2), (1-\rho^2)\sigma_1^2\right). \tag{5.18}$$

In the parent generation, the breeding values are distributed as $a \sim N(z_P, V_A)$, the environmental effects are distributed as $e \sim N(0, V_E)$, and these two are independent of each other. Thus, the phenotypic values $p = a + e$ are distributed as $p \sim N(z_P, V_A + V_E)$. Further, the covariance between breeding values and phenotypic values is $\text{Cov}(p, a) = \text{Cov}(a + e, a) = \text{Cov}(a, a) = V_A$, and thus their correlation is $V_A/\sqrt{V_A(V_A + V_E)} = h$, where $h^2 = V_A/(V_E + V_A)$ is the heritability of the trait. Now, applying Eq. (5.18) so that the parental breeding value a plays the role of x_1, and the parental phenotypic value p plays the role of x_2, we obtain

$$a|p \sim N(z_P + h^2(p - z_P), (1 - h^2)V_A). \tag{5.19}$$

We recall that the distribution of individuals in the parent generation with phenotype p follows $q_P(p)$, and that an individual with phenotype p leaves on average $Cf(p)$ offspring. By Eq. (5.19), the mean breeding value of those offspring is $z_p + h^2(p - z_P)$. Thus, the mean breeding value in the offspring generation will be

$$z_O = \frac{\int_{p=-\infty}^{\infty} \left[z_P + h^2(p - z_P)\right] Cf(p)q_P(p)dp}{\int_{p=-\infty}^{\infty} Cf(p)q_P(p)dp}. \tag{5.20}$$

Simplifying gives

$$\Delta z = z_O - z_P = h^2 S, \tag{5.21}$$

where

$$S = \frac{\int_{p=-\infty}^{\infty} p f(p) q_P(p) dp}{\int_{p=-\infty}^{\infty} f(p) q_P(p) dp} - z_P \qquad (5.22)$$

is called the selection differential, i.e. the mean trait value of selected individuals minus the mean trait value before selection (Falconer and Mackay, 1996). In other words, the selection differential S measures the association between trait values and fitness. Equation (5.21) is known as the breeder's equation (Lush, 1943). The h^2 part of the breeder's equation shows that the speed of evolutionary change due to selection is proportional to heritability. This is, as mentioned previously, because selection acts on phenotypes, only the heritable part of selection is carried over to the next generation.

To illustrate, let us consider the case of directional selection, with fitness increasing with increasing phenotypic value. Somewhat arbitrarily, we assume that the fitness function behaves as $f(p) = \exp(\alpha p)$, where the parameter $\alpha \geq 0$ controls the intensity of selection. If $\alpha = 0$, the model is neutral. If $\alpha = 1$, increasing the phenotypic value by one increases the relative fitness of the individual by factor $\exp(1) \approx 2.8$. As the phenotypic values are distributed in the parent generation as $p \sim N(z_P, V_A + V_E)$, Eq. (5.22) leads (after some algebra that the reader may wish to verify) to the selection differential $S = \alpha(V_A + V_E)$. Thus, the expected change over one generation in population mean phenotype is $\Delta z = h^2 S = \alpha V_A$. As is intuitive, the rate of evolution depends on the intensity of selection (α) as well as the amount of raw material that selection can operate upon (V_A).

Figure 5.8 shows how the breeding values and phenotypic values evolve over time in a simulated finite population exposed to mild ($\alpha = 0.1$) or strong ($\alpha = 0.3$) selection. In both cases, the amount of genetic variation decreases drastically over time. This is because we have not incorporated mutation, and because we excluded recombination by assuming a haploid clonal population. Therefore, the genotypes of the next generation are simply a subset of the genotypes of the present generation. Both drift and selection reduce genetic variation, and therefore it reduces faster with stronger selection. In contrast, phenotypic variation reduces only little, as it is also influenced by environmental variation. As illustrated by the lower panels of Figure 5.8, the breeder's equation (Eq. (5.21)) is successful in predicting how the population mean phenotype evolves over time as a response to selection.

Earlier we considered how a single trait responds to selection. However, phenotypes are characterized by a suite of multiple traits, and thus evolutionary processes are generally of multivariate nature. As an individual's fitness can be influenced simultaneously by many traits, modelling evolution separately for each of the traits can lead to a misleading picture. The multivariate breeder's equation reads (Lande and Arnold, 1983)

$$\Delta \mathbf{z} = \mathbf{G}\mathbf{P}^{-1}\mathbf{S}. \qquad (5.23)$$

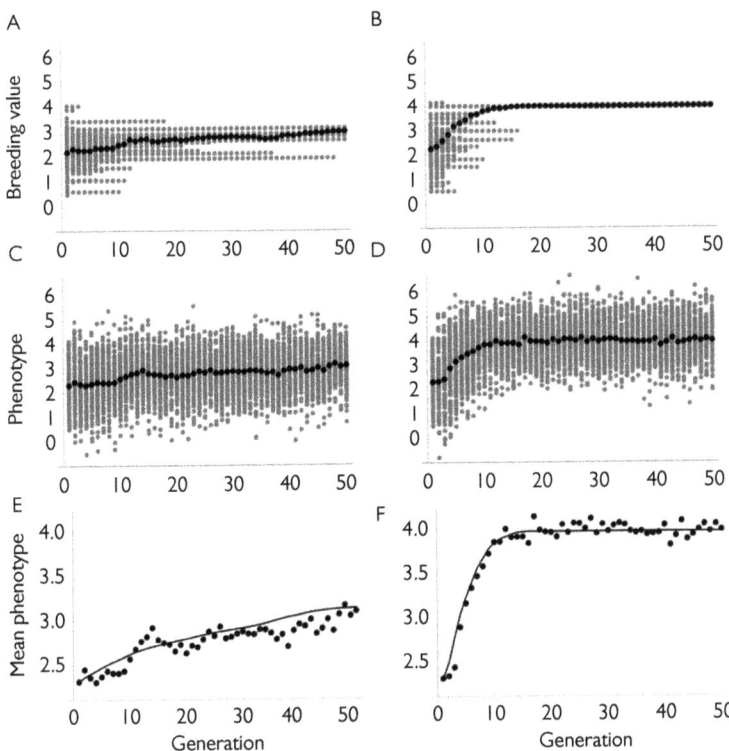

Figure 5.8 Univariate evolution by drift and selection. We consider a clonal population consisting of 100 individuals in each generation, with $V_A = V_E = 0.5$ and thus heritability $h^2 = 0.5$. We have ignored mutations, and thus the breeding values in the next generation are a subset of those of the parent generation. In all panels, grey dots refer to individuals and black dots to population means. The upper panels show the breeding values, and the middle panels the phenotypic values. The lower panels compare the evolution of population means to the cumulative prediction by the univariate breeder's equation (Eq. (5.21)). We have assumed directional selection by setting the fitness function to $\exp(\alpha p)$, with the strength of selection being $\alpha = 0.1$ in (A, C, E) and $\alpha = 0.3$ in (B, D, F).

Here **G** and **P** = **G** + **E** are the matrices of genetic and phenotypic variances and covariances, which we introduced in Eq. (5.10), and the superscript −1 refers to matrix inverse. Comparing Eq. (5.23) to Eq. (5.21) shows that the matrix **GP**$^{-1}$ plays the role of heritability in the multivariate case. In addition to measuring which fraction of the phenotypic variation is heritable, it also accounts for phenotypic and genetic correlations to measure heritability along different directions of the trait space.

Figure 5.9 extends the univariate example of Figure 5.8 to the bivariate case with two traits. We have assumed that, in addition to trait 1 that we considered in Figure 5.8, there is another trait 2, and that the two traits are genetically correlated with correlation coefficient 0.7. As we have assumed that the environmental variances

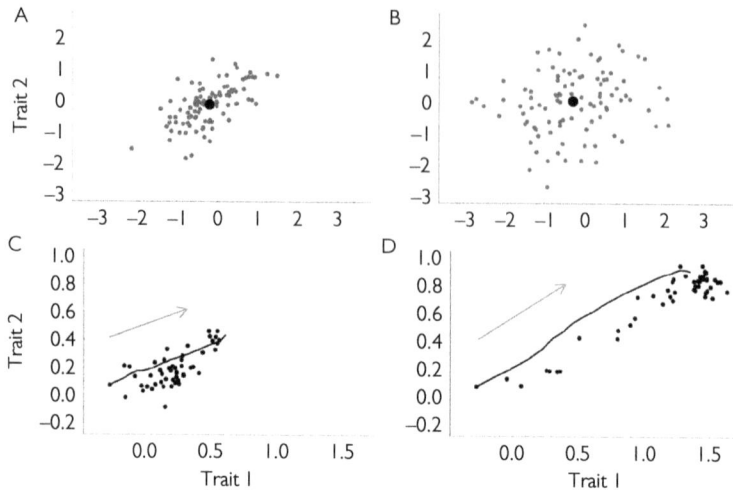

Figure 5.9 Bivariate evolution by drift and selection. We consider here the same populations as in Figure 5.8, but in addition to the trait considered there (called here trait 1) we consider another trait (called trait 2). Trait 2 is not under selection, but genetically correlated (with correlation coefficient 0.7) with trait 1. Panels A and B show respectively the distributions of breeding values and phenotypic values in the first generation, with grey dots corresponding to individuals and black dots to population means. The lower panels compare the evolution of population means to the cumulative prediction by the multivariate breeder's equation (Eq. (5.23)), with the strength of selection being $\alpha = 0.1$ in panel C and $\alpha = 0.3$ in panel D. The grey arrows indicate the direction of the evolutionary process.

are not correlated between the two traits, the breeding values are more correlated than the phenotypic values. We have assumed that selection operates on trait 1 only, while trait 2 is selectively neutral. As shown by the lower panels of Figure 5.9, the population mean value of trait 1 increases as a response to selection, as it did in the univariate case of Figure 5.8. But so does also trait 2, even if this trait is selectively neutral. This is because trait 2 is positively correlated with trait 1, and thus it is selected as a co-product of the selection operating on trait 1, as predicted by Eq. (5.23). This example illustrates how the structure of the **G**-matrix can influence the direction and speed of multivariate trait evolution (Griswold et al., 2007).

5.3.3 Population divergence due to drift and selection

In Section 5.2, we discussed how local populations diverge from each other due to genetic drift only. We now extend the discussion by adding selection in the form of local adaptation. In the context of the nine-spined stickleback example introduced in Box 5.1, postglacial land uplift subdivided a part of the ancestral marine population into a number of lake populations, some of which were later further subdivided into pond populations. To mimic such a scenario in a simplified setting, we

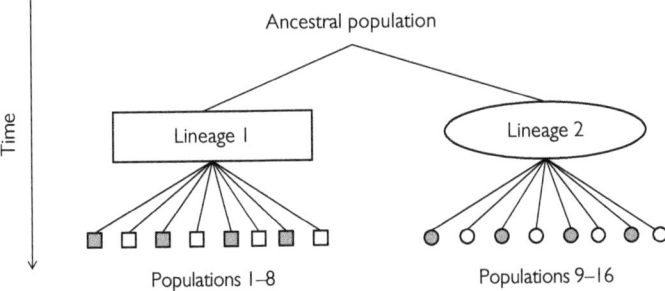

Figure 5.10 A scenario of population divergence simulated in Figure 5.11. We assume that the ancestral population first splits into two lineages for the duration of 20 generations. The lineages then split into eight populations each, which evolve further for the duration of 10 generations. Each generation of each lineage and each population consists of 100 individuals, which produce 1,000 propagules under the assumption of random mating. Out of the 1,000 propagules, 100 are selected to form the next generation. We assume $n = 100$ loci influencing two traits without any genetic or phenotypic correlation. In the scenario with local adaptation, the grey and white populations are assumed to be exposed to different environmental conditions, as described in the legend to Figure 5.11.

assume in Figure 5.10 that an ancestral population splits into two lineages that evolve independently from each other during a number of generations. During this period of evolution, we assume no selection, only drift. Later, the lineages split further into a set of 8 local populations each, which evolve independently of each other for another number of generations.

In Scenario A, simulated in Figure 5.11A, evolution remains neutral also after the lineages split into local populations. Thus, this scenario is fully compatible with pure genetic drift. In the trait space, the populations originating from lineage 1 form one group, and populations from lineage 2 form another group (Figure 5.11A). This is because populations within each lineage share part of their evolutionary history, making them more related to each other than populations between the lineages (Figure 5.12A). As discussed in the previous section, related populations are expected to resemble each other under neutrality (Eq. (5.14)).

In Scenario B simulated in Figure 5.11B, local adaptation comes into play. We have assumed here that there are two different kinds of environments, which favour different kinds of phenotypes. In the context of the nine-spined sticklebacks (Box 5.1), these could be environments with and without predatory fish. The black and grey crosses in Figure 5.11B show the fitness optima for the two population types, i.e. the phenotypic values that maximize fitness. As shown by the colour coding in Figure 5.10, we assume that half of the populations from each lineage inhabit each environmental type. As was the case with Scenario A, the shared evolutionary history makes the populations within the lineages more related to each other than populations between the lineages (Figure 5.12B). But now the phase of local adaptation distorts the neutral expectation by which related populations should resemble each

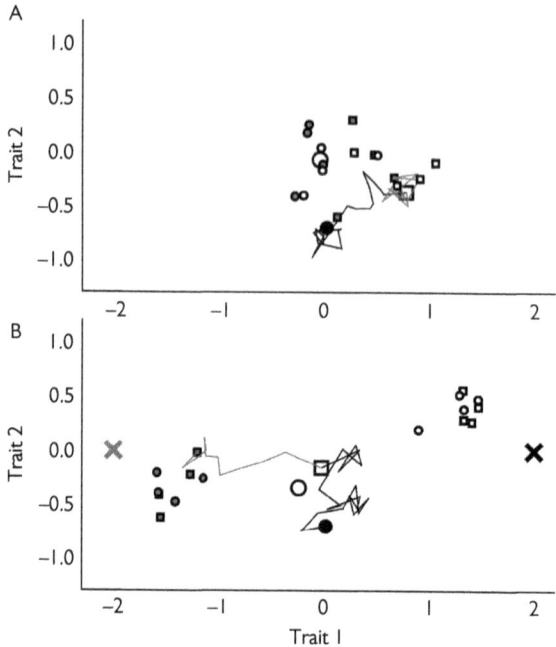

Figure 5.11 Population divergence by drift and selection under the scenario illustrated in Figure 5.10. All symbols refer to mean population phenotypes. The black dot is the ancestral population, from which lineages 1 (large square) and 2 (large circle) evolve for 20 generations under neutral dynamics. These split then further into 8 populations (shown by the small squares and circles) that evolve for an additional 10 generations. The lines exemplify the random walk performed by population 1 of lineage 1 from the ancestral generation to the present generation. In (A) there is no selection during any part of the process, whereas in (B) we have incorporated local adaptation during the last 10 generation. In the latter case, the optimal phenotypes are shown by the crosses, the black and grey colours corresponding to local populations depicted with empty symbols and grey symbols, respectively. The fitness functions are set to the density functions of bivariate normal distributions, with means set to the optima, and variance-covariance matrices to $5\mathbf{I}$, where $\mathbf{I}$ is the identity matrix.

other: the populations group in Figure 5.11B based on the local environments they are exposed to, rather than based on the lineages they originate from.

Figure 5.11 illustrates how local adaptation can influence greatly the evolution of those traits that are under selection. Does selection have consequences also for neutral population genetic structure? As we have already discussed, populations within a lineage are more related to each other than populations belonging to two different lineages (Figure 5.12). Similarly, within a lineage, individuals that belong to the same population are more related than individuals that belong to different populations (Figure 5.12). These relatedness patterns seem very similar for the cases without and with selection, but they are not identical. To quantify the difference, we apply one

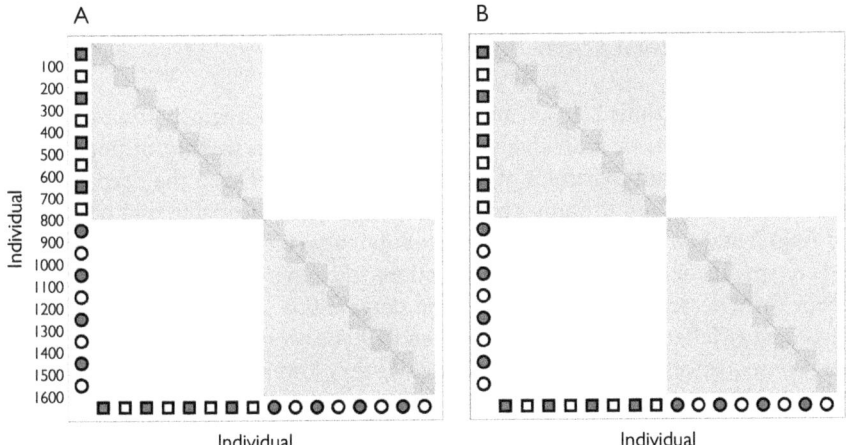

Figure 5.12 Coancestries generated by the population scenarios illustrated in Figures 5.10 and 5.11. The matrices show the coancestry coefficients among all 1,600 individuals (16 populations with 100 individuals each) in the final generation. As in Figures 5.10 and 5.11, the circles and squares indicate the lineages from which the local populations originate. We computed the coancestry coefficients recursively by applying Eq. (5.3) to the pedigree generated by the simulation.

widely used measure for population structure, namely F_{ST} (Weir and Cockerham, 1984), defined by

$$F_{ST} = \frac{f_1 - f_2}{1 - f_2}. \tag{5.24}$$

Here f_1 is the probability of identity for two alleles sampled from two individuals belonging to the same population, whereas f_2 is the probability of identity for two alleles sampled from two individuals belonging to different populations. Averaging the coancestry coefficients shown in Figure 5.12 within and among populations, we obtain for the neutral case $f_1 = 0.140, f_2 = 0.050$, and thus $F_{ST} = 0.095$. For the case with selection we obtain $f_1 = 0.146, f_2 = 0.051$, and thus $F_{ST} = 0.099$. These numbers illustrate that selection has slightly increased within-population relatedness. This is because with selection some of the individuals have contributed disproportionally to the next generation, thereby decreasing the effective population size. While F_{ST} is a convenient summary measure to characterize the overall level of population divergence, we note that it fails to characterize population structure in more detail. For example, the above values of F_{ST} do not tell whether the populations belong to two different lineages. This kind of information is retained in population-to-population matrix of coancestry coefficients $\theta_{XY}^{\mathcal{P}}$ (Eq. (5.11)).

We will return to the data shown in Figures 5.11 and 5.12 in Section 5.5, where we apply statistical methods with the aim of asking whether these data have been shaped by selection.

5.4 Evolutionary dynamics under habitat loss and fragmentation

As discussed in Sections 2.4, 3.4, and 4.4, habitat loss and fragmentation can influence the movements of individuals, the dynamics and persistence of populations, and the assembly and dynamics of species communities. Given the deep interplay between ecological and evolutionary dynamics, it is not surprising that habitat loss and fragmentation can also have many evolutionary consequences. Fragmentation leads to smaller and more isolated populations, thus increasing the role of random genetic drift, decreasing gene flow between populations, and increasing the level of inbreeding (McRae, 2006; Charlesworth and Charlesworth, 2012). These have adverse consequences for ecological dynamics, as they lower the fitness of populations and make them less capable of adapting to new situations (van der Werf et al., 2009). The interplay between ecological and evolutionary dynamics in extinction processes has been called the extinction vortex (Fagan and Holmes, 2006; Blomqvist et al., 2010). However, evolutionary processes can also help species by making them better adapted to survive in a world dominated by anthropogenic influences. The term evolutionary rescue is used to describe a situation in which a species would otherwise go extinct, but natural selection modifies the species traits so that it avoids extinction (Bell, 2012; Ferriere and Legendre, 2013).

Among the many kinds of evolutionary consequences that habitat loss and fragmentation can have, we consider in this section the evolution of dispersal, which is one of the classical problems in evolutionary biology (Roff, 1975; Hamilton and May, 1977; Mcpeek and Holt, 1992; Dieckmann et al., 1999; Sutherland et al., 2013). For metapopulations inhabiting fragmented landscapes, a high enough dispersal ability is needed to enable recolonizations of extinct habitat patches (Section 3.4), and thus one could expect that fragmentation leads to increased selection for dispersal (e.g. Hanski and Mononen, 2011). However, dispersal can be costly (Bonte et al., 2011), and in very isolated habitats, dispersing individuals may be lost in the matrix without possibility to reproduce. Therefore, one could expect selection to favour reduced dispersal (Dieckmann et al., 1999). In this section, we employ evolutionary models to understand how these counteracting forces play out under different scenarios of habitat loss and fragmentation.

5.4.1 Evolution of dispersal in the Hamilton–May model under adaptive dynamics

In adaptive dynamics (Geritz et al., 1998; Abrams, 2005; Waxman and Gavrilets, 2005), also called evolutionary invasion analysis (Hurford et al., 2010), ecological and evolutionary time-scales are separated. This is done by assuming that at any given time, the resident population is monomorphic, i.e. all individuals have the same genotype. The outcome of the evolutionary play is addressed by asking which kinds of mutants are able to invade the resident population.

5.4 Evolutionary dynamics under habitat loss and fragmentation

The classical Hamilton and May (1977) model of dispersal evolution is an example of the adaptive dynamics approach. This model is spatially implicit and consists of a set of breeding sites, each of which can be occupied by one individual. The individuals are assumed to produce a large number m of propagules, out of which a fraction v disperses, and thus the fraction $1 - v$ stays in the natal site. Dispersal is risky, as only a fraction p of the dispersing propagules survives the dispersal period. The surviving propagules settle randomly among the sites. In each site, limited availability of resources allows only one individual to survive. The surviving individual is selected randomly among all the propagules that compete for the site, including the dispersed and non-dispersed ones.

The analysis of the Hamilton and May (1977) model can be greatly simplified by assuming that the number of propagules m that each individual produces is very large, so that $m \to \infty$. One consequence of this is that all sites are always occupied. As a related consequence, one may ignore demographic stochasticity in the distribution of the dispersed propagules, as we will do later.

To apply adaptive dynamics, we assume that the system consists of two types of individuals. We assume that individuals of type 1 disperse the fraction v_1 of their offspring, whereas individuals of type 2 disperse the fraction v_2 of their offspring. As each site is occupied by one individual only, the two types cannot coexist locally. We denote the fraction of sites occupied by individuals of type 1 by q_1, so that the fraction of sites occupied by individuals of type 2 is $q_2 = 1 - q_1$. We assume complete heritability of the trait v, i.e. that type 1 individuals always produce type 1 individuals, and type 2 individuals always produce type 2 individuals.

To derive an equation for evolutionary dynamics, we consider a focal site that is occupied by a type 2 individual, and compute the probability that the site will be occupied by a type 2 individual also in the next generation. The type 2 individual that was originally in the focal site will leave the fraction $1 - v_2$ of its m propagules there. Out of the other patches, the fraction q_2 consists of type 2 individuals, each of which produces mv_2p propagules that successfully disperse to the other sites. Thus, the total number of type 2 propagules in the focal patch is $m(1 - v_2) + mq_2v_2p$. As type 1 propagules can arrive to that site only by dispersal, their number is mq_1v_1p. Thus, a site that consisted originally of a type 2 individual will consist of a type 2 individual in the next generation with probability

$$k_{22} = \frac{1 - v_2 + q_2v_2p}{1 - v_2 + q_2v_2p + q_1v_1p}. \tag{5.25}$$

With the same logic, a site that consisted originally of a type 1 individual will consist of a type 2 individual in the next generation with probability

$$k_{21} = \frac{q_2v_2p}{1 - v_1 + q_1v_1p + q_2v_2p}. \tag{5.26}$$

The fraction of type 2 individuals in the next generation is $k_{22}q_2 + k_{21}q_1$. Thus, the growth rate g of the type 2 population (i.e. the number of sites occupied by

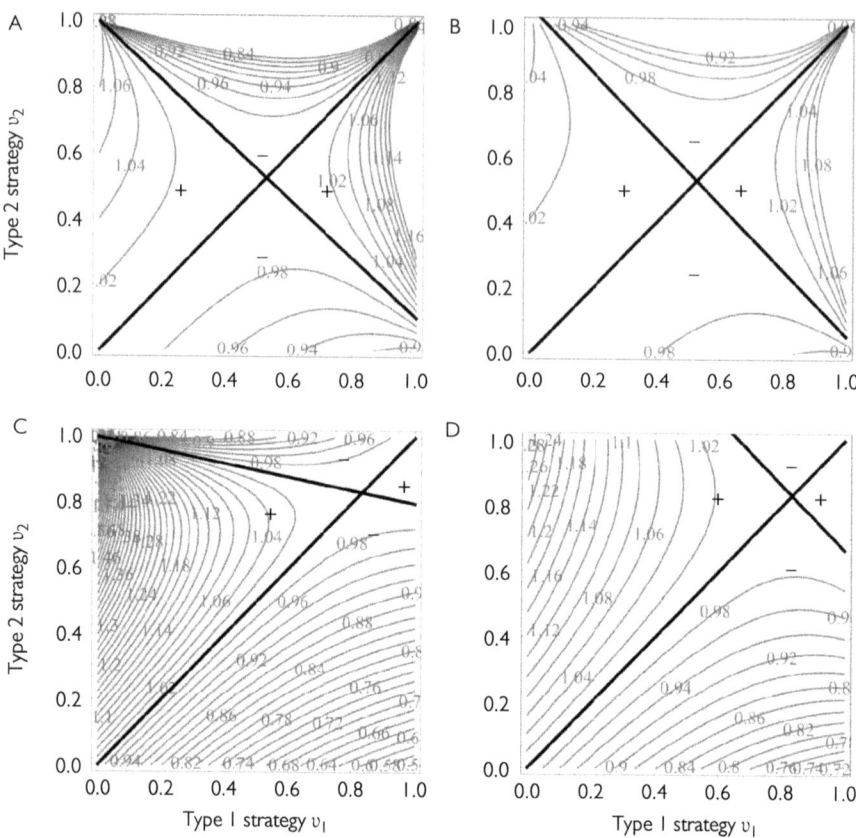

Figure 5.13 Pairwise invasibility plots for the Hamilton and May (1977) model for evolution of dispersal. The contour lines show the growth rate of type 2 individuals (Eq. (5.27)) in a population that consists of a mixture of type 1 and type 2 individuals. The x- and y-axes show the dispersal strategies v_1 and v_2 for the two types of individuals. The black lines separate the areas in which the density of the type 2 individuals increases or decreases, indicated by a plus or a minus sign, respectively. In (A, B), we have assumed that the probability of surviving the dispersal period is $p = 0.1$, whereas in (C, D) we have assumed that $p = 0.8$. In (A, C) we have assumed that the type 2 individuals are initially at low density ($q_2 \to 0$), whereas in (B, D) they occupy initially half of the sites ($q_2 = 0.5$).

type 2 individuals in the next generation divided by the number of sites they occupy in the present generation) is

$$g_2 = \frac{k_{22}q_2 + k_{21}q_1}{q_2}. \tag{5.27}$$

Figure 5.13 illustrates how the growth rate of type 2 individuals depends on the strategies v_1 and v_2, on their present frequencies q_1 and q_2, and on the mortality parameter p. The black lines and the plus and minus signs divide the space into regions

5.4 Evolutionary dynamics under habitat loss and fragmentation • 197

where the frequency of type 2 individuals increases or decreases. Let us start making sense of this figure by looking at Figure 5.13A, where we have assumed that only the fraction $p = 0.1$ of the propagules survive dispersal. We have further assumed that the type 2 population is initially rare. This corresponds to the modelling framework of adaptive dynamics (Geritz et al., 1998; Dieckmann et al., 1999; Dercole and Rinaldi, 2008; Kisdi and Geritz, 2010), where type 1 individuals are called residents and type 2 individuals are called mutants. Figure 5.13A is called a pairwise invasibility plot (PIP; Dieckmann et al., 1999; Kisdi and Geritz, 2010). The mutant is assumed to challenge a resident, which has reached its population dynamical equilibrium. In the present model, this simply means that the residents occupy all patches. Then mutants, here individuals of type 2, arrive to the system at low density. The question is whether the mutants are able to invade the system and replace the resident. This question can be addressed by looking at the contour lines in Figure 5.13A, as they show the invasion fitness of the mutant, i.e. the expected number of offspring that a mutant individual will contribute to the next generation. In the adaptive dynamics literature, the invasion fitness of a mutant is denoted by $S_r(m)$, where m is the strategy of the mutant (here $m = v_2$), and r is the strategy of the resident (here $r = v_1$).

Let us still consider Figure 5.13A and assume that the resident has strategy $v_1 = 0.2$. Then assume that one of the residents mutates. Assuming that the mutations are small, the mutant strategy will be close to the resident strategy, e.g. $v_2 = 0.19$ or $v_2 = 0.21$. If the mutant's strategy is $v_2 = 0.19$, the resident–mutant combination belongs to the negative growth rate region in Figure 5.13A, meaning that the mutant's density will decrease, and thus is not able to invade the system. If the mutant's strategy is $v_2 = 0.21$, the resident–mutant combination belongs to the positive growth rate region in Figure 5.13A, meaning that the mutant will increase in density. In many population models (such as the present one, as discussed later), the fact that the mutant is able to grow at low density means that it will eventually replace the resident. Thus, it becomes the new resident, and now the resident follows the strategy $v_1 = 0.21$. Eventually, a new mutant will appear, which again is able, or not able, to invade the system, depending on how its strategy v_2 compares with the strategy v_1 of the resident.

The point where the two black lines cross in Figure 5.13A is of particular importance. Let us denote the strategy of the resident (type 1 individual) at this point by $v_1 = v^*$. Assume that the mutant (type 2 individual) follows any other strategy $v_2 \neq v^*$ than the resident. As all locations below and above the point where the two lines cross belong to the region of negative mutant growth rate, the mutant will not be able to invade the system. Thus, the strategy v^* is resistant to invasion by any kind of mutants. That is why it is called the evolutionary stable strategy (ESS), and it is the expected endpoint of evolution.

In the present model the ESS can be computed analytically, and it is given by (Hamilton and May, 1977)

$$v^* = \frac{1}{2-p}. \tag{5.28}$$

What are the evolutionary insights that one may derive from Eq. (5.28)? Let us first assume that there is no dispersal mortality, i.e. that $p = 1$. In this case the ESS strategy is $v^* = 1$, and thus it is beneficial to disperse all propagules randomly among all the sites and leave none in the natal site. This is because the propagules that remain in the natal site have no advantage, but instead they have a disadvantage, as they would be competing for the site with their siblings. With increasing dispersal mortality, it is beneficial to leave more of the propagules in the natal site (compare the lower and upper panels in Figure 5.13), as now dispersal bears a cost. But even if dispersal mortality is very high, it pays off to disperse at least half of the offspring: at the limit $p \to 0$, Eq. (5.28) yields $v^* = 1/2$. This is again the case because it is better to compete against non-relatives than against relatives. If keeping all propagules in the natal sites, the propagules would compete with their siblings, and from the mother's point of view it would be irrelevant which one would win. Only those propagules that are dispersed have the chance of spreading the mother' genes also to other sites. This exemplifies the concept of inclusive fitness (Queller, 1992; Taylor, 1996). For the mother it is not only relevant how many offspring she will produce, but also how many offspring her offspring will produce, to all forthcoming generations.

But what about if the mutants are initially not rare, so that instead of a single mutant the resident population is simultaneously challenged by a large number of mutants? The right-hand panels in Figure 5.13 make this assumption, but are otherwise identical to the left-hand panels of the same figure. While the growth rates of the mutants are different between the left- and right-hand panels, the lines cross at the same points. Indeed, one can show that in this model the strategy given by Eq. (5.28) is resistant to mutants that start from any initial density (Hamilton and May, 1977).

5.4.2 Evolution of dispersal in the plant population model under quantitative genetics

Adaptive dynamics provides a mathematically convenient framework for evolutionary analyses, but at the same time it makes a large number of simplifying assumptions. We next relax some of these assumptions by continuing with evolution of dispersal, but considering the individual-based plant population model of Chapter 3, which we now supplement with quantitative genetics. We assume that dispersal, or more precisely the length scale of the dispersal kernel, is influenced by a clonally inherited haploid gene. In the absence of mutational effects, the breeding value of an offspring (a_O) would equal that of the parent (a_P). We account for mutations by letting $a_O = a_P + m$, where the mutational effect m is distributed as $N(0, V_M)$, where V_M denotes mutational variance. Following the animal model introduced in Section 5.2, we could assume that the phenotype (the length scale of the dispersal kernel) is a sum of genetic and environmental effects, $L = a + e$, where $e \sim N(0, V_E)$, with V_E denoting environmental variance. However, to simplify the analysis, we ignore environmental effects and thus assume full heritability. Unlike in the examples of Section 5.3, we do not make explicit assumptions on how the fitness of the individuals depends on the phenotype. This is because the model is of eco-evolutionary nature, and thus the fitness consequences of the trait values arise through ecological dynamics. Here, the phenotype influences where the offspring are dispersed, which in turn influences their survival and reproduction.

5.4 Evolutionary dynamics under habitat loss and fragmentation • 199

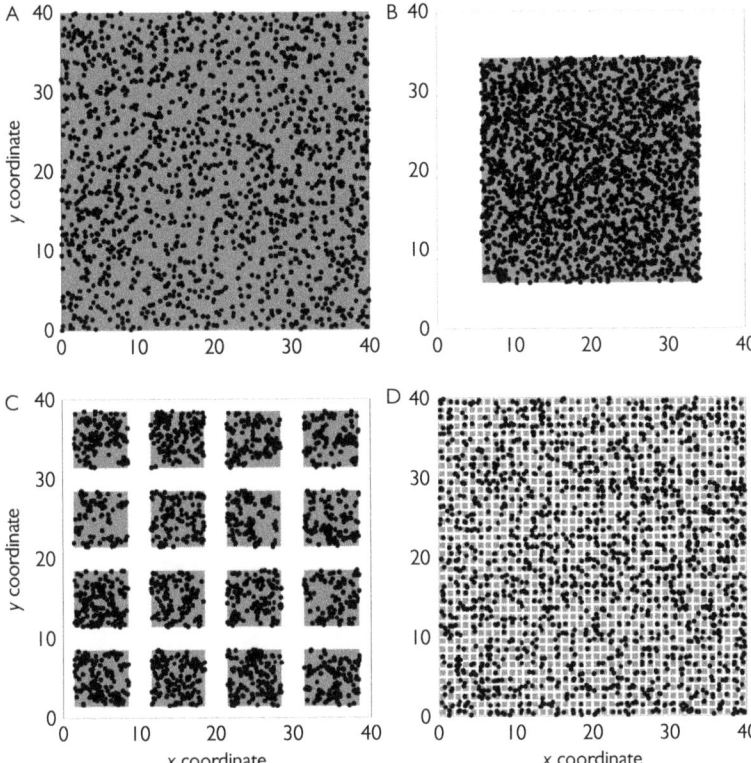

Figure 5.14 Simulating the evolution of dispersal in the plant population model under habitat loss and fragmentation. We assume either that the entire domain consists of habitat (A), or that 50% of the landscape consists of habitat, the habitat being present in one contiguous block (B), in few large patches (C), or in many small patches (D). The dots show the distribution of individuals in the final generation. The distributions of the evolved dispersal distances are shown in Figure 5.15. The parameter values of the plant population model (introduced in Chapter 3) are set to $f = 4$ and $m = 1$, except that in (A) we have set $f = 2$ so that the mean-field model is the same for all panels. Both the competition and dispersal kernels follow bivariate normal distributions with variance-covariance matrix $L^2\mathbf{I}$, where the length scale parameter is set to $L = 1$ for competition, and allowed to evolve for dispersal. Initial population density was set to the mean-field density $f - m$ within the patches, and to 0 in the matrix. The initial distribution of genotypes was assumed to be uniform in the range $L \in (0,5)$. Mutational variance was set to $V_M = 0.05$, and mutations were accepted only if the dispersal distance remained in the range $L \in (0,5)$. The dynamics were simulated until time $t = 40$.

To examine the evolutionary consequences of habitat loss and fragmentation, we ran individual-based simulations of the model in the four landscapes shown in Figure 5.14. In Figure 5.14A, the full landscape consists of habitat. As we assumed that there is no trade-off between dispersal ability and other life-history traits, the ESS is to disperse the offspring as far as possible. The argument here is the same

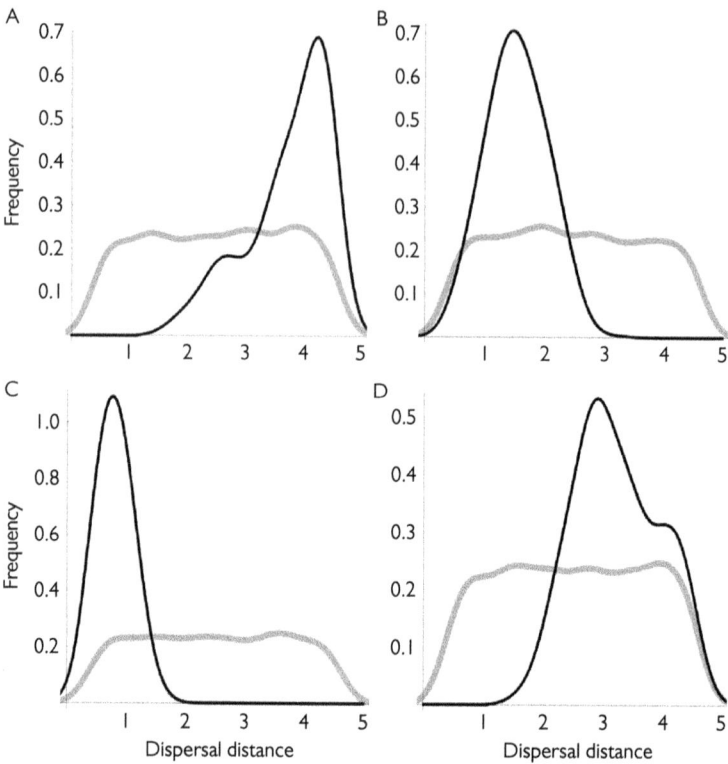

Figure 5.15 The influence of habitat loss and fragmentation for the evolution of dispersal in the plant population model. (A–D) show the initial ($t = 0$, grey lines) and evolved ($t = 40$, black lines) distributions of dispersal distances in scenarios of Figure 5.14A–D, respectively.

as in the Hamilton–May model without mortality: short dispersal bears the cost of increased competition and in particular increased kin competition. However, the stationary distribution of dispersal distances in Figure 5.15A does not involve only the ESS, i.e. the largest possible dispersal distance, but a range of distances. This is due to randomness associated with genetic drift and mutation. Beyond a sufficiently large dispersal distance, the benefit of dispersing even further becomes negligible, in which case the strength of selection is weak.

We then assumed that half of the habitat has been lost, and that the species is not able to establish in the unsuitable matrix. In Figure 5.14B, the remaining habitat is present in one contiguous block. Now individuals that live near the habitat edge risk dispersing their offspring to the unsuitable matrix, which bears a strong fitness cost. For this reason, there is selection for decreased dispersal distances (Figure 5.15B), which is counteracted by the benefits of long-distance dispersal discussed in the previous paragraph. This situation thus resembles the case of the Hamilton–May model with dispersal mortality. In Figure 5.14C, the remaining habitat is present as a set of habitat patches. As the habitat patches are now smaller than the large block of contiguous habitat in Figure 5.14B, a larger proportion of the individuals is located near the edge, and thus dispersal is even more strongly selected against (Figure 5.15C).

In Figure 5.14D, we have further decreased the size of the habitat patches, so that now the landscape consists of many small patches. By the reasoning just discussed, one could think that this situation selects for even shorter dispersal distances. But this is not the case: this landscape selects for longer dispersal distances than the landscapes of Figures 5.15B and C. This is because of two reasons. First, the habitat patches are now so small that they cannot sustain viable local populations. Therefore, a mother that would disperse her offspring only within a patch would not be able to transfer her genes to the next generation. Second, at any point in time, some of the habitat patches will be empty, as seen from Figure 5.14D. This gives an extra reward for dispersal, as it means that propagules dispersing far away from their mother can colonize areas where they escape conspecific competition. As the individuals are more uniformly distributed in the homogeneous landscape of Figure 5.14A, there dispersal does not provide such opportunity.

5.5 Statistical approaches to genetics and evolutionary ecology

Many statistical methods in population genetics and evolutionary biology can be traced back to Ronald Fisher (1890–1962), who made a number of fundamental contributions to statistics (e.g. to ANOVA and maximum likelihood-based methods) inspired by his work in population genetics. With the recent advances in molecular biology, especially with high-throughput sequencing (e.g. Loman et al., 2012), the access to genetic information has rapidly increased, resulting in an equally rapid increase in the need for new kinds of bioinformatics and statistical methods. In this section, we illustrate some statistical methods that link directly to the processes discussed in Sections 5.2–5.4. We start by examining how the relatedness among populations can be inferred from neutral markers. We then ask how the heritability of a trait can be estimated by examining the extent to which related individuals resemble each other. We then combine genotypic and phenotypic data with the aim of identifying which loci influence a given trait. Finally, we attempt to infer whether populations have diverged due to neutral or selective processes, using either genotypic data only or a combination of genotypic and phenotypic data.

5.5.1 Inferring population structure from neutral markers

Analyses of genetic population structure are the starting point for addressing many kinds of questions in evolutionary ecology. The aim of such analyses is to quantify the extent to which different individuals or different populations are related to each other. Knowledge about individual-level relatedness helps to delineate populations and to assign parents or parental populations to individuals (e.g. Dayanandan et al., 1999; Jones and Ardren, 2003). Knowledge about population-level relatedness informs about past levels of gene flow among populations, e.g. whether the populations are reproductively isolated from each other, or if they act as one panmictic unit (e.g. Cheng et al., 2013; Liberal et al., 2014). As we discussed in Section 5.2, relatedness between the individuals i and j can be measured by the coancestry coefficient θ_{ij}, whereas relatedness between the populations X and Y can be measured by the population-level coancestry coefficient $\theta^{\mathcal{P}}_{XY}$.

Related individuals share more genes than unrelated individuals, and thus relatedness can be inferred by assessing the similarity of genotypes. As not only drift but also selection influences genotypic similarity, relatedness should be estimated from allelic variation in neutral loci, i.e. using neutral markers. In Figure 5.16 we have used neutral genetic data to estimate the population-to-population coancestry coefficients $\theta^{\mathcal{P}}_{XY}$ among the 16 populations involved in the scenarios simulated in Figures 5.10–5.12. To do so, we have simulated the single nucleotide polymorphism (SNP; e.g. Morin et al., 2004) genotyping of 20 individuals from each of the 16 populations for 200 bi-allelic neutral loci. To estimate the matrix of coancestry coefficients $\theta^{\mathcal{P}}_{XY}$ from these data, we applied the method of Karhunen and Ovaskainen (2012). In brief, this model-based approach includes parameters that relate to the effective population sizes within each lineage, thus modelling the amount of genetic

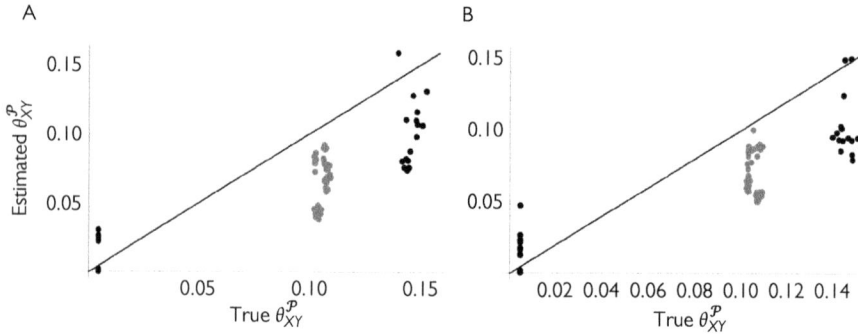

Figure 5.16 Estimating population structure from neutral loci. We estimated population-level coancestry coefficients for $\theta^{\mathcal{P}}_{XY}$ for the 16 populations simulated in Section 5.3. The y-axis shows the posterior mean estimates based on the admixture F-model of Karhunen and Ovaskainen (2012), estimated using the R-package rafm (Karhunen et al., 2013). The x-axis shows the true value, computed by averaging the individual-level coancestry coefficients (shown in Figure 5.12) to the population level using Eq. (5.11). The lines depict the identity $y = x$. The data consist of 200 neutral SNPs (bi-allelic markers) genotyped for 20 individuals from each population. (A) is based on the scenario without selection (illustrated in Figures 5.11A and 5.12A) and (B) on the scenario with selection (illustrated in Figures 5.11B and 5.12B). The black dots near the origin correspond to population pairs that belong to different lineages, the grey dots to population pairs that belong to the same lineages, and the black dots with high relatedness values to the within-population coancestry coefficients $\theta^{\mathcal{P}}_{XX}$.

drift, as well as admixture coefficients that model the amount of gene flow among the lineages. As illustrated by Figure 5.16, the estimation method correctly identifies that populations that belong to the same lineage are more related to each other than populations that belong to different lineages. It also captures a signal of the more recent evolutionary history by assigning higher values for within-population relatedness than for relatedness among populations from the same lineages.

5.5.2 Estimating additive genetic variance and heritability

Heritability is one of the most central concepts in evolutionary ecology. We recall that (narrow sense) heritability is defined as the fraction of phenotypic variance that is due to genetic effects, $h^2 = V_A/(V_A+V_E)$, where V_A is the amount of additive genetic variation and V_E is the amount of environmental variation. The extent to which a trait is heritable influences how similar related individuals are to each other compared to the similarity between unrelated individuals. One simple method for examining this is the parent–offspring regression. In Figure 5.17, we have applied parent–offspring regression to a simulated breeding study, where the parental generation comes from one of the populations that we simulated in Section 5.3. For each breeding pair, we have measured the trait value for 5 offspring. Figure 5.17 shows the mean phenotypic values of offspring against that of their parents. The regression slope β fitted to the data gives an estimate of heritability. To see why this is the case, we recall that the

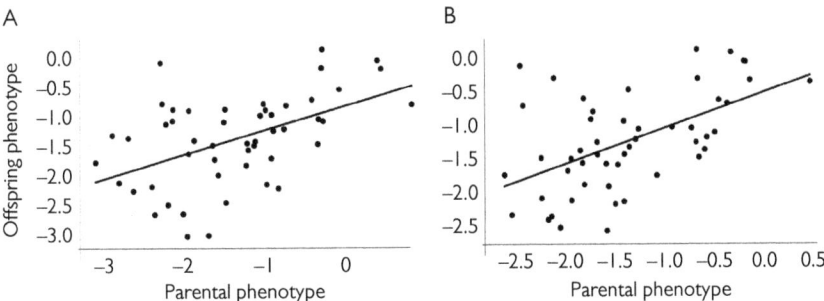

Figure 5.17 Estimating heritability and the amount of genetic variance. We consider trait 1 of population 1 of lineage 1 from the scenario without selection illustrated in Figures 5.10–5.12. We formed 50 breeding pairs from the final generation, and let each pair produce 5 offspring. In the figure the dots show the mean offspring phenotype with respect to the mean parental phenotype. The two panels are based on replicating the same breeding experiment twice. The slope of the regression line gives an estimate of heritability h^2, and its maximum likelihood estimate (95% confidence interval) is 0.41 (0.20 . . . 0.62) in (A) and 0.53 (0.29 . . . 0.78) in (B). The variance among trait values in the parental generation gives an estimate of the phenotypic variance, $V_p = 1.72$. The estimate of additive genetic variance (based on $V_A = h^2 V_P$) is 0.71 for (A) and 0.92 for (B). In this example, the true values are known as the data are generated by an individual-based simulation, and they are $h^2 = 0.43$ and $V_A = 0.75$.

regression slope between the response variables Y and the explanatory variable X can be written as

$$\beta = \frac{\text{Cov}(X, Y)}{\text{Var}(X)}. \tag{5.29}$$

In the context of Figure 5.17, X is the mean parental phenotype and Y is the offspring phenotype. As discussed in Section 5.2, the covariance in breeding values between individuals i and j is $\text{Cov}(a_i, a_j) = 2\theta_{ij} V_A$. We denote by subscripts O, S, and D the offspring, the sire, and the dam, respectively. We recall from Section 5.2 that within-individual coancestry is $\theta_{SS} = \theta_{DD} = \theta_{OO} = 1/2$, and coancestry between parent and offspring is $\theta_{SO} = \theta_{DO} = 1/4$. The fact that $\beta = h^2$ then follows from

$$\text{Cov}(X, Y) = \text{Cov}\left(\frac{a_S}{2} + \frac{e_S}{2} + \frac{a_D}{2} + \frac{e_D}{2}, a_O + e_O\right) = \frac{V_A}{2}, \tag{5.30}$$

and

$$\text{Var}(X) = \text{Var}\left(\frac{a_S}{2} + \frac{e_S}{2} + \frac{a_D}{2} + \frac{e_D}{2}\right) = \frac{V_A}{2} + \frac{V_E}{2}. \tag{5.31}$$

The heritability estimates obtained in Figure 5.17 are in line with the underlying true values, which are known because we used simulated data. However, as illustrated by the wide confidence intervals of the heritability estimates, as well as by the large difference between the two independent breeding experiments (Figure 5.17), obtaining precise estimates for heritability requires a large amount of data.

5.5.3 Using association analysis to detect loci behind quantitative traits

The aim of quantitative trait loci (QTL) mapping (e.g. Yi and Xu, 2008; Tétard-Jones et al., 2011) is to identify which loci or genomic regions influence a quantitative trait of interest. In particular, genome-wide association analyses scan the entire genome in an attempt to identify those SNPs associated with the trait of interest. Such analyses have become widely used especially in the context of human diseases, and in the context of animal and plant breeding for traits of economic importance (e.g. Sachidanandam et al., 2001; Abecasis et al., 2002; Gabriel et al., 2002), but also in evolutionary ecology (e.g. Johstron et al., 2011).

One way to perform QTL mapping is to conduct a series of linear regressions. In each of the regressions, the sampling units are the individuals for which the SNPs and the trait have been measured, the trait is the response variable, and one of the SNPs is the explanatory variable. In Figure 5.18, the y-axis shows the statistical significance for such regressions, transformed so that a large value corresponds to a high level of statistical significance. As we have conducted a large number of tests, we encounter the problem of multiple testing (Benjamini and Hochberg, 1995), due to which a large number of loci not associated with the trait (the grey dots) exceed the usual significance threshold $p = 0.05$ (the dashed line). To account for multiple testing, one needs to apply a more stringent significance threshold (the continuous line).

5.5 Statistical approaches to genetics and evolutionary ecology • 205

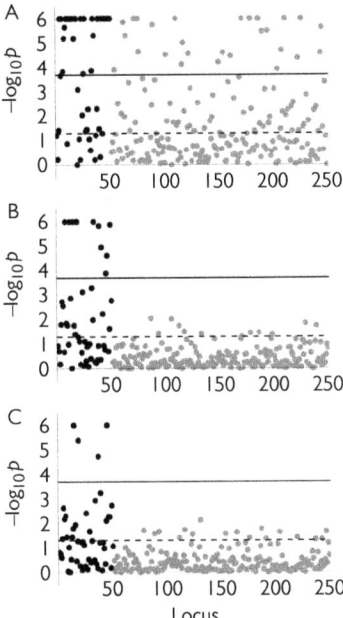

Figure 5.18 Illustration of quantitative trait loci (QTL) mapping. The x-axis shows the locus identifier, with the first 50 loci (shown by black dots) being the SNPs which influence the quantitative trait in reality, whereas the remaining 200 SNPs (shown by grey dots) do not influence the quantitative trait. The y-axis shows the $-\log_{10} p$ value of the linear regression, where phenotype is explained by gene frequencies in the focal locus, truncated to the value 6 if larger. The dashed line corresponds to $p = 0.05$ and the continuous line to the Bonferroni corrected value of $p/250 = 0.0002$ that accounts for multiple testing. The data involve the trait 1 values for all the 100 individuals from each of the 8 populations of lineage 1, from the scenario with selection illustrated in Figures 5.10–5.12. In (A) we have sampled the individuals from the last generation of the simulations and ignored their population structure. The analyses of (B) are based on the same number (800) of unrelated individuals. (C) is otherwise identical to (A), but we have accounted for population structure by subtracting the population mean value from each phenotype.

In the ideal case, all the causal loci (the black dots in Figure 5.18) would be above the significance threshold, while none of the non-causal loci (the grey dots) would do so. The reason why many of the black dots do not exceed the threshold is simply an issue of statistical power. While all those loci influence the trait in question, some of them may have only a small effect in the sense that their allelic effects vary only little. Further, for some loci there may be not much variation in the genotype. In particular, if one of the allelic variants has become fixed, it is impossible to pinpoint whether the locus influences the phenotype.

The reason for the many false positives (grey dots that are above the significance threshold corrected for multiple testing in Figure 5.18A) is that the data points employed in the regression model are not independent. This is because the individuals are not independent, as both their genotypes and their phenotypes are influenced by their shared evolutionary histories. In the ideal case, the individuals used for QTL mapping would be completely independent of each other. Figure 5.18B confirms that in such a case we would correctly identify some of the causal loci, but at the same time avoid false positives. Unfortunately, in many cases it can be very difficult to sample independent individuals. However, more sophisticated versions of QTL mapping account for population structure, such as population-based QTL mapping and family-based association mapping (e.g. Yu et al., 2006; McMullen et al., 2009). As an example, in Figure 5.18C we have repeated the analysis of Figure 5.18A, but controlled for population structure by subtracting from each trait value the population mean. The figure illustrates that accounting for population structured enabled us to avoid false positives.

5.5.4 Detecting loci under selection from genotypic data

We now turn to the question of how to identify signals of selection from data. We start by using genotypic data only, after which we will combine genotypic and phenotypic data. By selection we refer here specifically to local adaptation, and thus our interest is in asking whether the populations simulated in Section 5.3 have differentiated from each other due to selection in addition to the influence of genetic drift. With genotypic data only, we cannot ask which phenotypic traits are influenced by local adaptation, but we can try to identify loci that influence a trait that has been under selection.

Under local adaptation, the allelic frequencies of loci under selection are expected to be different between populations that have been exposed to different kinds of environments. In Figure 5.19 we have quantified the degree to which the populations differ from each other in their allelic frequencies using F_{ST}, which is a commonly used measure for the amount of among-population variation relative to within-population variation (Slatkin, 1985; Whitlock and McCauley, 1999). The neutral loci (shown by the grey dots) offer a reference level to which the loci influencing the quantitative trait (the black dots) can be compared. In the scenarios without selection (the left-hand panels in Figure 5.19), the distributions of the F_{ST} values are similar for the neutral loci and the quantitative trait loci, as is expected. In the scenarios with selection (the right-hand panels of Figure 5.19), some of the selected loci attain higher F_{ST} values than the neutral loci, suggesting that they may have been under local adaptation. However, the pattern is quite mild in Figure 5.19B, in which we have computed F_{ST} by comparing all the populations. This is because selection has not made all the populations to diverge from each other, only those that inhabit different kinds of environments. Let us assume that we would know that the populations come from two different kinds of environments, and that we would expect that environmental variation triggers local adaptation. For example, in the case of nine-spined sticklebacks (Box. 5.1), we could have sampled small ponds and marine environments, which we would expect to differ in their predatory pressure. In this case, we could

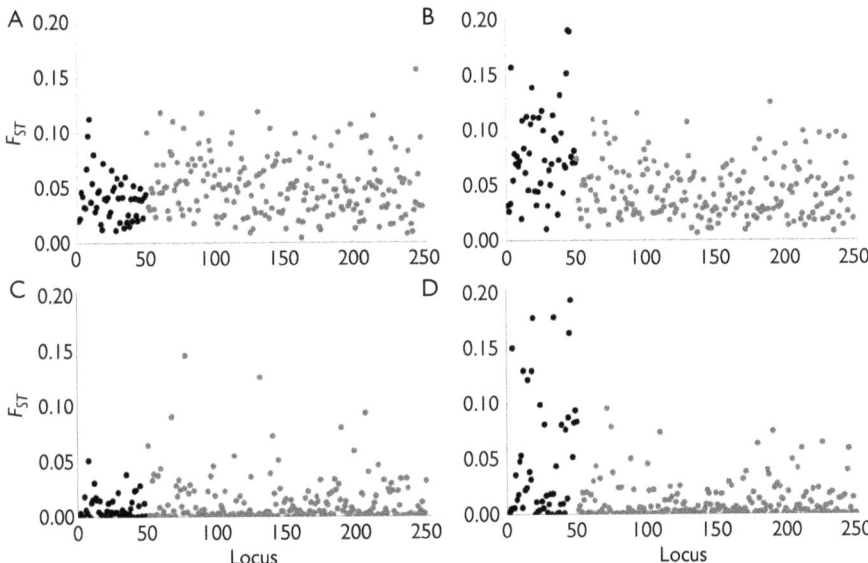

Figure 5.19 Detecting loci under selection from genotypic data. The x-axis shows the locus identifier, with the first 50 loci (shown by black dots) being the SNPs which influence the quantitative trait, whereas the remaining 200 SNPs (shown by grey dots) do not influence the quantitative trait and are thus neutral. We calculated locus-specific F_{ST} values as $(f_1 - f_2)/(1 - f_2)$, where f_1 and f_2 are the probabilities that two randomly selected alleles are identical if they represent individuals that belong to the same (f_1) or different (f_2) populations. In all panels, the data originate from the 8 populations of lineage 1 from the scenarios illustrated in Figures 5.10–5.12. In (A, B), we have computed F_{ST} by comparing within-population variation to between-population variation among the 8 populations. In (C, D), we have computed F_{ST} by comparing variation within environmental types to the variation between environmental types. (A, C) correspond to the scenario without selection, whereas (B, D) correspond to the scenario with selection.

pool the populations that inhabit the same kind of environment. As illustrated in Figure 5.19D, such a test is more powerful in identifying the loci that have been under local adaptation.

5.5.5 Detecting traits under selection from genotypic and phenotypic data

Let us then consider the question that is maybe the most interesting one for evolutionary ecologists working with natural populations: is there evidence that a particular trait (or a combination of traits) has undergone local adaptation?

Local adaptation is expected to make populations diverge from each other in terms of the mean trait values, as we illustrated in Figure 5.11B. But isolated populations are expected to diverge also just through random genetic drift, as we illustrated in Figure 5.11A. Thus, how can we tell whether the pattern of population divergence differs from that expected by genetic drift only? To resolve this question, we recall from Eq. (5.15) that with drift only, and with the assumption of normally distributed traits, the population mean trait values are distributed as

$$a^{\mathcal{P}} \sim N\left(0, 2V_A \theta^{\mathcal{P}}\right). \tag{5.32}$$

Here the population-level coancestry matrix $\theta^{\mathcal{P}}$ measures how related the populations are to each other. In Figure 5.16, we already estimated the coancestry matrix $\theta^{\mathcal{P}}$ from neutral marker data. Thus, given estimates of the population mean traits $a^{\mathcal{P}}$ and the amount of (ancestral) additive variance V_A, we can use Eq. (5.32) to compare the observed distribution of population means to their expected distribution. The S-test (S standing for selection) of Ovaskainen et al. (2011) uses the Mahalanobis (1936) distance to ask whether the population means have diverged according to Eq. (5.32), or whether there is evidence for a deviation from this null expectation. A value of the S-test close to zero yields evidence for stabilizing selection, meaning that the traits have diverged less than expected by drift only. A value of the S-test close to 1 yields evidence for diversifying selection, meaning that the traits have diverged more than expected by drift only. If the data are perfectly in line with the null hypothesis of genetic drift, the expected value of the S-test is 0.5.

To perform the S-test, we have assumed that we have access to the same neutral marker data that were used to estimate $\theta^{\mathcal{P}}$ in Figure 5.16. To obtain estimates of $a^{\mathcal{P}}$ and V_A, we have performed a simulated breeding design in which individuals sampled from the natural populations were used as the parental generation. With such data, the population mean phenotypes are informative about $a^{\mathcal{P}}$, whereas the resemblance of siblings contains information about heritability and thus about V_A (see Figure 5.17). To account for parameter uncertainty in $a^{\mathcal{P}}$, $\theta^{\mathcal{P}}$, and V_A, we have applied a Bayesian estimation scheme implemented in the R-package driftsel (Karhunen et al., 2013). The S-test confirms (Table 5.1) that the data generated

Table 5.1 Detecting signals of selection from a combination of genotypic and phenotypic data.

Case	Traits	S-test	H-test
without selection	1	0.45	0.42
	2	0.44	0.35
	1 and 2	0.45	0.32
with selection	1	>0.99	>0.99
	2	0.53	0.52
	1 and 2	>0.99	0.99

We consider the 16 populations simulated in Section 5.3 and illustrated in Figures 5.10–5.12. The neutral marker data are the same as used in Figure 5.16. The quantitative trait data are based on a similar breeding design as used in Figure 5.17, with 5 offspring measured for 10 breeding pairs for each of the 16 populations. Both the S-test and the H-test were computed with the R-package driftsel (Karhunen et al., 2013).

with drift only are indeed compatible with Eq. (5.32) and its multi-trait analogue, which we used to analyse both traits at the same time. When the data were simulated according to the local adaptation scenario, the S-test provides evidence for selection for trait 1 but not for trait 2 (Table 5.1). This is in line with the fact that in the simulated scenarios we assumed diversifying selection to operate on trait 1, whereas we assumed stabilizing selection for trait 2 (see the locations of the crosses in Figure 5.11B).

While the S-test pinpoints a signal of local adaptation, it does not address which environmental variables have influenced trait divergence. This can be done by the H-test (H standing for habitat), which accounts for environmental information by asking whether populations from similar habitats share more similar trait values than expected by genetic drift only (Karhunen et al., 2014). Thus, in the present case the H-test will ask whether populations occupying the two environmental types have evolved dissimilar trait values. The ingredients needed for applying the H-test are the same genotypic and phenotypic data needed for the S-test, as well as environmental covariate data. A value of the H-test close to 1 indicates that the populations from similar environments have more similar trait values than would be expected by chance. This is indeed the case for trait 1 in the scenario with selection, but not in the scenario without selections (Table 5.1). This result is in line with Figure 5.11B, in which the filled and empty symbols form two distinct clusters, and thus the trait values are distributed according to habitat similarity rather than to population relatedness.

5.6 Perspectives

5.6.1 Mathematical approaches to modelling genetics and evolution

Evolution is a complex process, shaped by the interplay between the evolutionary forces of selection, drift, gene flow, recombination, and mutation. We have illustrated in this chapter that one can take very different approaches to modelling evolution, from simple analytical formulae such as the breeder's equation, to running individual-based simulations, to statistical approaches such as the animal model. These approaches have important differences in how the evolutionary forces are—or are not—incorporated in them.

Individual-based models include the evolutionary forces simply by construction. For example, in our evolutionary plant model, genetic drift was included because the genes in the offspring generation are a sample of the genes present in the parental generation. Recombination was included because we modelled the genotype explicitly through allelic states. Selection was included without the need to specify any fitness function, because in this and other eco-evolutionary models, the evolving trait has fitness consequences through its influence on demographic dynamics. In our evolutionary plant model, the dispersal kernel of the parent determined how likely its offspring were to be dispersed to environments that favour survival and reproduction.

In less mechanistic models, some of the evolutionary forces are not included by construction. Consequently, either they need to be modelled explicitly or they are ignored. For example, the animal model does not have an explicit genotype–phenotype map, but the breeding values of individuals are modelled statistically. Consequently, the effect of recombination needs to be incorporated by assuming that the breeding value of an individual is a random variable, with the expectation set to the mean of the parents, and the variance set to half of the genetic variance in the population (Hartl and Clark, 1997). One simplification associated with many applications of the animal model is that the amount of additive genetic variation is assumed to remain constant at the population level (e.g. Hanski et al., 2011). However, as illustrated by individual-based simulations (Figure 5.6), not only the individual breeding values but also the amount of additive variation is a random variable that evolves over time. This phenomenon is known as the evolution of the **G**-matrix, which is an active area of research (Roff, 2000; Steppan et al., 2002; Griswold et al., 2007; Arnold et al., 2008).

At one extreme are the models of adaptive dynamics, such as the Hamilton and May (1977) model considered here; ecological dynamics are assumed to take place at a much faster rate than evolutionary dynamics, so that the feedback between these two can be ignored. As a consequence, at any given time the resident population can be assumed to be monomorphic, consisting just of a single genotype that has become fixed in the population. In these models, evolutionary changes take place only when the resident population is successfully challenged by a mutant. While empirical data may or may not support this scenario, the simplicity of the adaptive dynamics approach makes it a powerful tool for understanding many evolutionary phenomena.

One approach to modelling evolution that we did not consider here is the use of random walk models or their diffusion approximations. In such models, the response variable is the mean trait value of the population (Whitlock, 1995). Recalling from Chapter 2 that also animal movements can be modelled with random walk, we may draw an analogy between the movements of a single individual and the evolution of a population mean trait value. In this analogy, genetic drift corresponds to the random component of movement, modelled mathematically by the second derivative over space (for animal movement) or over the traits (for evolution). Selection in turn corresponds to a biased component in movement, modelled by first derivative over space (for animal movement) or over the traits (for evolution). In case of directional selection (e.g. the larger the phenotype, the larger the fitness), the analogy in animal movement would be the tendency to move to a particular direction, e.g. a bird flying all the time to the north towards the breeding grounds during their spring migration. In case of stabilizing selection (e.g. the closer the phenotype to an optimal value, the higher the fitness), the analogy in animal movement would be the tendency to return to a particular location, e.g. to perform home-range movements. While animals move in real landscapes, population means can be viewed to evolve in 'fitness landscapes' or 'adaptive landscapes' (Wright, 1932; Vincent, 1988; Mustonen and Lässig, 2009; Stoltzfus and Yampolsky, 2009; Schoville et al., 2012).

It is evident that all of our evolutionary models were great simplifications of reality. To start with, phenotypes are not influenced only by the genes that code for proteins, but also by gene regulation, i.e. the level at which gene is expressed under different stages of development, under different parts of the organisms, and under different environmental conditions (Hartl, 2014). Such interactions between the regulatory and protein coding regions of the genome are one of the reasons for epistatic effects, which we have ignored here. Further, there is an increasing amount of evidence for epigenetic effects, i.e. heritable phenotypic variation that is not related to variation in DNA sequences (Goldberg et al., 2007), providing new challenges for the modelling of genetics and evolution.

5.6.2 Some insights derived from evolutionary models on dispersal evolution

We used two contrasting modelling frameworks to illustrate how species may cope with habitat fragmentation through an evolutionary response. Among the many traits that may show an evolutionary response, we considered dispersal, more specifically either dispersal propensity (as in the Hamilton–May model) or dispersal distance (as in the plant population model). While in the Hamilton–May model the dispersal strategy boils down to either disperse or not, in the plant population model all propagules disperse, so that in a fragmented landscape the likelihood of changing the site is controlled by the interplay between dispersal distance and landscape structure. In spite of the structural differences between these two model types, both yielded qualitatively similar predictions. Without any cost, evolution led to high dispersal propensity (the Hamilton–May model) or large dispersal distances (the plant population model), as limited dispersal bears the cost of increased competition with kin. When increasing the cost of dispersal, both models led to decreased dispersal. While the dispersal cost was implemented as an explicit parameter in the Hamilton–May model, in the plant population model it operated through the cost of losing habitat association, i.e. the aggregation of kin around high-quality habitats (Bolker, 2003). As illustrated by the plant population model, a highly fragmented landscape structure may lead to colonization–extinction dynamics favouring dispersal through the availability of empty patches. Such an extension was actually discussed already by Hamilton and May (1977).

In our models, we ignored the large variety of costs that dispersal may bear, and which may be levied before, during, or after dispersal (Bonte et al., 2011). Such costs may constrain the range of viable dispersal distances, and thus lead to the evolution of intermediate dispersal distances (Skelsey et al., 2012). Predictive modelling of dispersal evolution thus requires knowledge about the trade-offs that involve dispersal traits (Zheng et al., 2009a).

Instead of using the extended plant population model to evolutionary dynamics, we could equally well have extended the butterfly metapopulation model to assess the evolutionary consequences of habitat loss and fragmentation. This would have resulted in an eco-evolutionary model similar to that of Hanski and Mononen (2011), who used their model to study the interplay between habitat structure, population dynamics, and dispersal evolution. Hanski and Mononen (2011) reached essentially the same conclusion as we did here: habitat loss and fragmentation can favour either increased or decreased dispersal, depending on the parameter regime.

Since the pioneering work by Hamilton and May (1977), a large array of models (e.g. Roff, 1975; Comins et al., 1980; Mcpeek and Holt, 1992; Dieckmann et al., 1999; Levin et al., 2003; Ronce, 2007; Bolker, 2009; Cantrell et al., 2009; Gibbs et al., 2010; North et al., 2011) has been constructed to study the evolution of dispersal. One consistent insight from these models is that dispersal is favoured in dynamic landscapes where existing habitat patches are subject to decay and new patches appear in unpredictable locations.

5.6.3 The many uses of genetic data

Rapid developments in sequencing technology have led to a revolution in evolutionary ecology by allowing one to acquire vast amounts of sequence data, facilitating many kinds of analyses at the levels of genes, individuals, populations, species, and higher taxonomical units. This had led to the parallel development of statistical approaches and bioinformatics pipelines. In our statistical section, we presented five examples that illustrated some of the very basic approaches for using genetic data to learn about ecological and evolutionary processes.

As the first example, we showed how the relatedness among populations can be inferred from neutral markers. Neutral markers are routinely used for the construction of pedigrees by assigning individuals to putative parents or source populations, and to the delineation of individuals into populations (e.g. Cardon and Palmer, 2003; Cheng et al., 2013; He et al., 2014).

As the second example, we showed how the degree of heritability and the amounts of genetic and environmental variance can be estimated using parent–offspring regression. The parent–offspring regression is one special case of the methods examining how closely related individuals resemble each other. A much more general statistical approach is to fit the animal model to data. The animal model accounts not only for relatedness among parents and their offspring, but for the level of relatedness between any pair of individuals. As one example, the driftsel analyses (Karhunen et al., 2013) that we applied to analyse population divergence were based on fitting the animal model simultaneously to data from all populations. These analyses accounted both for population-to-population relatedness generated by the shared evolutionary histories and for individual-to-individual relatedness generated by the breeding design. The R-package MCMCglmm by Hadfield (2010) implements a more general version of the animal model than what we considered here, including the possibility to estimate non-additive genetic components.

As the third example, we attempted to identify those loci that influence some quantitative trait of interest. We approached this question by performing QTL analyses, in which we applied the standard linear regression model independently for each locus. More sophisticated analyses, such as Bayesian LASSO (Yi and Xu, 2008), allow one to bring prior information to such analyses, in particular the expectation that only a small number of the candidate genetic markers are likely to influence any given trait.

As the fourth example, we used a F_{ST}-based approach to find signatures of selection from genomic data. As reviewed by Lachance and Tishkoff (2013), also many other kinds of approaches are possible, the pros and cons of which depend on the nature of the available data and the specific questions to be addressed. Some of these approaches are not based on allele frequencies but on haplotypes, i.e. collections of alleles that are typically inherited together because they are located in linked genes (e.g. Sabeti et al., 2002). Inferring signals of selection from genetic data is challenging due to the possibility of false positives, i.e. loci not involved in selective processes but which show up in the analyses for some other reason. Thus, it is safest to consider that the outcome of such analyses is a list of candidate genes, the validity of which is to be confirmed with other methods (Lachance and Tishkoff, 2013).

As the fifth example, we looked for signals of selection using a combination of neutral marker data and quantitative trait data. While we used a model-based approach behind the S-test, a more traditional approach to address the same question would have been to employ a $F_{ST} - Q_{ST}$ test (e.g. Martin et al., 2008; Whitlock and Guillaume, 2009; Edelaar et al., 2011). The neutral expectation of the S-test corresponds to $F_{ST} = Q_{ST}$, whereas $F_{ST} < Q_{ST}$ indicates diversifying selection and $F_{ST} > Q_{ST}$ indicates stabilizing selection. The ingredients needed for the $F_{ST} - Q_{ST}$ test can be derived from the ingredients used in the S-test. To see this, we note that F_{ST} is a summary statistic of the population-to-population relatedness matrix $\theta^{\mathcal{P}}$, as F_{ST} can be computed by comparing the diagonal elements of the matrix (which measure within-population relatedness) to its off-diagonal elements (which measure among-population relatedness). Similarly, Q_{ST} is a measure of population differentiation derived from quantitative trait data, and it can be computed by comparing the level of population differentiation (variance among the population means $a^{\mathcal{P}}$) to the amount of genetic variation within populations (V_A). The S-test has more statistical power than the $F_{ST} - Q_{ST}$ test because it employs the full distributions of trait and neutral marker values, not just the summary statistics F_{ST} and Q_{ST} (Ovaskainen et al., 2011).

In this chapter, we did not even touch on many of the rapidly developing areas fundamental to current genetic and evolutionary research. For example, genome assembly (e.g. Flicek and Birney, 2009; Li et al., 2010) refers to the process in which pieces of DNA sequence reads are put together by comparing matching patterns at overlapping regions, like when assembling a jigsaw puzzle. Gene annotation focuses on finding genes from the genome and predicting what those genes do, e.g. based on finding similar sequences that have already been annotated for related organisms (e.g. Curwen et al., 2004; Wu et al., 2009). The construction of genetic and linkage maps in turn refers to determining in which chromosomes and where within those chromosome the genes are located, and how frequently pairs of genes are inherited together (e.g. Rastas et al., 2013).

Appendix A

Mathematical methods

A.1 A very brief tutorial to linear algebra

In this section, we cover the very basics of matrix and vector algebra. Matrices and vectors are arrays of numbers, and they are usually denoted by bold letters, capitalized for matrices and non-capitalized for vectors. So let us define the example matrices **A**, **B**, and **C**, and the vectors **x** and **y** by

$$\mathbf{A} = \begin{pmatrix} a_{11} & a_{12} \\ a_{21} & a_{22} \end{pmatrix} = \begin{pmatrix} 2 & 1 \\ 0 & 5 \end{pmatrix},$$

$$\mathbf{B} = \begin{pmatrix} 4 & 0 \\ -3 & 2 \end{pmatrix}, \quad \mathbf{C} = \begin{pmatrix} 2 & 0 & -1 \\ 1 & 2 & 1 \end{pmatrix}, \quad \mathbf{x} = \begin{pmatrix} -1 \\ 2 \end{pmatrix}, \quad \mathbf{y} = (5). \quad (A.1)$$

The matrices **A** and **B** have 2 columns and 2 rows, so they are a 2×2 matrices, whereas the matrix **C** has 2 rows and 3 columns, so it is a 2×3 matrix. Vectors are just special cases of matrices of which either of the dimensions is 1. Row vectors consist of a single row and thus have dimension $1 \times n$, whereas column vectors have a single column and thus have dimension $n \times 1$. For example, the column vector **x** is a 2×1 matrix. Numbers can also be considered as special cases of matrices, with dimension 1×1, as is the case with **y** in Eq. (A.1).

Two matrices that have the same dimensions (same numbers of rows and columns) are added by simply summing their elements, yielding e.g.

$$\mathbf{A} + \mathbf{B} = \begin{pmatrix} 6 & 1 \\ -3 & 7 \end{pmatrix}. \quad (A.2)$$

Matrices are multiplied by a number by multiplying every element by that number, so that e.g.

$$5\mathbf{C} = \begin{pmatrix} 10 & 0 & -5 \\ 5 & 10 & 0 \end{pmatrix}. \quad (A.3)$$

Two matrices can also be multiplied by each other, but matrix multiplication is different from element-wise multiplication. To illustrate, multiplying the matrix **A** by the matrix **C** yields **D** = **AC**, with

$$\mathbf{D} = \begin{pmatrix} 5 & 2 & -1 \\ 5 & 10 & 5 \end{pmatrix}. \quad (A.4)$$

For example, the element (1, 3) of the matrix **D** has value −1. This can be derived by taking the row 1 of matrix **A** (with values 2 and 1) and the column 3 of matrix **C** (with values −1 and 1), multiplying these values element by element (giving −2 and 1), and then summing the values. More generally, assume that the dimensions of the matrix **A** are $n \times k$, and that the dimensions of the matrix **C** are $k \times m$. These matrices can be multiplied together because the 'inner dimensions' match: the second dimension of the first matrix and the first dimension of the second matrix are both k. The dimensions of the product matrix $\mathbf{D} = \mathbf{AC}$ will be given by the 'outer dimensions', i.e. the first dimension of the first matrix and the second dimension of the second matrix. Thus, the dimensions of **D** are $n \times m$. The element $\mathbf{D}_{ij}$ is computed from the elements in the row i of matrix **A** and the column j of matrix **C**. Both of these have k numbers, which are multiplied element by element, and added together. As an equation, $d_{ij} = a_{i1}c_{1j} + a_{i2}c_{2j} + \cdots + a_{ik}c_{kj} = \sum_{l=1}^{k} a_{il}c_{lj}$.

Unlike with numbers, with matrices it does not generally hold that $\mathbf{AC} = \mathbf{CA}$, and therefore it does matter in which order matrices are multiplied by each other. For example, with the example matrices given in Eq. (A.1), we computed the product **AC**, but it is not even possible to compute the product **CA** because the inner dimensions of the matrices do not match in this order.

One often needed matrix operation is that of the transpose, denoted by the superscript T. Transposing a matrix simply means switching the roles of rows and columns, so that e.g. the transpose of the matrix **C** would be

$$\mathbf{C}^T = \begin{pmatrix} 2 & 1 \\ 0 & 2 \\ -1 & 1 \end{pmatrix}. \quad (A.5)$$

One common application of matrix algebra is solving a system of linear equations. Thus, we wish to find the solution **x** to the equation $\mathbf{Ax} = \mathbf{b}$, where the matrix **A** and the vector **b** are given. For example, let **A** be as given in Eq. (A.1), an let $\mathbf{b} = (0\ 10)^T$. Note that as shown in Eq. (A.5) we have used the transpose T to convert the 1×2 row vector $(0\ 10)$ into the 2×1 column vector $(0\ 10)^T$. The matrix equation $\mathbf{Ax} = \mathbf{b}$ corresponds to the system of the two linear equations

$$\begin{cases} 2x_1 + x_2 = 0, \\ 5x_2 = 10, \end{cases} \quad (A.6)$$

where we wish to find the numbers x_1 and x_2 which solve the system. It is easy to see that the solution here is $x_1 = -1$ and $x_2 = 2$, or in the matrix form $\mathbf{x} = (x_1\ x_2)^T = (-1\ 2)^T$. In matrix form, the solution can be written as the product $\mathbf{x} = \mathbf{A}^{-1}\mathbf{b}$, where $\mathbf{A}^{-1}$ is called the inverse of the matrix **A**.

Another commonly needed matrix operation is that of finding eigenvalues and eigenvectors. These can be computed only for square matrices, i.e. $n \times n$ matrices with an equal number of rows and columns. For such a matrix **A**, a number λ and a $n \times 1$ column vector **x** are said to form an eigenvalue–eigenvector pair if $\mathbf{Ax} = \lambda \mathbf{x}$. For example, for the matrix **A** given in Eq. (A.1), $\lambda = 2$ and $\mathbf{x} = (1\ 0)^T$ form an eigenvalue–eigenvector pair. To see this, we apply matrix multiplication to see that $\mathbf{Ax} = (2\ 0)^T$, which is the same as $\lambda \mathbf{x} = 2\mathbf{x} = (2\ 0)^T$. A $n \times n$ matrix generally has n eigenvalue–eigenvector pairs. The reader may wish to verify that the second such pair for the example matrix **A** is given by $\lambda = 5$ and $\mathbf{x} = (1\ 3)^T$.

More precisely, solutions **x** to the equation $\mathbf{Ax} = \lambda \mathbf{x}$ are called the right eigenvectors. Vectors **y** that satisfy $\mathbf{y}^T \mathbf{A} = \lambda \mathbf{y}^T$ are correspondingly called left eigenvectors. The eigenvalues λ are the same in both cases but the left and right eigenvector are generally different. For example, the left eigenvector of the matrix **A** associated with the eigenvalue $\lambda = 2$ is $\mathbf{y} = (-3\ 1)^T$, whereas the left eigenvector associated with the eigenvalue $\lambda = 5$ is $\mathbf{y} = (0\ 1)^T$. The left eigenvectors of the matrix **A** are the right eigenvectors of the matrix $\mathbf{A}^T$. We will not go further here on how eigenvalues and eigenvectors can be computed—in practice this can be done with matrix algebra routines implemented in many programming languages.

Other commonly needed matrix operations include e.g. computing the determinant of a matrix, and various matrix decompositions such as the LU decomposition and the Cholesky decomposition. These are explained in many basic textbooks of linear algebra, such as Poole (2014).

A.2 A very brief tutorial to calculus

A.2.1 Derivatives, integrals, and convolutions

A derivative measures the rate of change. If $f(t)$ is a function of time t, the derivative $df(t)/dt$, also denoted by $f'(t)$, tells how fast the function changes (increases or decreases) as time increases. Similarly, if $f(x)$ is a function of one-dimensional space x, then $f'(x)$ tells how the function changes (increases or decreases) as one moves across the space from left to right, i.e. in increasing direction of the x coordinate. In Figure A.1, when moving from left to right, the function $f(x)$ (shown in Figure A.1A) first increases, and thus the derivative $f'(x)$ (shown in Figure A.1B) attains positive values in this part of the space. Then the function decreases (and thus the derivative is negative), then increases (and thus the derivative is positive), and finally decreases again (and thus the derivative is negative). In those locations where the function attains a local maximum or minimum, it does not increase nor decrease, and thus the derivative equals 0.

The second derivative $f''(x)$ is simply defined as the derivative of the first derivative. Thus, it measures how fast the first derivate increases or decreases, as can be seen by comparing Figures A.1B and C as we just did for Figures A.1A and B. A comparison between Figures A.1A and C illustrates how the second derivative $f''(x)$ measures the curvature of the function $f(x)$. If thinking of the function f as e.g. the population density, the second derivative is positive in locations where population density is lower than in the neighbouring locations, whereas the second derivative is negative in locations where population density is higher than in the neighbouring locations. For example, in the middle of Figure A.1A (at $x = 2$), the function $f(x)$ has a local minimum. At this location the function is convex, which corresponds to a positive second derivative, meaning that population density at $x = 2$ is lower than on nearby locations.

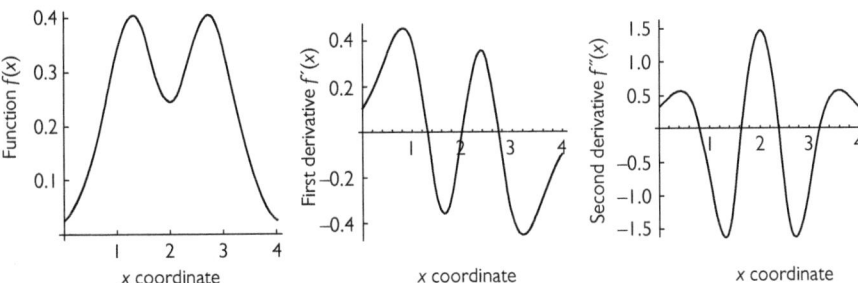

Figure A.1 Illustration of the concepts of derivative and integral. (A) shows an arbitrary function. Its first and second derivatives are shown in (B) and (C), respectively. Conversely, if thinking that (C) shows the function, (B) and (A) show its first and second integrals, respectively.

Similarly, the third derivative $f'''(x)$, also denoted by $f^{(3)}(x)$, is defined as the derivative of the second derivative. More generally, the derivative of order n is denoted by $f^{(n)}(x)$.

Integral is the inverse of derivative, so e.g. the integral of $f'''(x)$ is $f''(x)$, the integral of which is $f'(x)$, the integral of which is $f(x)$. We may write $f(x) = \int_{-\infty}^{x} f'(y)dy + c$. The upper limit of the integration is x and thus the value of a function at x equals its derivative integrated from some lower limit up to that point. We have set here the lower limit to $-\infty$, but it could have been set to any value. This is because knowing the derivative does not determine the function completely, only up to an integration constant, which we have denoted above by c. Changing the lower limit of the integration can be compensated for by changing the integration constant.

The area under a curve can be computed with an integral. For example, for the function f of Figure A.1A, it holds that $\int_0^4 f(x)dx = 1$, and thus the region between the x-axis and the curve has an area of 1.

In addition to derivation and integration, some ecological models involve a convolution term. Convolution does not apply to a single function but to a pair of functions. The convolution between the functions $f(x)$ and $g(x)$ is denoted by $(f * g)(x)$, and it is defined by

$$(f * g)(x) = \int_{-\infty}^{\infty} f(y)g(x-y)dy. \tag{A.7}$$

Figure A.2 gives examples of how the convolution of two functions behaves. If thinking of the function shown in Figure A.2A as the density (number per unit area) of individuals (say, trees) and the function of the Figure A.2B as the dispersal kernel (say, of seeds), the convolution shown in Figures A.2C describes the distribution of the seeds after dispersal. In this example the dispersal kernel g is symmetric around 0, in which case the convolution produces a smoothing (i.e. a local average) of the function f. Note that in this example the density of seeds is not highest where the density of trees is highest. In Figure A.2D we have taken the convolution of the function f with itself. As the function is not centred around 0, the convolution has also moved the function in space. As we will discuss in Appendix A.3, one application of convolutions relates to computing the distribution for the sum of two random variables.

A.2.2 Differential equations

Differential equations are routinely used in many kinds of models in ecology and evolution. As a very simple example, assume that the population consists initially (at time $t = 0$) of a single individual, and that the population increases at a constant rate 2, meaning that 2 new individuals arrive per time unit to the population. The mathematical description of this problem can be written as the differential equation $f'(t) = 2$ and the initial condition $f(0) = 1$. The solution to this problem is $f(t) = 1 + 2t$, and thus the population density increases linearly with time, with the slope 2. To see that this function indeed solves the problem, we first note that $f(0) = 1 + 2 \times 0 = 1$ and thus the proposed solution satisfies the initial condition. Further, by computing the derivative of the function $f(t) = 1 + 2t$ with respect to time t yields $f'(t) = 2$, and thus the proposed solution satisfies also the differential equation.

As a side note, this model predicts that e.g. at time $t = 0.25$ the population will consist of 1.5 individuals, which is clearly not feasible. This is because differential equations are deterministic models defined in a continuous state space (see Table 1.1), and consequently they are best suited for modelling large populations not influenced by stochastic fluctuations. If having initially 1,000 individuals instead of a single individual, there would be 1,500 individuals at time 0.25. While the model would still predict non-integer values, approximating 1,501 individuals by 1,500.5 individuals is more accurate than approximating 2 individuals by

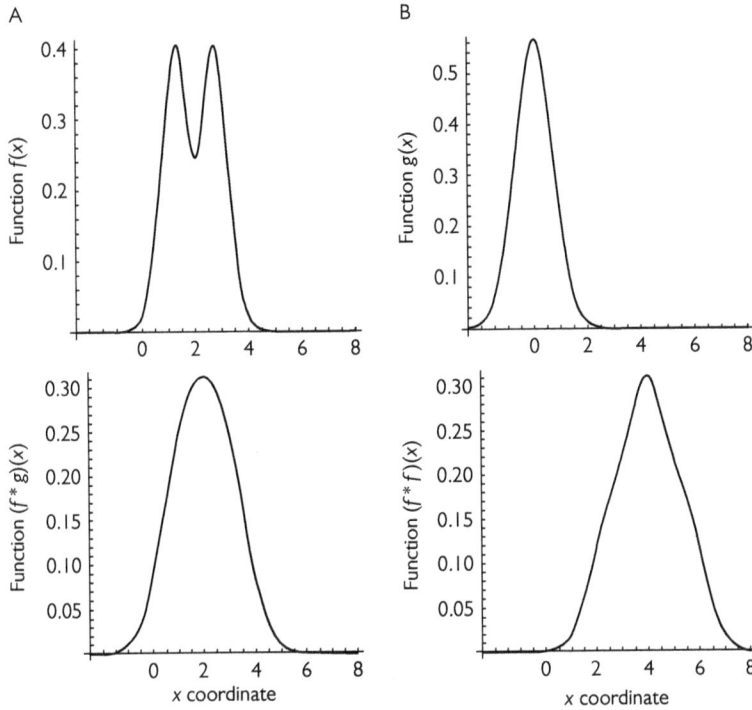

Figure A.2 Illustration of the concept of convolution. (A) and (B) show two arbitrary functions, denoted by $f(x)$ and $g(x)$, respectively. (C) shows the convolution $(f * g)(x)$ and (D) the convolution $(f * f)(x)$.

1.5 individuals. Another option is to consider the differential equation as modelling the expected value of a stochastic process: if starting with a single individual at time $t = 0.0$, then at $t = 0.25$ the population may consist of 1, sometimes of 2 (or 3 or more) individuals, with 1.5 being the expected value. In Appendix A.3 we will discuss Markov processes which can be seen as stochastic counterparts of differential equations.

In the example just described, we assumed that the arrival rate of new individuals is independent of the present size of the population. Let us then assume that new individuals appear because the individuals that make up the present population produce offspring. If the per capita rate (number per unit time) by which the present individuals produce offspring is 2, then the growth rate of the population is $2f(t)$, where $f(t)$ is the current size of the population. This reasoning leads to the differential equation $f'(t) = 2f(t)$. The solution to this problem is $f(t) = e^{2t}$. To see that this is indeed the solution, one needs to check that $f(t) = e^{2t}$ satisfies the initial condition as well as the differential equation. Clearly, $f(0) = 1$, and thus the initial condition holds. Further, $f'(t) = 2e^{2t}$, and thus also the differential equation $f'(t) = 2f(t)$ holds. Note that now the solution $f(t) = e^{2t}$ does not correspond to linear growth but to exponential growth.

Differential equations can also involve higher derivatives. For example, consider the second-order differential equation $f''(t) = 2f'(t) + 3f(t)$, where the highest order derivative (here, second derivative) is put to the left-hand side (this is what is being modelled), and the right-hand side involves lower order derivatives (here, the function itself and the first derivative). A second-order differential equation needs to be supplemented with two initial conditions, one for the function and one for its derivative, e.g. $f(0) = 1$ and $f'(0) = 7$. As an exercise, the reader may wish to verify that the solution to this problem is $f(t) = 2e^{3t} - e^{-t}$. To do so, one needs to check that the proposed function $f(t)$ satisfies both of the initial conditions as well as the differential equation. For the latter, simply compute the value of the left-hand side $f''(t)$ and the value of the right-hand side $2f'(t) + 3f(t)$, and note that they are indeed the same.

A.2.3 Systems of differential equations

Many ecological models are written as coupled systems of differential equations. Examples involve e.g. multi-species models (say, predator–prey model, with one equation for each species) and metapopulation models (with one equation for each patch). The mathematically simplest version is that of a linear system of autonomous differential equations. With the help of the matrix notation introduced in Appendix A.1, such a system can be written as $y'(t) = Ay(t)$, where the variable t is usually time. The equation needs to be supplemented with an initial condition, which can be written as $y(t) = y_0$. This is a linear system because $Ay(t)$ is a linear function of y. It is an autonomous system because $Ay(t)$ depends on the variable t only through $y(t)$, i.e. the matrix A is independent of time.

Autonomous linear systems of differential equations can be solved with a standard procedure. This consists of making an 'ansatz' (an educated guess which will later be verified to hold) that the solution is of the form $y(t) = xe^{\lambda t}$. So let us substitute this ansatz to the differential equation $y'(t) = Ay(t)$ and see what happens. The derivative of the ansatz is given by $y'(t) = \lambda xe^{\lambda t}$, and the matrix product is given by $Ay(t) = Axe^{\lambda t}$. Equating these two gives $\lambda xe^{\lambda t} = Axe^{\lambda t}$. This will be true for all times t if and only if $Ax = \lambda x$. Thus, we have shown that the ansatz $y(t) = xe^{\lambda t}$ is a solution to the differential equation, if (λ, x) is an eigenvalue–eigenvector pair of the matrix A (see Appendix A.1). In general, if A is a $n \times n$ matrix, it will have n eigenvalue–eigenvector pairs, which we may denote by (λ_i, x_i) for $i = 1, \ldots, n$. As the system of differential equations is linear, not only all the n functions $y_i(t) = x_i e^{\lambda_i t}$ satisfy the differential equations, but also any linear combination $y(t) = \sum_{i=1}^{n} c_i x_i e^{\lambda_i t}$. The initial condition $y(0) = y_0$ is needed to determine the weights c_i of the linear combination. As for $t = 0$ it holds that $e^{\lambda_i t} = 1$, we obtain $y(0) = \sum_{i=1}^{n} c_i x_i$. Denoting by X the $n \times n$ matrix which has the eigenvectors x_i as its columns, we can write the initial condition in matrix notation as $Xc = y_0$, where c is the $n \times 1$ column vector with elements c_i. The solution can be written $c = X^{-1} y_0$, where X^{-1} is the inverse of the matrix X (see Appendix A.1).

As an example, let A be the 2×2 matrix

$$A = \begin{pmatrix} -1 & 1 \\ -3 & 1 \end{pmatrix}, \tag{A.8}$$

and let $y_0 = (-1, 1)^T$. Denoting the two components of the vector $y(t)$ by $y_1(t)$ and $y_2(t)$, the system of differential equations can also be written as (see Appendix A.1)

$$\begin{cases} y_1'(t) = -y_1(t) + y_2(t), \\ y_2'(t) = -3y_1(t) + y_2(t), \end{cases} \tag{A.9}$$

with the initial condition $y_1(0) = -1$, $y_2(t) = 1$. The eigenvalues of the matrix A are $\lambda_1 = \sqrt{2}i$, $\lambda_2 = -\sqrt{2}i$, where i is the imaginary unit with $i^2 = -1$. It may seem confusing that the eigenvalues are complex numbers, as the original system of differential equations had nothing to do

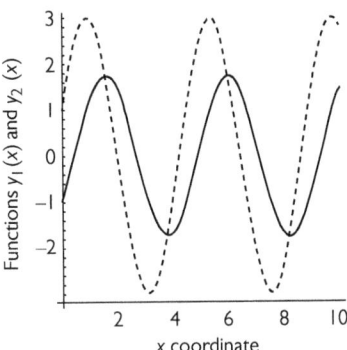

Figure A.3 The solution (Eq. (A.10)) to the system of differential equations given by Eq. (A.9). The function $y_1(t)$ is shown by continuous line and the function $y_2(t)$ by dashed line.

with complex numbers! But there is nothing wrong here, the complexity of the eigenvalues just indicates that the solution will oscillate in time (this relates to what is called Euler's formula: $e^{ia} = \cos a + i \sin a$). After computing the eigenvalue–eigenvector pairs and solving $c = \mathbf{X}^{-1}\mathbf{y}_0$, the solution can be presented as

$$\begin{cases} y_1(t) = \sqrt{2} \sin\left(\sqrt{2}t\right) - \cos\left(\sqrt{2}t\right), \\ y_2(t) = 2\sqrt{2} \sin\left(\sqrt{2}t\right) + \cos\left(\sqrt{2}t\right). \end{cases} \quad (A.10)$$

Reassuringly, the imaginary parts have cancelled out from the solution, and thus we needed complex numbers only as an intermediate step. The dynamics of this system are illustrated in Figure A.3. They resemble predatory–prey dynamics discussed in Section 4.2.

Systems of differential equations used to model ecological or evolutionary dynamics will often be non-linear and non-autonomous. For the general case, a system of first-order differential equations can be written as $\mathbf{y}'(t) = \mathbf{f}(\mathbf{y}(t), t)$, where $\mathbf{f}$ is some vector-valued function which may depend in any non-linear manner on the current state of the system $\mathbf{y}(t)$, as well as on time t. Unlike for linear and autonomous systems, there is no general recipe on how to solve this equation, and one often needs to resort to numerical solutions. However, specific properties of the system (most importantly, local stability of equilibriums solutions) can often be found analytically by linearizing the system. We refer the interested reader to the many textbooks covering these topics (e.g. Swokowski 2000).

A.2.4 Partial differential equations

Sometimes differential equations involve derivatives both with respect to time and space. In such a case, they are called partial differential equations, to be separated from ordinary differential equations which involve derivatives only with respect to one variable. As an example, consider the diffusion equation (Eq. (2.2)) of Chapter 2, simplified here to a single spatial dimension as

$$\frac{\partial f(x,t)}{\partial t} = D \frac{\partial^2 f(x,t)}{\partial x^2}. \quad (A.11)$$

The state variable in this equation, denoted by $f(x, t)$, is a function of both time t and spatial location x. In the context of Chapter 2, $f(x, t)$ would model the probability density of an individual's location. The symbol ∂ that appears on both sides of the equation stands for a partial derivative. In the left-hand side the partial derivative is written as $\partial/\partial t$, and it measures the rate of change with respect to time t. Thus, the left-hand side of the equation asks how fast the function f increases or decreases with time. The right-hand side of the equation gives the answer: the rate of change is proportional to the second spatial derivative. Note that in the right-hand side, the partial derivative is taken not with respect to time t but with respect to the spatial variable x, and it is not the first but the second partial derivative, as indicated by the superscripts 2 in $\partial^2/\partial x^2$. Based on the interpretation of the second derivative, Eq. (A.11) says that the function f will increase in locations where it currently has a smaller value than in neighbouring locations, and that it will decrease in locations where it currently has a higher value than in neighbouring locations. Therefore, the function becomes smoother than it was originally, as peaks become flattened out and dips become filled.

A partial differential equation also needs to be supplemented by an initial condition $f(x, 0) = f_0(x)$. Unlike with ordinary differential equations, the initial condition is not a single value, but a function $f_0(x)$ which describes the initial state of the function f in all spatial locations x.

A.2.5 Difference equations

The difference between differential and difference equations is that difference equations are defined for discrete time, $t = 0, 1, 2, \ldots$, while differential equations are defined with respect to continuous time. For example, a discrete-time analogy for the differential equation $f'(t) = 2f(t)$ would be the difference equation $f(t + 1) = 2f(t)$. This equation tells that every time-step the number of individuals in the population doubles. Starting again with a single individual, $f(0) = 1$, the solution to this problem is $f(t) = 2^t$. Like the solution of the corresponding differential equation, this also corresponds to exponential growth. As in this book we mainly focus on continuous-time models, we have focused on differential equations, but very similar mathematical tools exist also for difference equations. Often the same biological problem can be modelled almost identically by a differential or a difference equation, and thus the choice between these two is in many cases somewhat arbitrary. However, in other cases these two approaches can give qualitatively different answers.

A.3 A very brief tutorial to random variables

Differential equations and difference equations (Section A.2) are examples of deterministic models. This means that 'running' the same model repeatedly always leads to the same solution. Many, if not all, ecological and evolutionary phenomena are, however, not of deterministic but of stochastic nature, meaning that running exactly the same model twice (e.g. repeating the same experiment twice) usually results in different outcomes. Such random variation can be captured by stochastic processes. A central concept behind stochastic processes is that of a random variable, a concept that is also needed for understanding the mathematical foundations of statistics that we discuss in Appendix B.

A.3.1 Discrete valued random variables

Random variables are usually denoted by capital letters such as X. An example of a random variable X could be the outcome of a toss of a coin or a die. In the case of the die, there are six possible outcomes, $X \in (x_1, x_2, \ldots, x_6)$, which simply correspond to the number of dots (called pips) on the six faces of the die. Thus, in this example we may write $x_1 = 1, \ldots, x_6 = 6$, or in a more general notation $x_k = k$, where $k = 1, 2, 3, 4, 5, 6$. Assuming a fair (i.e. unbiased) die, the probabilities of each of these outcomes are identical, so that $p_k = 1/6$ for $k = 1, \ldots, 6$. X is a discrete valued random variable, as the outcomes form a discrete set. The probability distribution of X can be described simply by listing the probabilities of all possible outcomes, as we have just done.

One of the most basic properties of a random variable is its expectation, denoted by $E(X)$. The expected value is defined as the mean value of the outcomes, weighted by their probabilities, $E(X) = \sum_k p_k x_k$. For the die example, the expected result, also called the mean or the average, is $E(X) = \frac{1}{6} + \frac{2}{6} + \cdots + \frac{6}{6} = \frac{7}{2} = 3.5$. If throwing the die 1,000 times, the sum of the values should be close to 3,500. In addition to the expectation, another routinely used statistic is variance, which measures the amount of variability among the outcomes. Variance is defined as $\mathrm{Var}(X) = E\left[(X - E(X))^2\right] = \sum_k p_k (x_k - E(X))^2$. In this formula, $X - E(X)$ asks how far the random variable X is from its expectation $E(X)$. Variance is thus defined as the expectation of the squared difference between the random variable and its expectation. In the case of the die example, we obtain

$$\mathrm{Var}(X) = \frac{(1-3.5)^2}{6} + \frac{(2-3.5)^2}{6} + \cdots + \frac{(6-3.5)^2}{6} = \frac{35}{12} \approx 2.92.$$

Random variables can have infinitely many possible outcomes. As one example, let us consider a random variable for which the possible outcomes are all the non-negative integers $k = 0, 1, 2, \ldots$. Let us assume that the first outcome ($k = 0$) takes place every second time, i.e. with probability $p_0 = 1/2$. Let us assume that the second outcome ($k = 1$) takes place in every second case, conditional on $k = 0$ not taking place, and thus with probability $p_1 = 1/2 \times 1/2 = 1/4$. Assuming that the third possible outcome ($k = 2$) would again hold every second time, conditional on $k = 0$ or $k = 1$ not taking place, we have $p_2 = 1/2 \times 1/4 = 1/8$. Continuing similarly, we obtain for the general case $p_k = 1/2^{k+1}$. The probabilities p_k are illustrated in Figure A.4A. They describe the probability density function of a random variable, which follows the geometric distribution with parameter value $1/2$. For any discrete probability distribution such as the geometric distribution, summing up the probabilities of all possible outcomes gives one, i.e. $\sum_k p_k = 1$.

Applying the formulae of expectation and variance described two paragraphs earlier, we find that these are for the geometrically distributed (with parameter $1/2$) random variable $E(X) = \sum_k p_k x_k = \sum_k k/2^{k+1} = 1$, and $\mathrm{Var}(X) = \sum_k p_k (x_k - E(X))^2 = \sum_k (k-1)^2/2^{k+1} = 2$. We note that for this distribution, we were able to compute the values of the infinite sums exactly, but in a more general case one may need to compute them numerically.

A.3.2 Continuous valued random variables

Let us then move to random variables with a continuous state-space. An archetypal example of such a random variable is given by the normal distribution $N(\mu, \sigma^2)$ with mean μ and variance σ^2. The notation $X \sim N(\mu, \sigma^2)$ is to be read as 'the random variable X is distributed according to the normal distribution with mean μ and variance σ^2'. Let us consider for simplicity the so-called standard normal distribution $N(0, 1)$, with mean 0 and variance 1. The random variable X can obtain any real value. Thus, if we sample once from the distribution, we may obtain $X = 0.2311$, whereas the next sample may be $X = -24.5921$. While any value is possible, not all values are equally likely. The relative probabilities of different values are described by the probability density function of the distribution. In the case of $N(0, 1)$, the

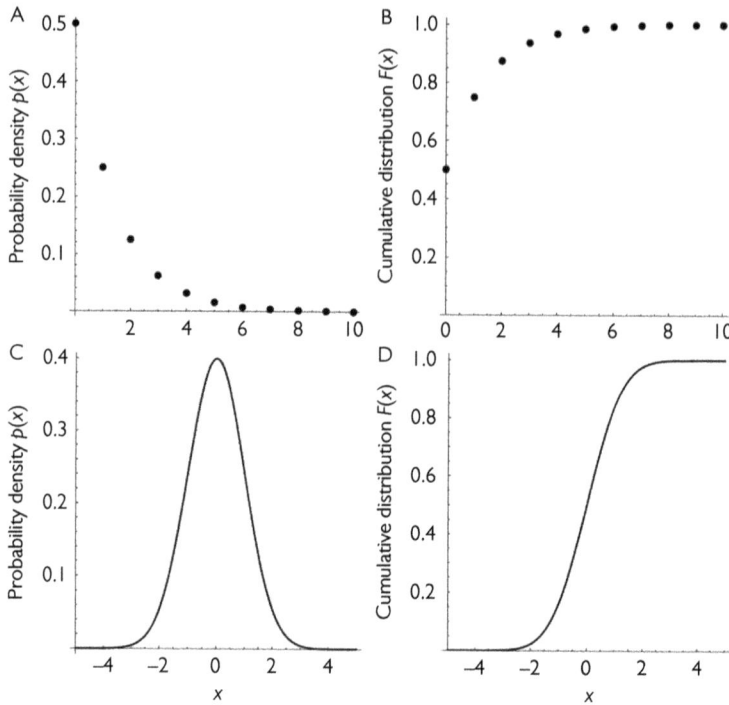

Figure A.4 Representing random variables through probability distributions. (A, B) illustrate a discrete valued random variable (geometric distribution with parameter 1/2), whereas (C, D) illustrate a continuous valued random variable (normal distribution with mean 0 and variance 1). (A, C) show probability density functions and (B, D) cumulative distribution functions.

probability density function (illustrated in Figure A.4C) is $p(x) = e^{-x^2/2}/\sqrt{2\pi}$. The value of the probability density at the most likely value of $x = 0$ is $p(0) = 1/\sqrt{2\pi} \approx 0.40$, whereas a value as far from 0 as -24.5921 is extremely unlikely, $p(-24.5921) \approx 1.9 \times 10^{-132}$.

In the case of a discrete valued random variable, the values of the probability density function are probabilities which sum to 1. But in the case of a continuous random variable, there is a continuum of possible values. Thus, we cannot say that the most likely value $x = 0$ is sampled with probability 0.4. Even if the sampled value could be $x = 0$, mathematically the probability of sampling this value is 0, as is the probability of sampling any specific single value. With continuous valued random variables, we may ask with what probability the random variable will fall within a specific range. The answer to this question is obtained by integrating the probability density function over that range. For example, as the area under the probability density curve within the range $-1 < x < 1$ is $\int_{-1}^{1} p(x)dx = \int_{-1}^{1} (e^{-x^2/2}/\sqrt{2\pi})dx \approx 0.683$, the fraction of cases where the random variable sampled from $N(0, 1)$ will have its value within this range is 68.3%.

The probability by which the random variable will obtain a value that is at most a given threshold x is called the cumulative distribution function, often denoted by $F(x)$.

For continuous random variables, $F(x)$ is defined by $F(x) = \int_{-\infty}^{x} p(y) dy$, whereas for discrete random variables it is defined by $F(x) = \sum_{y=-\infty}^{x} p_y$. Note that inside the integral (or sum) we have used another name for the variable (y) because x is reserved for the integration (summation) limit.

In the case of discrete valued random variables, the probabilities must sum to 1. Similarly, in the case of a continuous valued random variable, the probability density function needs to integrate to 1, i.e. $\int_{-\infty}^{\infty} p(x) dx = 1$. These statements can be expressed with the help of the cumulative distribution function as $F(\infty) = 1$. Figure A.4 illustrates how the cumulative distribution functions of the geometric and the normal distributions converge to 1 as x increases. Note that the probability density function of the standard normal distribution involves a factor $1/\sqrt{2\pi}$. This factor is called the normalizing constant, as it has been chosen so that $F(\infty) = 1$.

The expected value of a continuous valued random variable is defined like that for a discrete valued random variable, except replacing the summation by an integral. Thus, $E(X) = \int_{-\infty}^{\infty} p(x) x dx$, where each possible value x of the random variable X is weighted by its probability density $p(x)$. For $X \sim N(0, 1)$, it holds that $E(X) = \int_{-\infty}^{\infty} (e^{-x^2/2}/\sqrt{2\pi}) x dx = 0$. Similarly, the variance of a continuous valued random variable is defined by $Var(X) = \int_{-\infty}^{\infty} p(x)(x - E(X))^2 dx$, yielding for the $X \sim N(0, 1)$ $Var(X) = \int_{-\infty}^{\infty} (e^{-x^2/2}/\sqrt{2\pi}) x^2 dx = 1$. Note that we derived that the mean and variance of a $N(0, 1)$ random variable are indeed 0 and 1.

A.3.3 Joint distribution of two or more random variables

Often it is interesting to consider two or more random variables at the same time. Denoting two random variables by X and Y, we may consider their joint distribution, i.e. the probabilities (or values of the probability density) of all possible combinations of their outcomes (x, y). Covariance and correlation measure the degree by which the two random variables depend on each other. The covariance is defined as $Cov(X, Y) = E[(X - E(X))(Y - E(Y))]$, and correlation is obtained by scaling the covariance by the variances, $Cor(X, Y) = Cov(X, Y)/\sqrt{Var(X) Var(Y)}$. While covariance can obtain any value from $-\infty$ to $+\infty$, correlation is restricted to the range from -1 to 1. Two variables are statistically independent if their covariance, and thus correlation, is 0.

To illustrate the concept of covariance, let us consider the multivariate normal distribution. The notation $X \sim N(\mu, \Sigma)$ means that the $n \times 1$ random vector X follows a multivariate normal distribution with mean μ (a $n \times 1$ vector) and variance-covariance matrix Σ (a $n \times n$ matrix). As an example, let us consider the bivariate case $(n = 2)$ with

$$\mu = \begin{pmatrix} -2 \\ 1 \end{pmatrix}, \quad \Sigma = \begin{pmatrix} 4 & 7 \\ 7 & 25 \end{pmatrix}. \tag{A.12}$$

In this example, the vector X has two elements, $X = (X_1 X_2)^T$. The first one of these is distributed as $N(-2, 4)$ and the second one as $N(1, 25)$, where the means and the variances of the univariate normal distributions are obtained from the mean vector μ and the diagonal elements of the variance-covariance matrix Σ. The covariance 7 between the two random variables can be translated into the correlation of $7/\sqrt{4 \times 25} = 0.7$. The dots in Figure A.5A illustrate 200 random deviates sampled from this distribution, and the ellipses the 50 and 95% quantiles of the probability distribution. The mean of the distribution is centred at μ, and the positive correlation assumed between the two variables is evident in the figure. Figure A.5B illustrates otherwise the same case but the covariance has been changed to -7, leading to a negative correlation of -0.7.

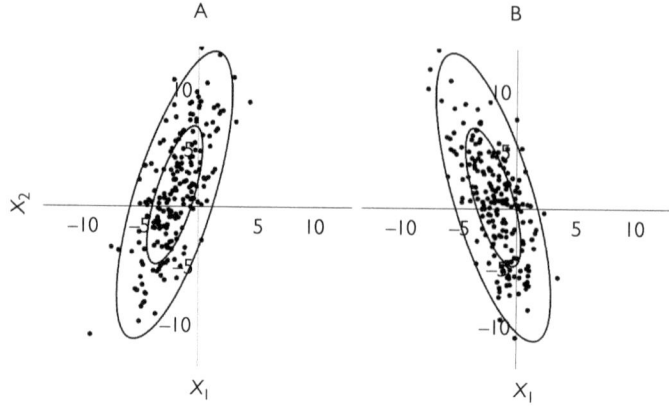

Figure A.5 Illustration of a bivariate probability distribution. In (A), the dots show 200 random deviates sampled from the bivariate normal distribution of Eq. (A.12), whereas the lines show the 50 and 95% quantiles of this distribution. (B) is otherwise identical but the covariance has been changed to −7.

A.3.4 Sums of random variables

Expectations, variances, and covariances have many useful properties. For example, one may wish to study the properties of a random variable Z obtained as a weighted sum $Z = aX + bY$ of two random variables X and Y, where the weights a and b are real numbers. The expectation and variance of Z can be computed from those of the original variables as $E(Z) = aE(X) + bE(Y)$, and $Var(Z) = a^2 Var(X) + b^2 Var(Y) + 2ab Cov(X, Y)$. Similarly, if $Z = aX + bY$ and $W = cP + dQ$, then

$$\text{Cov}(Z, W) = ac\,\text{Cov}(X, P) + ad\,\text{Cov}(X, Q) + bc\,\text{Cov}(Y, P) + bd\,\text{Cov}(Y, Q). \quad (A.13)$$

If the random variables X and Y are independent of each other, the covariance is 0, and thus the formula for the variance simplifies to

$$\text{Var}(aX + bY) = a^2 \text{Var}(X) + b^2 \text{Var}(Y). \quad (A.14)$$

In addition to expectation and variance, it is possible to compute the entire probability distribution for the (possibly weighted) sum of two random variables. To illustrate, let us consider the example of throwing the die (see Section A.3.1) but now doing it twice, with the first and second outcomes called X and Y. The sum of these two outcomes $Z = X + Y$ may obtain any value between 2 and 12. For example, the value of 5 will be obtained by the (X, Y) combinations $(1, 4), (2, 3), (3, 2)$ and $(4, 1)$. As each of these takes place with probability 1/36, we obtain $\Pr(Z = 5) = 4/36 = 1/9$. More generally, if we denote by p^X and p^Y the probability distributions of the random variables X and Y (which in the die example are identically distributed, but that more generally could be different), we have

$$\Pr(Z = z) = \sum_k p^X_k p^Y_{z-k}. \quad (A.15)$$

Equation (A.15) is simply based on going through all the cases for which the sum of the two random variables obtains the specific value z. This happens if the first variable X obtains any value k, and the second variable Y obtains the value $z - k$, which makes the sum equal z, as we just illustrated for the die example for $z = 5$.

We note that Eq. (A.15) holds only if the random variables X and Y are independent of each other. To see this, let us not throw the second die at all, but decide to set the value of the second die to the value of the first die. In this case, the outcome of the second die Y is still a random variable, as it obtains each of the values from 1 to 6 with probability 1/6. However, it is not independent from the outcome of the first die X. Now the possible outcomes of Z are 2, 4, 6, 8, 10, and 12, and each of these takes place with probability 1/6.

Equation (A.15) holds also for continuous valued random variables if replacing the sum by an integral. Thus assuming that X and Y are independent continuous valued random variables, the probability density function of $Z = X + Y$ is given by

$$p^Z(z) = \int_{-\infty}^{\infty} p^X(x) p^Y(z - x)\, dx = \left(p^X * p^Y\right)(z), \tag{A.16}$$

where the symbol $*$ denotes convolution (see Section A.2). As an example, let $X \sim \text{Uniform}(0, 1)$ be an uniformly distributed random variable in the range from 0 to 1. Let also the random variable $Y \sim \text{Uniform}(0, 1)$ follow the same distribution. Evidently, the sum $Z = X + Y$ may obtain any value from 0 to 2. At first sight one might expect that Z could obtain any value from 0 to 2 equally likely, i.e. that $Z \sim \text{Uniform}(0, 2)$. This would actually be the case if we would make the two random variables fully correlated by sampling only X and then setting $Y = X$, as we did in the case of the die example where we set the value of the second die to that of the first die.

To compute the distribution of Z when X and Y are independent, let us apply Eq. (A.16). With $X \sim \text{Uniform}(0, 1)$, the probability density function is $p^X(x) = 1$ if $0 < x < 1$ and $p^X(x) = 0$ otherwise. As Y is identically distributed, it has the same probability density function. Thus we obtain

$$p^Z(z) = \int_{-\infty}^{\infty} p^X(x) p^Y(z - x)\, dx = \int_0^1 p^Y(z - x)\, dx = \int_{\max(z-1,0)}^{\min(z,1)} 1\, dx. \tag{A.17}$$

In Eq. (A.17) we have first utilized the fact that $p^X(x) = 1$ for $0 < x < 1$ and $p^X(x) = 0$ otherwise. We have then utilized the fact that $0 < z - x < 1$ if and only if $z - 1 < x < z$. As the integrand is just the constant 1, the value of the integral is the length of the integration interval, and therefore $p^Z(z) = \min(z, 1) - \max(z - 1, 0)$. Thus, if $0 < z < 1$, we have $p^Z(z) = z$. If $1 < z < 2$, we have $p^Z(z) = 1 - (z - 1) = 2 - z$. This probability density looks like a triangle: the most likely value for the sum is $Z = 1$, and the value of the probability density decreases linearly to 0 if approaching the extremes of 0 or 2. Informally, there are more combinations of X and Y which produce the sum of $Z = 1$ than e.g. the sum of $Z = 0.1$. Note that this is analogous to the earlier discussion on the sum of two throws of the die, for which case e.g. $\Pr(Z = 5) = 4/36$ is greater than $\Pr(Z = 2) = 1/36$.

A special case often needed in the context of hierarchical generalized linear models (Section B.1) is the sum of two normally distributed variables. Assume that $X \sim N(\mu_X, \sigma_X^2)$ and that $Y \sim N(\mu_Y, \sigma_Y^2)$, and let $Z = X + Y$. Assuming that X and Y are independent, it holds that $Z \sim N(\mu_X + \mu_Y, \sigma_X^2 + \sigma_Y^2)$. Thus, the sum of two normally distributed random variables is also normally distributed, the expected value being the sum of the expectations, and the variance being the sum of the variances.

A.3.5 An application of random variables to quantitative genetics

As an illustration of how to work with random variables, we follow here Appendix A of Ovaskainen et al. (2011) to derive Eq. (5.4) in the main text. Our starting point is Eq. (5.1), which models the breeding value of an individual i as

$$a_i = \sum_{j=1}^{n} \sum_{k=1}^{2} \sum_{u=1}^{m_j} x_{ijku} v_{ju}. \quad (A.18)$$

We consider the allelic states x_{ijku} as random variables (even if they are denoted by non-capitalized letters), and the allelic effects v_{ju} as fixed. By the covariance formula for sum of random variables (Eq. (A.13)), we obtain

$$\text{Cov}(a_{i_1}, a_{i_2}) = \sum_{j=1}^{n} \sum_{k_1,k_2=1}^{2} \sum_{u_1,u_2=1}^{m_j} v_{ju_1} v_{ju_2} \text{Cov}(x_{i_1 jk_1 u_1}, x_{i_2 jk_2 u_2}). \quad (A.19)$$

We then apply the definition of covariance to write

$$\text{Cov}(x_{i_1 jk_1 u_1}, x_{i_2 jk_2 u_2}) = \text{E}(x_{i_1 jk_1 u_1} x_{i_2 jk_2 u_2}) - \text{E}(x_{i_1 jk_1 u_1}) \text{E}(x_{i_2 jk_2 u_2}). \quad (A.20)$$

We denote the frequency of the allele u of gene j in the ancestral population by p_{ju}. As we assume that the genes under consideration are neutral, the expected allele frequency for any individual is equal to that in the ancestral population, and thus $\text{E}(x_{ijku}) = p_{ju}$. Recall that x_{ijku} is an indicator function taking the value 1 for one of the allelic variants u and 0 for all other allelic variants. Thus, while one might at first think that $\text{E}(x_{ijku} x_{ijku}) = p_{ju}^2$, it actually holds that $\text{E}(x_{ijku} x_{ijku}) = p_{ju}$. For the same reason, it holds that $\text{E}(x_{ijku_1} x_{ijku_2}) = 0$ for $u_1 \neq u_2$. The value of the product $x_{i_1 jk_1 u} x_{i_2 jk_2 u}$ equals 1 if in gene j the allele k_1 of the individual i_1 is of the same type u as the allele k_2 of the individual i_2, whereas the product equals 0 if this is not the case. This means that the probability by which the allele k_1 of the individual i_1 is of the same type u as the allele k_2 of the individual i_2 is given by $\text{E}(x_{i_1 jk_1 u} x_{i_2 jk_2 u})$. Therefore, the probability that two randomly chosen alleles in a gene j are of the same type u for the individuals i_1 and i_2 is given by $\sum_{k_1,k_2=1}^{2} \text{E}(x_{i_1 jk_1 u} x_{i_2 jk_2 u})/4$. This probability can be decomposed into two components. First, the alleles may be identical by descent, the ancestral allele being of type u. Given the definition of coancestry coefficients (see Section 5.2), this happens with probability $\theta_{i_1 i_2} p_{ju}$. The second option is that the alleles are not identical by descent, but that the alleles in the ancestral generation were both of type u. As the probability of the latter case is $(1 - \theta_{i_1 i_2}) p_{ju}^2$, we obtain

$$\frac{1}{4} \sum_{k_1,k_2=1}^{2} \text{E}(x_{i_1 jk_1 u} x_{i_2 jk_2 u}) = \theta_{i_1 i_2} p_{ju} + (1 - \theta_{i_1 i_2}) p_{ju}^2. \quad (A.21)$$

As the individuals i_1 and i_2 can have different allelic variants only if the alleles are not identical by descent (recall from Section 5.2 that we have ignored mutations), we obtain for $u_1 \neq u_2$

$$\frac{1}{4} \sum_{k_1,k_2=1}^{2} \text{E}(x_{i_1 jk_1 u_1} x_{i_2 jk_2 u_2}) = (1 - \theta_{i_1 i_2}) p_{ju_1} p_{ju_2}. \quad (A.22)$$

Equation (A.22) can be written as

$$\sum_{k_1,k_2=1}^{2} \text{Cov}(x_{i_1 jk_1 u_1}, x_{i_2 jk_2 u_2}) = \left(\sum_{k_1,k_2=1}^{2} \text{E}(x_{i_1 jk_1 u_1} x_{i_2 jk_2 u_2})\right) - 4 p_{ju_1} p_{ju_2}. \quad (A.23)$$

To simplify Eq. (A.23), we denote by $h_{ju_1 u_2} = \delta_{u_1 u_2} p_{ju_1} - p_{ju_1} p_{ju_2}$, where $\delta_{u_1 u_2}$ is Kronecker's delta, which obtains the value of 1 if $u_1 = u_2$ and the value of 0 if this is not the case. In this notation, we have

$$\sum_{k_1,k_2=1}^{2} \text{Cov}(x_{i_1 jk_1 u_1}, x_{i_2 jk_2 u_2}) = 4 \theta_{i_1 i_2} h_{ju_1 u_2}. \quad (A.24)$$

Substituting this expression into Eq. (A.19) yields

$$\text{Cov}(a_{i_1}, a_{i_2}) = 4\theta_{i_1 i_2} \sum_{j=1}^{n} \sum_{u_1, u_2 = 1}^{m_j} h_{ju_1 u_2} v_{ju_1} v_{ju_2}, \qquad (A.25)$$

which equals Eq. (5.4), if we measure the amount of additive variation V_A present in the ancestral generation by

$$V_A = 2 \sum_{j=1}^{n} \sum_{u_1, u_2 = 1}^{m_j} h_{ju_1 u_2} v_{ju_1} v_{ju_2}. \qquad (A.26)$$

A.4 A very brief tutorial to stochastic processes

After the discussion of random variables, we are ready to move to stochastic processes. A stochastic process is a random variable that depends on time. Thus, instead of a single random variable X, we consider a collection of random variables $X(t)$ indexed by time t. Time can be continuous or discrete, and the state-space (i.e. the set of values which the random variable can attain) can be continuous or discrete as well.

A.4.1 Markov chains

Let us start from the discrete time setting ($t = 0, 1, \ldots$), and consider the simplest possible example in which the state space involves only two states, so that the random variable $X(t)$ has either the value 1 or the value 2. A realization of such a stochastic process may look like $(1, 1, 1, 2, 2, 1, 2, 2, 1, 2, 1, \ldots)$, where we have showed the values of the random variables $X(t)$ at times $t = 0, 1, 2, \ldots$.

The simplest and most important class of this kind of models is given by Markov chains. The key feature of a Markov chain is that the probability distribution of the random variable $X(t)$ is allowed to depend only on the state of the system in the previous step, i.e. on the value of $X(t-1)$. Markov chains can be described with the help of a transition matrix $\mathbf{P}$. The elements of this matrix, denoted by p_{ij}, are the probabilities that the process will next move to a state j if it is currently in state i. As an example, let us consider the transition matrix

$$\mathbf{P} = \begin{pmatrix} p_{11} & p_{12} \\ p_{21} & p_{22} \end{pmatrix} = \begin{pmatrix} 0.8 & 0.2 \\ 0.6 & 0.4 \end{pmatrix}. \qquad (A.27)$$

If the system is presently in state 1, it may either stay in that state (with probability $p_{11} = 0.8$) or move to state 2 (with probability $p_{12} = 0.2$). If the system is currently in state 2, it will stay in that state with probability $p_{22} = 0.4$ and move to state 1 with probability $p_{21} = 0.6$. Figure A.6A shows a simulation of this Markov chain model. As the transition matrix is set up so that state 1 is favoured, it is intuitive that the system spends more time in that state than in state 2.

Markov chains are mathematically convenient since much of their behaviour can be derived with the help of matrix algebra. Let us denote by $\mathbf{z}(t) = (z_1(t), z_2(t))$ a row vector that describes the probability distribution of the systems state at time t, coded so that $z_1(t)$ is the probability that the system is in state 1 and $z_2(t) = 1 - z_1(t)$ is the probability that the system is in state 2. At each time step, the probability distribution $\mathbf{z}(t)$ changes according to the rules of the transition matrix: the probability that the system will be in state 1 at time $t+1$ is the probability that it was at time t in state 1 and stays there, plus the probability that it was in state 2 and will move to state 1. As an equation, $z_1(t+1) = z_1(t) p_{11} + z_2(t) p_{21}$. This equation and the analogous equation for $z_2(t+1)$ can be combined in the more compact matrix notation (see Appendix A.1) as $\mathbf{z}(t+1) = \mathbf{z}(t)\mathbf{P}$. Thus, the probability distribution $\mathbf{z}(t+1)$ is obtained by multiplying the probability distribution $\mathbf{z}(t)$ by the transition matrix $\mathbf{P}$ from the right.

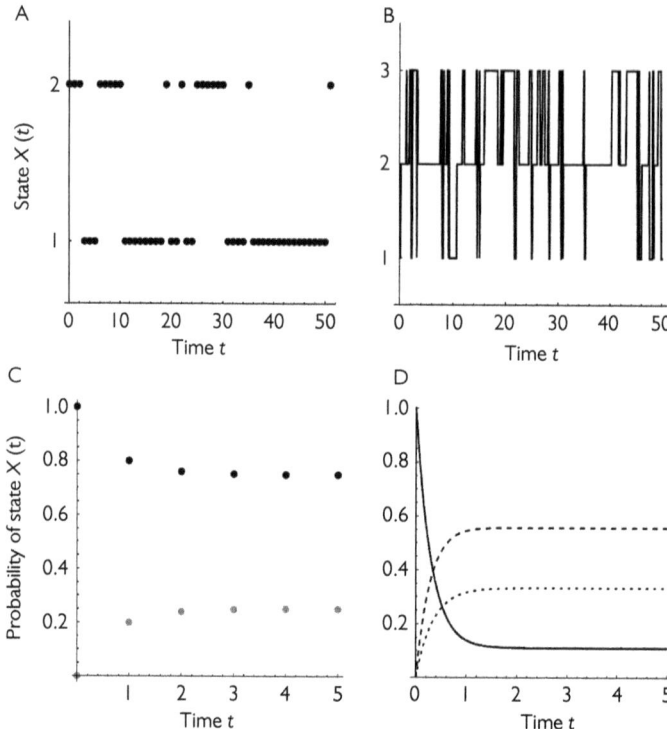

Figure A.6 Illustrations of Markov chains and Markov processes. The upper panels show stochastic simulations of the discrete time Markov chain (A) and the continuous time Markov process (B) of Section A.4. (C, D) show how the probability of being in a given state evolves in time. In (C), the Markov chain is initiated from state 1 at time $t = 0$, and the black (respectively, grey) dots show how the probability of being in state 1 (respectively, in state 2) evolves over time. In (D), the Markov process is initiated from state 1 at time $t = 0$, and the lines show how the probabilities of being in state 1 (continuous line), state 2 (dashed line), and state 3 (dotted line) evolve over time.

As an example, let us assume that the system is initially in state 1 and thus $\mathbf{z}(1) = (1, 0)$. Then we may compute that $\mathbf{z}(1) = (0.8, 0.2)$, $\mathbf{z}(2) = (0.76, 0.24)$, and $\mathbf{z}(3) = (0.752, 0.248)$, which numbers are illustrated in Figure A.6C. We may use matrix algebra to jump directly over multiple time steps. For example, $\mathbf{z}(t + 2) = \mathbf{z}(t + 1)\mathbf{P} = \mathbf{z}(t)\mathbf{P}^2$, where $\mathbf{P}^2$ means that the matrix $\mathbf{P}$ is raised to the second power and thus multiplied by itself in the sense of matrix multiplication. More generally, $\mathbf{z}(t + k) = \mathbf{z}(t)\mathbf{P}^k$, and thus $\mathbf{P}^k$ gives the transition matrix describing how the probability distribution changes over k time steps.

The long-term behaviour of a Markov chain is characterized by its stationary state. In the example in the previous paragraph, the probability distribution $\mathbf{z}(t)$ approaches $\mathbf{z}^* = (0.75, 0.25)$ as time t increases. Mathematically, this can be written as $\lim_{t \to \infty} \mathbf{z}(t) = \mathbf{z}^*$. Thus, if sampling the system long enough after the process started, the probability to find it in state 1 will be 0.75 and the probability to find it in state 2 will be 0.25. This means that the system will spend 75% of its time in state 1 and 25% of its time in state 2. The stationary state z^* can be found analytically by noting that it must satisfy the equation $z^* = z^*P$. As discussed in Appendix A.1, this means that z^* is a left eigenvector of P associated with the eigenvalue 1. Solving the eigenvector problem with standard matrix algebra tools yields $z^* = (0.75, 0.25)$.

Note that the probabilities $z(1)$, $z(2)$, and $z(3)$ listed earlier approach, but do not equal, the stationary state z^* because it matters from which state the system is initiated. During the initial transient period, the system 'forgets' from which state it was initiated and it converges to the stationary state that is independent of the initial condition. In other words,

$$\lim_{k \to \infty} P^k = \begin{pmatrix} 0.75 & 0.25 \\ 0.75 & 0.25 \end{pmatrix}, \qquad (A.28)$$

showing that whichever was the initial condition, the system will after a sufficiently long time be in state 1 or in state 2 with probabilities 0.75 and 0.25, respectively.

A.4.2 Markov processes

The continuous time analogue of a Markov chain is a Markov process. While in a Markov chain the system stays a fixed time period in each state before moving to another state (or possibly staying in the same one), in a Markov process the durations of the time steps are exponentially distributed random variables. A Markov process is determined by specifying its transition rates.

Before continuing with the Markov process, let us discuss the relationship between rates and probabilities. Unlike probabilities, rates are not constrained to be between 0 and 1 but they can have any positive values. The probability p by which an event with a rate q takes place during a short time interval dt is given by $p = qdt$. This equation holds only for very small time intervals, so that technically it holds only for $dt \to 0$. To see this, consider e.g. the rate $q = 2$, and let the duration of the time interval be $dt = 1$. The formula $p = qdt$ would predict that the probability by which the transition takes place is 2, which clearly cannot be valid. For an arbitrary time interval Δt of any duration, the probability of the event happening is $p = 1 - e^{-q\Delta t}$, which gives the probability of $p = 0.85$ for $q = 2$ and $dt = 1$. If assuming a short time interval, the linear relationship $q\Delta t$ becomes an increasingly good approximation. For example, if $q = 2$ and $\Delta t = 0.01$, the approximation $q\Delta t = 0.02$ is very close to the exact value of $p = 1 - e^{-q\Delta t} = 0.0198$.

Let us then illustrate a Markov process with an example system with three states. We let the transition rate of moving from state 1 to 2 be $q_{12} = 2$, and set the other rates to $q_{13} = 1$, $q_{21} = 0$, $q_{23} = 1$, $q_{31} = 1$, and $q_{32} = 1$. Markov processes can be simulated with the help of the Gillespie (1977) algorithm, the description of which also illustrates mathematical properties of Markov processes. Let us assume that the process is currently in state 1. As the process leaves this state with rates $q_{12} = 2$ (to state 2) and $q_{13} = 1$ (to state 3), the total rate at which the process leaves the state 1 is $q_{12} + q_{13} = 3$. Mathematical theory of Markov processes now tells us that the time that the system stays in state 1 is exponentially distributed with parameter equalling the total rate 3. The expected value of the exponential distribution is the inverse of the parameter (in this case $1/3$), so that a large rate means a short waiting time. When simulating the process, we may thus randomize the duration Δt that the system spends in the present state using an exponentially distributed random number. The next decision to be made in the simulation is to which state the system moves when it leaves the present state 1. This is determined by the relative rates. By dividing the rates $q_{12} = 2$ and $q_{13} = 1$ by their sum, we see that the system moves to state 2 with probability $2/3$ and to the state 2 with probability $1/3$. Figure A.6B illustrates a simulation of the Markov process just described.

Also Markov processes can be studied analytically with the help of matrix algebra. The analogue of the transition probability matrix **P** is the transition rate matrix **Q**. In the example in the previous paragraph, we have

$$\mathbf{Q} = \begin{pmatrix} -q_{12} - q_{13} & q_{12} & q_{13} \\ q_{21} & -q_{21} - q_{23} & q_{23} \\ q_{31} & q_{32} & -q_{31} - q_{32} \end{pmatrix} = \begin{pmatrix} -3 & 2 & 1 \\ 0 & -1 & 1 \\ 1 & 1 & -2 \end{pmatrix}. \quad (A.29)$$

Here the off-diagonal elements have been set to the transition rates, whereas the diagonal elements have been set so that the row sums equal to 0. This is in contrast with the transition probability matrix **P**, in which the row sums equal to 1.

While in the case of the discrete time Markov chain the state of the system evolves according to the difference equation $\mathbf{z}(t + 1) = \mathbf{z}(t)\mathbf{P}$, in the case of the continuous time Markov process it evolves according to the system of differential equations

$$\frac{d\mathbf{z}(t)}{dt} = \mathbf{z}(t)\mathbf{Q}. \quad (A.30)$$

Figure A.6D illustrates the solution to this system of differential equations, thus showing how the probability of being in each of the three states behaves if starting initially from state 1. After long enough time, the system converges to the stationary state. At the stationary state the probability distribution does not change anymore, and thus its derivative is 0, meaning that the stationary probability distribution $\mathbf{z}^*$ must satisfy the equation $\mathbf{0} = \mathbf{z}^*\mathbf{Q}$. Thus, $\mathbf{z}^*$ is the left eigenvector of the matrix **Q** corresponding to the eigenvalue 0. For our example, we obtain $\mathbf{z}^* = \frac{1}{9}(1\ 5\ 3)^T$, reflecting the amount of time that the system spends in the three states at the stationary state.

The transition rate matrix **Q** can be translated to a transition probability matrix with the help of the equation $\mathbf{P}(\Delta t) = e^{\Delta t \mathbf{Q}}$, where Δt is the time interval over which the transition probability matrix is to be computed. Here $e^{\Delta t \mathbf{Q}}$ is the matrix exponential of the matrix $\Delta t \mathbf{Q}$. It is defined, analogously to the Taylor expansion of the usual exponential function, as the power series $e^{\Delta t \mathbf{Q}} = \mathbf{I} + \frac{1}{2!}\Delta t^2 \mathbf{Q}^2 + \frac{1}{3!}\Delta t^3 \mathbf{Q}^3 + \ldots$, where **I** stands for the identity matrix. We note that $\lim_{\Delta t \to 0} \mathbf{P}(\Delta t) = \mathbf{I}$, simply confirming that over a vanishingly short time the system will stay where it was originally.

Sometimes Markov processes or Markov chains have an absorbing state. For example, consider the transition rate matrix

$$\mathbf{Q} = \begin{pmatrix} -3 & 2 & 1 \\ 0 & 0 & 0 \\ 1 & 1 & -2 \end{pmatrix}. \quad (A.31)$$

With this transition rate matrix, the system will never leave from state 2. Whether the system is initiated from state 1, 2, or 3, it will sooner or later enter state 2. For this reason, state 2 is called an absorbing state, and the stationary distribution is concentrated in this state, $\mathbf{z}^* = (010)^T$. In population models without the possibility of immigration from outside, population extinction will represent an absorbing state. In such a case, it is often of interest to study the quasi-stationary distribution, which describes the probability distribution after long enough time so that the initial condition has become irrelevant, but before the system has reached the absorbing state. Further discussion about quasi-stationary distributions can be found e.g. from Darroch and Seneta (1967), and an application in the ecological context e.g. from Ovaskainen (2001).

Appendix B

Statistical methods

B.1 Generalized linear mixed models

The most widely used statistical modelling framework in ecology (or probably in any field of science) is that of generalized linear mixed models (GLMMs). Here we will only cover a small selection of topics related to GLMMs, more detailed treatments being given in e.g. Bolker et al. (2009) and Zuur et al. (2009). In this section, our focus is on model formulation, whereas the issues of parameter estimation is discussed in Section B.2. We assume here that the reader is familiar with the notation and basic properties of random variables and distributions, which we reviewed in Appendix A.3.

B.1.1 Linear models

The baseline linear model can be written as

$$y_i = \beta_0 + x_{i1}\beta_1 + x_{i2}\beta_2 + \cdots + x_{ik}\beta_k + \varepsilon_i. \tag{B.1}$$

Here y_i is the response variable (also called the dependent variable) related to the sample i, and $x_{i1}, \ldots, x_{ik}$ are the measured covariates (also called independent variables) by which we wish to explain variation in the response variable. The $\beta_1, \ldots, \beta_k$ are the regression coefficients that are to be estimated. They measure how the covariates influence the response variable. The term ε_i is the residual, and it measures the deviation between the measurement y_i and the model prediction.

In the linear model, the residuals are assumed to be normally distributed, $\varepsilon_i \sim N(0, \sigma^2)$, where the parameter σ^2 measures the amount of residual variance. The mean of residual variation is set to 0 without loss of generality, because the overall mean is captured by the parameter β_0, called the intercept of the model. The residual variance σ^2 measures how much un-modelled variation remains in the data after accounting for the effects of the covariates. In addition to normality of the residuals, the linear model assumes homoscedasticity and independence of residuals (Crawley 2007; Zuur et al. 2010), both important issues to which we will return in more detail later.

As an example, the response variable y_i could be the abundance of a species in sampling unit (e.g. a field site or experimental unit) i. The covariates $x_{i1}, \ldots, x_{ik}$ represent properties of the sampling unit, e.g. habitat type and amount of food resources available for the species. In this example, a word version of the model, often used to communicate statistical models (Zuur et al. 2009), would read ABUNDANCE = HABITAT + FOOD. While a word version of a model is a good starting point for communication, we encourage the reader to write also an equation version. As we will illustrate later, the equation version makes the 'hidden' assumptions behind the model explicit, which is important especially in the case of more complex models than the baseline linear model.

To simplify the notation, let us rewrite the model just described as $y_i = L_i + \varepsilon_i$, where L_i is called the linear predictor, defined as

$$L_i = x_{i1}\beta_1 + \cdots + x_{ik}\beta_k = \sum_{j=1}^{k} x_{ij}\beta_j. \tag{B.2}$$

Comparing Eqs. (B.1) and (B.2), the reader may wonder what happened to the intercept β_0. For notational simplicity, we have included that in the covariates rather than treating it separately. This can be always done by setting the first covariate to $x_{1i} = 1$ for all sampling units i, in which case the parameter β_1 becomes the intercept of the model, and the parameters $\beta_2, \beta_3, \ldots, \beta_k$ measure the effects of the $k - 1$ covariates.

The linear model can involve both continuous explanatory variables (called covariates) and categorical explanatory variables (called factors). To illustrate, let us assume that in the model ABUNDANCE = HABITAT + FOOD, habitat is a factor with three categories (e.g. forest, bog, and grassland), whereas the availability of food resources is a continuous covariate (e.g. biomass of the food resources). We can normalize the effect of habitat being forest to 0, and thus let β_2 and β_3 model the effects of the habitat being bog or grassland, respectively, relative to the effect of the habitat being forest. Thus, we let the indicator variables x_{i2} and x_{i3} have values $x_{i2} = x_{i3} = 0$ if the sampling unit i consists of forest, $x_{i2} = 1, x_{i3} = 0$ if the unit i consists of bog, and $x_{i2} = 0, x_{i3} = 1$ if the unit i consists of grassland. As the amount of food is a continuous covariate, the value of x_{i4} is simply set to the amount of food available in sampling unit i. Thus, the model has in total four degrees of freedom, i.e. four regression parameters that are to be estimated. Out of these, β_1 corresponds to the intercept, β_2 and β_3 to a factor with three categories, and β_4 to a continuous covariate. Generally, a factor with m categories has $m - 1$ degrees of freedom, because the effect of one category can be normalized to be 0. To illustrate why this is the case, assume that there would be only one category, e.g. all habitats would consist of forest. In this case, the mean effect of habitat being forest would be captured by the intercept of the model, and thus it would not be meaningful to involve a separate parameter modelling the effect of the habitat type.

The notation introduced in the previous paragraphs includes the possibility of having not only covariates and factors, but also interactions among any kinds of explanatory variables. For example, an interaction between habitat and food can be modelled through the explanatory variables $x_{i5} = x_{i2}x_{i4}$ and $x_{i6} = x_{i3}x_{i4}$. In a word equation, such an interaction would be written as HABITAT * FOOD. The inclusion of an interaction term enables one to ask whether the influence of the amount of food resources on species abundance is different in different habitat types.

The linear model serves as a baseline that we will extend later in many ways. First, we consider different link functions and error distributions to allow non-normally distributed residuals. Second, we include random effects, or equivalently relax the assumption of independent residuals, as this makes it possible to consider e.g. spatial, temporal, or spatiotemporal models, or to account for relatedness among individuals. Third, we consider multivariate response variables to model e.g. multiple traits or multiple species. Fourth, we make the model hierarchical by modelling the regression parameters β with the help of higher level parameters.

B.1.2 Link functions and error distributions

To motivate the need for link functions and error distributions, consider the case of species distribution modelling (Section 3.5). Often the data available for such an exercise are of presence–absence nature, i.e. the response variable $y_i = 1$ if the species occurs in sampling unit i and $y_i = 0$ if the species does not occur there. In this case, the assumption of normally distributed residuals is clearly problematic. A widely applied approach to binary data is logistic regression, defined by

$$P[y_i = 1] = \text{logit}^{-1}(L_i). \tag{B.3}$$

Here $P[y_i = 1]$ denotes the probability that $y_i = 1$, i.e. that the species is found from the sampling unit i. As the response variable y_i can attain only the values 0 or 1, it always holds that $P[y_i = 0] = 1 - P[y_i = 1]$, and thus there is no need to specify a separate model for species absence. In the logistic regression model, the link function is the logit function, defined by

$$\text{logit}(p) = \log\left(\frac{p}{1-p}\right). \tag{B.4}$$

Here $0 \leq p \leq 1$ is a probability, whereas the linear predictor $L = \text{logit}(p)$ can attain any real value, $-\infty < L < \infty$. Thus, the inverse of the logit,

$$\text{logit}^{-1}(L) = \frac{1}{1+\exp(-L)}, \tag{B.5}$$

converts the linear predictor L into a probability, i.e. into a number between 0 and 1 (Figure B.1).

The reader may be wondering what happened with the residual ε_i, i.e. why the logistic regression model does not read as $P[y_i = 1] = \text{logit}^{-1}(L_i + \varepsilon_i)$. The reason here is that with binary outcomes, all variation in the data is captured already by the randomness of the Bernoulli random variable, i.e. whether the realized outcome is 0 or 1. Consequently, independently distributed residuals would not be identifiable in the model, and hence their inclusion is not feasible. However, as will be discussed later, the inclusion of a structured residual is possible.

The logistic link function is not the only possibility for converting a linear predictor L_i into a probability. Another commonly used approach is that of the probit regression. In this case, we write

$$P[y_i = 1] = \Phi(L_i), \tag{B.6}$$

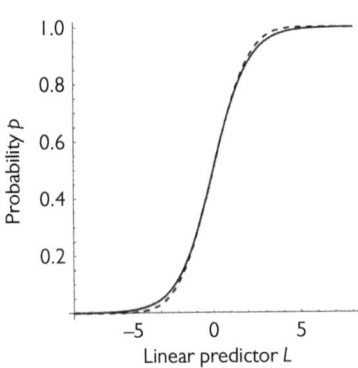

Figure B.1 A comparison of the logit and probit link functions. The continuous line shows the inverse logit link function (Eq. (B.5)), whereas the dashed line shows the probit link function (Eq. (B.6)). To illustrate the similarity in the shape, the probit link function has been scaled by setting the variance of the underlying normal distribution to $\sigma^2 = 8/\pi$.

where Φ is the cumulative distribution function of the standard normal distribution $N(0, 1)$, i.e. the normal distribution with zero mean and unit variance (see Section A.3).

Figure B.1 compares the logit and probit link functions against each other, with the probit link function scaled so that they match as well as possible. Both have the property that they map values from the real axis ($-\infty < L < \infty$) to probabilities ($0 \leq p \leq 1$), and thus both are appropriate for Bernoulli random variables. Which of these two should one select? Usually neither of them is better justified from the biological point of view. In theory, model selection (Appendix B.2) could indicate which one fits the data best, and thus one could let the data decide the selection among these two. In practice, especially with limited data, both are likely to fit the data equally well, and thus the choice is somewhat arbitrary. An advantage of logistic regression is that it has been the very widely applied in ecological research, facilitating comparison with other studies. An advantage of probit regression comes from the mathematical side. Namely, Eq. (B.6) can be written equivalently by letting $z_i = L_i + \varepsilon_i$, where $\varepsilon_i \sim N(0, 1)$, and by letting $y_i = 1$ if $z_i > 0$ and $y_i = 0$ if $z_i \leq 0$. To see that this formulation is equivalent with Eq. (B.6), we note that $z_i \leq 0$ if and only if $\varepsilon_i \leq -L_i$. By the definition of the cumulative distribution function (see Appendix A.3), the probability by which this happens is $\Phi(-L_i)$. Thus $P[y_i = 1] = 1 - P[y_i = 0] = 1 - \Phi(-L_i) = \Phi(L_i)$, where the last equality follows because the probability density of the standard normal distribution is symmetric around 0. As we will illustrate later, this alternative formulation makes the probit model a convenient starting point for modelling covariance structures, needed e.g. in spatial models or models of co-occurrence among species.

In addition to binary data, other cases that call for relaxing the assumption of normality include response variables that are restricted to positive values ($y_i > 0$; e.g. species abundance measured by biomass) and response variables that are counts ($y_i = 0, 1, 2, \ldots$; e.g. species abundance measured by the number of individuals). In the case of a positive response variable, one possibility is to apply the log-link function, i.e. write the model as $\log(y_i) = L_i + \varepsilon_i$. If the response variable is positive ($y_i > 0$), then its logarithm can attain any real value ($-\infty < \log(y_i) < \infty$), and thus after the log-transformation the application of the baseline linear model becomes feasible. Predictions for the original response variable are obtained by back-transforming with the inverse function of the logarithm, i.e. with the exponential function and therefore $y_i = \exp(L_i + \varepsilon_i)$.

In case of count data, the response variable is not only non-negative, but also restricted to integer values. In this case, a commonly applied approach is that of Poisson regression, defined as $y_i \sim \text{Poisson}(\exp(L_i))$. Here the expected value of the response variable y_i is modelled by $\exp(L_i)$, and thus it is constrained to positive values. The response variable is assumed to be distributed around its expectation according to the Poisson distribution, and thus it is constrained to integer values. As it is the case with logistic and probit regressions, Poisson regression does not have an explicit residual term, but all randomness in the response variable is captured by the inherent randomness of the Poisson distribution. However, the amount of variation predicted by the Poisson distribution is often smaller than the variation observed in the data. If that is the case, other approaches should be applied, such as over-dispersed Poisson or the negative-binomial model. A more complete treatment of link functions and error structures can be found from many textbooks of statistics, such as Crawley (2007).

B.1.3 Relaxing the assumption of independent residuals

The three assumptions of the basic linear model are that the residuals are (i) normally distributed, (ii) independent, and (iii) homoscedastic. The assumption of normality is clearly visible in the formula $\varepsilon_i \sim N(0, \sigma^2)$, and we already discussed earlier how this assumption can be relaxed. The assumption of homoscedasticity is also seen from this formula, as it means that the variance σ^2 is assumed to be the same for all sampling units, i.e. it does not depend on the index i. The assumption of independence is not seen from $\varepsilon_i \sim N(0, \sigma^2)$ because this formulation as such does not involve any statement about possible covariance between two sampling units (see Appendix A.3). One way to spell out the assumption of independence is to use the statistical jargon of 'i.i.d.', which stands for independent and identically distributed random variables. Alternatively, the assumption of independence can be spelled out by specifying that the covariance of residuals equals 0 for different sampling units $i \neq j$, i.e. that $\text{Cov}(\varepsilon_i, \varepsilon_j) = 0$.

At this point, it is convenient to move to a matrix notation (see Appendix A.1). To do so, we index the sampling units by $i = 1, \ldots, n$, and collect the response variables y_i, the residuals ε_i, and the linear predictors L_i into the vectors $\mathbf{y}$, $\boldsymbol{\varepsilon}$, and $\mathbf{L}$, respectively, all of dimension $n \times 1$. Further, we collect the covariates x_{il} into the $n \times k$ matrix $\mathbf{X}$, called the design matrix, and all the regression coefficients β_l into the vector $\boldsymbol{\beta}$ of dimension $k \times 1$. Utilizing matrix multiplication, we have $\mathbf{L} = \mathbf{X}\boldsymbol{\beta}$. Therefore, we can write the model as $\mathbf{y} = \mathbf{X}\boldsymbol{\beta} + \boldsymbol{\varepsilon}$ with $\boldsymbol{\varepsilon} \sim N(\mathbf{0}, \sigma^2 \mathbf{I}_n)$, where $N(\boldsymbol{\mu}, \boldsymbol{\Sigma})$ denotes the multivariate normal distribution with mean $\boldsymbol{\mu}$ (a vector with n elements) and variance-covariance matrix $\boldsymbol{\Sigma}$ (a matrix with $n \times n$ elements). The mean of the residual distribution is the zero vector $\boldsymbol{\mu} = \mathbf{0}$, i.e. a vector with value 0 for each of the n elements. The variance-covariance matrix of the residual distribution is set to $\boldsymbol{\Sigma} = \sigma^2 \mathbf{I}_n$, where $\mathbf{I}_n$ stands for the identity matrix of dimensions $n \times n$. The identity matrix has 1s at the diagonal elements ($I_{ii} = 1$) and 0s at the off-diagonal elements ($I_{ij} = 0$ for $i \neq j$). Thus, $\text{Cov}(\varepsilon_i, \varepsilon_i) = \sigma^2$, and thus residual variance is σ^2, as it should. Further, $\text{Cov}(\varepsilon_i, \varepsilon_j) = 0$ for $i \neq j$, and thus the residuals are independent among the sampling units. Given all this, the notation $\boldsymbol{\varepsilon} \sim N(\mathbf{0}, \sigma^2 \mathbf{I}_n)$ makes the assumptions of independence, normality, and homoscedasticity explicit. By doing so, it provides a convenient starting point for relaxing any of the assumptions, most importantly the one on independence. As we next discuss, this may be done by considering a more general error structure $\boldsymbol{\varepsilon} \sim N(\mathbf{0}, \boldsymbol{\Sigma})$, where the variance-covariance matrix $\boldsymbol{\Sigma}$ may have non-zero elements also at the off-diagonals.

As discussed in the context of species distribution modelling (Section 3.5), one motivation for relaxing the assumption of independence is that the data may come from a spatial setting. Such data often show spatial autocorrelation, meaning that data points acquired close to each other are more similar than data points acquired far away from each other. In such a case, the covariance between the residuals for the sampling units i and j can be assumed to depend on their distance d_{ij}. Thus, we may set $\Sigma_{ij} = f(d_{ij})$, where the function f is some decreasing function of distance. For example, we may assume that spatial autocorrelation decreases exponentially with increasing distance, so that $f(d) = \sigma^2 \exp(-\alpha d_{ij})$, where σ^2 is the same variance parameter as in the case of independent residuals. The additional parameter α is a scale parameter, $1/\alpha$ giving the distance at which spatial autocorrelation decays to $\exp(-1) \approx 0.368$. As correlation is defined as covariance divided by the square root of product of variances (see Appendix A.3), we obtain

$$\text{Cor}(\varepsilon_i, \varepsilon_j) = \frac{\text{Cov}(\varepsilon_i, \varepsilon_j)}{\sqrt{\text{Cov}(\varepsilon_i, \varepsilon_i)\text{Cov}(\varepsilon_j, \varepsilon_j)}} = \frac{\sigma^2 \exp(-\alpha d_{ij})}{\sqrt{\sigma^2 \exp(0) \sigma^2 \exp(0)}} = \exp(-\alpha d_{ij}). \tag{B.7}$$

This model thus assumes that residual correlation decays with increasing distance from 1 (for two sampling units located very near to each other) to 0 (for two sampling units located very far apart). The baseline model with independent residuals can be obtained as the special case $\alpha \to \infty$, as at this limit the length scale of spatial autocorrelation $1/\alpha$ approaches 0 and thus different sampling units will become independent.

Why should one involve a spatially structured residual, i.e. what difference does it make compared to assuming independent residuals? The reasons for including a spatially structured residual are twofold. First, when assuming the independence of residuals, the fitting procedure (whether based on least-squares, maximum likelihood, or Bayesian approach, see Appendix B.2) will mistakenly assume that there are more independent pieces of evidence about the influence of the explanatory variables on the response variable than there actually are. Therefore, the level of statistical significance of the explanatory variables is likely to be overestimated. In other words, a model with independent residuals can result in a very small p-value (say, $p = 0.00001$) in a case where a model that accounts for a spatially structured residual would show a non-significant p-value (say, $p = 0.3$). Second, estimation of the spatial scale of autocorrelation ($1/\alpha$) can be of interest by itself, as we discuss in the context of species distribution modelling (Section 3.5).

Let us then briefly consider the analysis of (non-spatial) time-series data, so that the observations y_t are indexed by time t. In this case, observations that are made close to each other in time can be expected to be positively correlated. Thus, in analogy with models with spatial autocorrelation, in a model with temporal autocorrelation one may assume that $\text{Cov}(\varepsilon_{t_1}, \varepsilon_{t_2}) = f(|t_1 - t_2|)$, where f is a decreasing function. One way of constructing temporal autocorrelation for regularly sampled data is to use the autoregressive (AR) process. For example, in the AR(1) model $\varepsilon_t = \phi\varepsilon_{t-1} + \omega_t$, where the parameter $|\phi| < 1$ measures the degree of temporal dependency, and $\omega_t \sim N(0, \sigma^2)$ models independent noise. This formulation is equivalent to assuming $\text{Cor}(\varepsilon_{t_1}, \varepsilon_{t_2}) = \phi^{|t_1 - t_2|}$ and thus leads to an exponentially decaying temporal autocorrelation. With spatiotemporal data the residual can be assumed to be structured in both space and time, e.g. by assuming the AR(1) process in time and that the noise term ω_t has a spatial covariance structure over space.

Other commonly used forms of spatial and temporal autocorrelation than the exponential kernel include e.g. Gaussian kernels, Matérn kernels, and kernels with a nugget. We address the interested reader to e.g. Crawley (2007) and Dormann et al. (2007) and references therein for more detailed treatments of spatial, temporal, and spatiotemporal models.

As a technical remark, we note that not all matrices can be used as a variance-covariance matrix Σ, as in the mathematical jargon the matrix needs to be symmetric ($\Sigma_{ij} = \Sigma_{ji}$) and positive definite. Technically, the assumption of positive definiteness requires that all eigenvalues of the matrix (see Appendix A.1) are positive. The condition of positive definiteness is needed to avoid unfeasible combinations of correlations. For example, if the sampling units 1 and 2 are highly positively correlated, and if the sampling units 2 and 3 are highly positively correlated, then necessarily also the sampling units 1 and 3 must be positively correlated.

B.1.4 Random effects

We have thus far written the linear predictor as $L = X\beta$, where the design matrix X includes the measured values of the continuous covariates as well as the indicator variables relating to the levels of the categorical factors. In the statistical jargon, we have thus far specified the covariates and factors as fixed effects. As an alternative, we may specify factors as random effects. Models with both fixed and random effects are called mixed models.

To discuss the distinction between fixed and random factors, let us assume that in the model ABUNDANCE = HABITAT + FOOD discussed earlier, the n sampling units would be acquired from m distinct regions, to be called sites, so that each site includes several sampling units. Sampling units within a site are not likely to represent independent data points, as they are all influenced by more similar conditions than sampling units that belong to different sites.

As discussed earlier, one option to account for this would be to assume a spatially structured residual. However, if the data are acquired from a discrete set of sites, a simpler option is to consider the site as a random factor. To do so, we extend the model to ABUNDANCE = HABITAT + FOOD + SITE. In an equation form, this model may be written as

$$y_i = \sum_{j=1}^{k} x_{ij}\beta_j + \gamma_{s(i)} + \varepsilon_i, \tag{B.8}$$

where $s(i)$ is the index of the site ($s = 1, \ldots, m$) to which the sampling unit i belongs, and γ_s is the contribution of the site s to species abundance. We may further write $\gamma_{s(i)} = \sum_{s=1}^{m} z_{is}\gamma_s$, where the indicator variables z_{is} describe to which site each sampling unit belongs, so that $z_{is} = 1$ if the sampling unit i belongs to site s and $z_{is} = 0$ otherwise. In the matrix notation, we may write the model as

$$\mathbf{y} \sim N(\mathbf{X}\boldsymbol{\beta} + \mathbf{Z}\boldsymbol{\gamma}, \sigma^2 \mathbf{I}_n), \tag{B.9}$$

where $\mathbf{Z}$ is the $n \times m$ matrix with elements z_{is}. Note that we have assumed that the residuals are independent after accounting for the random effect of the site.

In Eq. (B.9), the random factor SITE does not differ from the fixed factor of HABITAT in any visible way, as both are written in terms of a design matrix ($\mathbf{X}$ or $\mathbf{Z}$) multiplied by a vector of effects to be estimated ($\boldsymbol{\beta}$ or $\boldsymbol{\gamma}$). To state that SITE is a random factor, we make the additional assumption that $\boldsymbol{\gamma} \sim N(\mathbf{0}, \sigma_S^2 \mathbf{I}_m)$, i.e. that the effects of the sites are normally distributed with mean 0 and variance σ_S^2, where we have included the subscript S to distinguish the site-level variance σ_S^2 from the residual variance σ^2.

In case of a random factor, the interest is not on the individual effects γ_s that each of the m sites makes to the linear predictor. This is because the sites are considered as a random sample among the larger population of all possible sites from which one could have sampled the data. Thus, the important parameter is the amount of variance σ_S^2 among the sites rather than the effects γ_s of an individual site s. In contrast, with the fixed factor HABITAT, the focus is specifically on how the habitat being forest, bog, or grassland influences the abundance of the species. In some cases, one could actually consider also HABITAT as a random factor. This could be the case if the three habitat types of forest, bog, and grassland would be just a sample from a still larger population of habitat types (e.g. rocky outcrops, ponds, streams) from which one could have sampled the data. If declaring HABITAT as a random factor, we would be interested in how variation among habitat types influences variation in species abundance more generally, not just on the influences of these three habitat types.

Random effects and structured residuals often provide alternative ways of formulating the very same model. To make this point explicit, let us continue with the model that includes site as a random factor, and derive the corresponding model where the random effect has been replaced by a structured residual. To do so, we write

$$y_i = L_i + e_i, \tag{B.10}$$

where the fixed effects are included in the linear predictor, $L_i = \sum_{l=1}^{k} x_{il}\beta_l$, and the structured residual $e_i = \sum_{s=1}^{m} z_{is}\gamma_s + \varepsilon_i$ includes both the random effect of the site and the independent residual. As we assumed that the effects of the sites (the terms γ_s) are normally distributed and that each sampling unit belongs to one and only one site, it is easy to see (see the discussion in Appendix A.3 about sums of random variables) that each of the e_i is normally distributed with variance $\sigma_S^2 + \sigma^2$, i.e. that $e_i \sim N(0, \sigma_S^2 + \sigma^2)$. Further, as we assumed that the effects of the sites are independent among the sites, the covariance is

$$\Sigma_{ij} = \text{Cov}(e_i, e_j) = \begin{cases} \sigma_S^2 + \sigma^2, \text{if } i = j \\ \sigma_S^2, \text{if } i \neq j \text{ and } s(i) = s(j), \\ 0, \text{otherwise} \end{cases} \quad (B.11)$$

Thus, we can write the very same model either as $y \sim N(X\beta + Zs, \sigma^2 I_n)$, or alternatively as $y \sim N(X\beta, \Sigma)$, where the variance-covariance matrix Σ includes both the random effect of the site and the residual variation among the sampling units. In both cases, the parameters to be estimated include the regression coefficients β, the variance among the sites σ_S^2, and the residual variance σ^2.

As another example of a model with random effects, let us consider the animal model described in the context of quantitative genetics (Chapter 5). This model can be written for individual i as

$$p_i = L_i + a_i + e_i, \quad (B.12)$$

where the response variable p_i is the measured phenotype, L_i is the linear predictor involving the overall mean and any fixed effects possibly associated with individual i (such as the influence of sex or age of the individual), a_i is the breeding value (i.e. the additive genetic contribution to the phenotype), and the residual e_i includes the environmental effects. Let us assume that the data come from a breeding design in which a number of dams (mothers) are crossed with a number of sires (fathers). We denote by $d(i)$ and $s(i)$ the dam and the sire of the individual i, respectively, and by $a_{d(i)}$ and $a_{s(i)}$ the breeding values of the sire and the dam. Then the basic assumptions of Mendelian inheritance (see Chapter 5) lead to

$$a_i = \frac{a_{d(i)} + a_{s(i)}}{2} + w_i, \quad (B.13)$$

i.e. the expected breeding value of the offspring is the average of those for the dam and the sire. The term w_i represents randomness in the inheritance process and results in variation among the breeding values of siblings. We denote the amount of additive variation in the population by V_A and assume that it is the same for the parent and the offspring generations. In this case, we have $a_i \sim N(0, V_A)$, $a_{d(i)} \sim N(0, V_A)$, and $a_{s(i)} \sim N(0, V_A)$. Similarly, we denote the amount of environmental variation by V_E, so that $e_i \sim N(0, V_E)$. Applying the formula for the variance of sum of random variables (see Section A.3), we see that $w_i \sim N(0, V_A/2)$. Thus, a solely mathematical argument shows that the amount of random variation between the breeding values of full siblings is half of the additive variance in the population.

As a result, the animal model can be written as

$$y \sim N\left(X\beta + Za, \left(\frac{V_A}{2} + V_E\right)I_n\right). \quad (B.14)$$

Here the vector $a \sim N(0, V_A I_m)$ involves the breeding values among the m individuals in the parent generation. The elements of the design matrix Z are set to $z_{il} = 1/2$ if the individual l from the parent generation is either the sire or the dam of the individual i from the offspring generation, and otherwise $z_{il} = 0$. The amount of residual variation $V_A/2 + V_E$ involves a contribution both from the additive genetic effects ($V_A/2$) and from the environmental effects (V_E). Note that the identity matrix I_n describes that residual variation is independent among the individuals in the offspring generation.

In the case of the random factor being the site, each row of the design matrix **Z** has one entry with a value of 1, the remaining entries having a value of 0. In the case of the animal model, each row of the design matrix **Z** has two entries with value 1/2, the remaining entries having a value of 0. Thus, in the latter case each sampling unit (i.e. individual) obtains a contribution to its random effect from two sources, namely the dam and the sire. In the general case, random effects may have an arbitrary design matrix.

We will next write the animal model mathematically equivalently as

$$\boldsymbol{y} \sim N(\mathbf{X}\boldsymbol{\beta}, \boldsymbol{\Sigma}), \tag{B.15}$$

and thus move the random effect of the breeding value to the structured residual $\boldsymbol{\Sigma}$. By deriving the amount of covariance between two individuals, like we did earlier for the covariance among two sampling sites, it is easy to show that $\boldsymbol{\Sigma} = V_A \mathbf{A} + V_E \mathbf{I}_n$. Here the matrix **A** is called the coancestry matrix (see Chapter 5). For a breeding design that carries over one generation, its elements are given by

$$A_{ij} = \begin{cases} 1, & \text{if } i = j \\ 1/2, & \text{if } i \neq j \text{ and } i \text{ and } j \text{ are full sibs (have the same sire and dam)} \\ 1/4, & \text{if } i \neq j \text{ and } i \text{ and } j \text{ are half sibs (have the same sire or dam but not both)} \\ 0, & \text{otherwise} \end{cases} \tag{B.16}$$

The formulations of Eq. (B.14) (which considers the sire and the dam as random effects) and Eq. (B.15) (which involves a residual structured by the coancestry matrix) are mathematically equivalent. The formulation based on the coancestry matrix is a convenient starting point if the breeding design is more complex, e.g. if the data involve several generations. This is because the entries of A_{ij} can be computed for all pairs of individuals using the recursive relationship given by Eq. (5.3).

B.1.5 Multivariate models

In all models discussed in the previous sections, the response variable y_i has been univariate, i.e. a single number per sampling unit i. In many applications, the response variable is multivariate, thus consisting of several numbers combined into a vector $\boldsymbol{y}_i$. We denote the length of the vector $\boldsymbol{y}_i$ by m, not to be confused with m standing for the number of random effects in the earlier discussion. Assuming that there are n sampling units, the response variables can be organized either as a $n \times m$ matrix or as a vector of length nm. The two examples of multivariate models considered in this book consist of the animal model with more than one trait (say, weight and body length of the individuals) and joint species distribution modelling with more than one species. In the first case, the motivation of applying a multivariate model instead of modelling each trait separately is the possibility of identifying genetic or environmental correlations among the traits. In the second case, the motivation of applying a multivariate model instead of modelling each species separately is the possibility of assessing community-level patterns, such as co-occurrence among pairs of species.

We start again from the baseline univariate linear model $y_i = L_i + \varepsilon_i$, with the linear predictor defined as $L_i = \sum_{l=1}^{k} x_{il} \beta_l$. To make the model multivariate, we simply add one more index j, and thus write $y_{ij} = L_{ij} + \varepsilon_{ij}$, where $L_{ij} = \sum_{l=1}^{k} x_{il} \beta_{lj}$. Here the index i refers to the sampling unit ($i = 1, \ldots, n$) and the index j refers to the component of the response variable ($j = 1, \ldots, m$). We will discuss first the animal model of quantitative genetics, and thus we call the index j the trait and the index i the individual. However, we ignore for a moment the breeding values of the individuals, which we will add later. Thus, we let L_{ij} simply be the linear predictor for the trait j of individual i, and it is obtained by summing the effect β_{lj} that each covariate $x_{il}(l = 1, \ldots, k)$, such as sex or age of the individual, has on the trait j.

As such, adding the index j for the trait does not yet indicate that the model is of multivariate nature, as we have not described how the different traits relate to each other. As a starting point, let us consider the case of independent traits. In this case we could simply apply the model separately for each trait, assuming that $\varepsilon_{ij} \sim N(0, \sigma_j^2)$, where σ_j^2 denotes the amount of residual variation associated to trait j. This would correspond to the assumption that the residuals are not correlated among the traits, i.e. that for two traits j_1 and j_2, the covariance $\text{Cov}(\varepsilon_{ij_1}, \varepsilon_{ij_2})$ is 0. To write the model in a multivariate notation, we let $\boldsymbol{\varepsilon}_i$ denote a vector of length m with elements ε_{ij}, so that it contains the residuals for all traits j of individual i. Similarly, we denote by $\boldsymbol{y}_i$ and by $\boldsymbol{L}_i$ vectors including the response variables and the linear predictors of all traits for individual i, so that the model reads in matrix notation as $\boldsymbol{y}_i = \boldsymbol{L}_i + \boldsymbol{\varepsilon}_i$. Now the earlier-stated assumption of independent residuals among the traits (see section B.1.3) can be explicitly spelled out as $\boldsymbol{\varepsilon}_i \sim N(\boldsymbol{0}, \boldsymbol{\Sigma})$, where the variance-covariance matrix $\boldsymbol{\Sigma}$ of the multivariate normal distribution is defined as the diagonal matrix

$$\boldsymbol{\Sigma} = \begin{pmatrix} \sigma_1^2 & \cdots & 0 \\ \vdots & \ddots & \vdots \\ 0 & \cdots & \sigma_m^2 \end{pmatrix}. \tag{B.17}$$

To include the possibility of correlations among the traits, we simply relax the assumption that $\boldsymbol{\Sigma}$ is a diagonal matrix and thus let it be any feasible variance-covariance matrix, i.e. any symmetric positive definite matrix. As an example, we consider two traits, called the weight and the length. We include as fixed effects only the overall mean (i.e. the intercept), and assume that these are 10 for weight and 20 for length. We further assume that the standard deviations of these traits are 2 and 5, respectively, so that $\sigma_1^2 = 4$ and $\sigma_2^2 = 25$. Let us further assume a positive correlation among these two traits, with correlation coefficient $\rho = 0.7$. Then the covariance between the two traits is $\rho\sigma_1\sigma_2 = 7$, and thus the matrix $\boldsymbol{\Sigma}$ is defined as

$$\boldsymbol{\Sigma} = \begin{pmatrix} 4 & 7 \\ 7 & 25 \end{pmatrix}. \tag{B.18}$$

Figure A.5A, which we used earlier to discuss the concept of correlation between two random variables, illustrates the distribution of trait values among individuals with these parameter values. In the figure, the positive correlation assumed between the two traits resulted in tall individuals to be also heavy, on average.

Let us return to the full animal model, which includes the separation of additive genetic and environmental effects. We thus model the multivariate phenotype $\boldsymbol{p}_i$, i.e. the vector including the values of the m traits measured for individual i, as

$$\boldsymbol{p}_i = \boldsymbol{L}_i + \boldsymbol{a}_i + \boldsymbol{e}_i. \tag{B.19}$$

Here $\boldsymbol{L}_i$ is the vector of fixed effects (modelled as described earlier), $\boldsymbol{a}_i$ is the vector of additive genetic effects, and $\boldsymbol{e}_i$ is the vector of environmental effects. As we did with the univariate case, we assume that the environmental effects are independent among the individuals, whereas the additive genetic effects are structured by the coancestry matrix. In the multivariate case, we assume that both the genetic additive and the environmental effects may be correlated among the traits. To do so in the notation of Chapter 5, we replace the variance parameters V_E and V_A by variance-covariance matrices $\mathbf{E}$ and $\mathbf{G}$.

Let us first consider the covariance in the environmental effects between the trait j_1 of the individual i_1 and the trait j_2 of the individual i_2. By the assumption of independence among individuals, the covariance is 0 if $i_1 \neq i_2$. Within the individual ($i_1 = i_2$), the covariance is given by the element (j_1, j_2) of the matrix $\mathbf{E}$,

$$\mathrm{Cov}(e_{ij_1}, e_{ij_2}) = \mathbf{E}_{j_1 j_2}. \tag{B.20}$$

As we discussed in the case of the univariate animal model, the additive effects are not independent among the individuals, but their covariance depends on the relatedness between the two individuals, measured by $A_{i_1 i_2}$. In the multivariate case, the covariance in additive effects between the trait j_1 of the individual i_1 and the trait j_2 of the individual i_2 is given by the product

$$\mathrm{Cov}(a_{i_1 j_1}, a_{i_2 j_2}) = A_{i_1 i_2} \mathbf{G}_{j_1 j_2}. \tag{B.21}$$

This can be written in matrix notation as

$$\mathrm{Cov}(\boldsymbol{e}_{i_1}, \boldsymbol{e}_{i_2}) = \delta_{i_1 i_2} \mathbf{E}, \tag{B.22}$$

$$\mathrm{Cov}(\boldsymbol{a}_{i_1}, \boldsymbol{a}_{i_2}) = A_{i_1 i_2} \mathbf{G}, \tag{B.23}$$

where $\delta_{i_1 i_2}$ is Kronecker delta, with value $\delta_{i_1 i_2} = 1$ if $i_1 = i_2$ and $\delta_{i_1 i_2} = 0$ if $i_1 \neq i_2$.

To write the very same model in still more compact notation, instead of thinking of a and e as $n \times m$ matrices, let us consider them as vectors of length nm. We order the elements of these vectors first according to the individual, and then according to the trait within individuals. In the case of two traits, the first element would correspond to trait 1 of individual 1, the second element to trait 2 of individual 1, the third element to trait 1 of individual 2, and so on. Organizing the response variables and the linear predictors accordingly, we may write the model as

$$\boldsymbol{p} = \boldsymbol{L} + \boldsymbol{a} + \boldsymbol{e}, \tag{B.24}$$

where $\boldsymbol{e} \sim N(0, \mathbf{I}_n \otimes \mathbf{E})$ and $\boldsymbol{a} \sim N(0, \mathbf{A} \otimes \mathbf{G})$. The operator $\otimes$ stands for the Kronecker product. It means that the elements of the two matrices are to be multiplied in all combinations so that the end result of combining a $n \times n$ matrix ($\mathbf{I}_n$ or $\mathbf{A}$) with a $m \times m$ matrix ($\mathbf{E}$ or $\mathbf{G}$) is a $(nm) \times (nm)$ matrix.

Let us then consider a multivariate version of a species distribution model, assuming that the data are of presence–absence nature. So now the response variable $\boldsymbol{y}_i$ is a vector of 0s and 1s, the number of elements in the vector being the number of species m. For example, $\boldsymbol{y}_i = (1, 0, 1)$ will correspond to the situation in which species 1 and 3 are found but species 2 is not found from the sampling unit i. One reason we may wish to apply a multivariate model instead of modelling each species separately is that it allows us to account for co-occurrences among the species, thus asking if some species pairs are found together more or less often than expected by the predictions based on the fixed effects. Reflecting the earlier discussion on multivariate models, the idea is to model species occurrences as $\boldsymbol{y}_i = \boldsymbol{L}_i + \boldsymbol{\varepsilon}_i$, where the correlations among the species are accounted through a covariance structure of the residuals, $\boldsymbol{e}_i \sim N(0, \boldsymbol{\Sigma})$. The reason this is not entirely straightforward is that now the data are Bernoulli distributed (0s and 1s) rather than normally distributed, and thus we need to apply either logistic or probit regression. As discussed earlier in the context of link functions and error distributions, these models do not have a normally distributed residual. But we recall that probit regression can be written as $z_i = L_i + \varepsilon_i$, where $\varepsilon_i \sim N(0, 1)$, and letting $y_i = 1$ if $z_i > 0$ and $y_i = 0$ if $z_i \leq 0$.

The trick here is that in this formulation the model involves a normally distributed 'residual' ε_i, through which we can introduce a correlation structure. Thus, let us add the index j to denote the species, write $z_{ij} = L_{ij} + \varepsilon_{ij}$, and let $y_{ij} = 1$ if $z_{ij} > 0$ and $y_{ij} = 0$ if $z_{ij} \leq 0$. If assuming independent occurrences among the species, we would let $\varepsilon_{ij} \sim N(0, 1)$ independently among the species j. In matrix notation, where the vector $\boldsymbol{\varepsilon}_i$ includes the residuals for all species for sampling unit i, this corresponds to the assumption of $\boldsymbol{\varepsilon}_i \sim N(\mathbf{0}, \mathbf{I}_m)$. To include a correlation structure, we may write $\boldsymbol{\varepsilon}_i \sim N(\mathbf{0}, \mathbf{R})$, where $\mathbf{R}$ is a correlation matrix, i.e. a symmetric positive-definite matrix with values at the diagonal constrained to be 1s. The reason $\mathbf{R}$ needs to be a correlation matrix instead of a variance-covariance matrix is that each ε_{ij} should be distributed as $\varepsilon_{ij} \sim N(0, 1)$, and thus the variances need to be constrained to 1.

To illustrate why the multivariate probit model is able to capture co-occurrence patterns, consider the case of two species. Let us assume that for a sampling unit i the linear predictor L_{ij} has the value 0 for both of the species, so that the marginal probabilities of occurrence are $P[y_{ij} = 1] = \Phi(L_{ij}) = 0.5$ for both species. Let us denote the cross-species correlation coefficient by ρ, so that the matrix $\mathbf{R}$ is given by

$$\mathbf{R} = \begin{pmatrix} 1 & \rho \\ \rho & 1 \end{pmatrix}. \tag{B.25}$$

If setting $\rho = 0$ and thus assuming independent occurrences, the four possible outcomes of the response variable have equal probabilities: $P[\mathbf{y}_i = (0, 0)] = P[\mathbf{y}_i = (0, 1)] = P[\mathbf{y}_i = (1, 0)] = P[\mathbf{y}_i = (1, 1)] = 0.25$. Assuming a positive correlation with $\rho = 0.9$ modifies these to $P[\mathbf{y}_i = (0, 0)] = P[\mathbf{y}_i = (1, 1)] = 0.43$ and $P[\mathbf{y}_i = (0, 1)] = P[\mathbf{y}_i = (1, 0)] = 0.07$. Thus, in this case both of the species still occur with probability 0.5, but now it is more likely that either both species are found from the sampling unit or neither of the species is found from the sampling unit. Symmetrically, assuming a negative correlation with $\rho = -0.9$ yields $P[\mathbf{y}_i = (0, 0)] = P[\mathbf{y}_i = (1, 1)] = 0.07$ and $P[\mathbf{y}_i = (0, 1)] = P[\mathbf{y}_i = (1, 0)] = 0.43$, and thus in this case the two species co-occur less often than random.

B.1.6 Hierarchical models

There is a large number of ways by which models can be hierarchical, i.e. that they include multiple levels nested within each other. To start with, most models with random effects can be considered hierarchical, at least in some sense. For example, the animal model we have been discussing has parameters that relate to the level of individuals (breeding values) and parameters that relate to the level of the population (additive genetic variance). For another example, the species abundance model with site as a random effect, also already discussed, has parameters that relate to the level of sampling units (the regression coefficients) and parameters that relate to collections of sampling units (random effect of the site). To make the hierarchical structure more evident, the site-level model could be extended to involve not only the random effect $\mathbf{s} \sim N(\mathbf{0}, \sigma_S^2 \mathbf{I}_m)$, but also fixed effects. This can be done by including site-level covariates in a design matrix $\mathbf{X}^S$, and denoting their effects by the vector of regression coefficients $\boldsymbol{\beta}^S$. The site-level part of the model reads then as $\mathbf{s} \sim N(\mathbf{X}^S \boldsymbol{\beta}^S, \sigma_S^2 \mathbf{I}_m)$.

Another way of developing a hierarchical structure is to assume that random variation among sites influences not only the overall abundance of the species (the intercept of the model), but also the effects of the covariates (the regression coefficients of the sampling unit level model). Such models are called random intercept–slope models, where the word slope refers to the graphical interpretation of a regression coefficient. To simplify the treatment, let us consider a linear model with intercept and only one covariate, and thus write $y_i = \beta_0 + x_{i1}\beta_1 + \varepsilon_i$.

Assuming again that the data have been acquired from a set of sites, we recall that a model with site as a random factor can be written as $y_i = \beta_0 + \gamma_{0s(i)} + x_{i1}\beta_1 + \varepsilon_i$. Here $s(i)$ is the index of the site s to which the sampling unit i belongs, and the effects of the sites are assumed to be distributed as $\boldsymbol{\gamma}_0 \sim N(\mathbf{0}, \sigma_S^2 \mathbf{I}_m)$. In this case, the site influences the intercept, which is on average β_0, but for the site s it is modified to $\beta_0 + \gamma_{0s}$, and can thus be larger or smaller than the average. To incorporate random variation in the slope, we model the slope for site s by $\beta_1 + \gamma_{1s}$, where γ_{1s} is the random contribution of the site s to the slope. Thus, the random intercept–slope model reads

$$y_i = (\beta_0 + \gamma_{0s(i)}) + x_{i1}(\beta_1 + \gamma_{1s(i)}) + \varepsilon_i. \tag{B.26}$$

While we assumed in the random intercept model that $\boldsymbol{\gamma} \sim N(\mathbf{0}, \sigma_S^2 \mathbf{I}_m)$, here we include in the vector $\boldsymbol{\gamma}$ the random effects for both the intercept (γ_{0s}) and the slope (γ_{1s}). To do so, we utilize the Kronecker product notation and write $\boldsymbol{\gamma} \sim N(\mathbf{0}, \mathbf{I}_m \otimes \boldsymbol{\Sigma})$, where the identity matrix $\mathbf{I}_m$ means that we assume independent effects among the sites. In the variance-covariance matrix

$$\boldsymbol{\Sigma} = \begin{pmatrix} \sigma_{s1}^2 & \rho\sigma_{s1}\sigma_{s2} \\ \rho\sigma_{s1}\sigma_{s2} & \sigma_{s2}^2 \end{pmatrix} \tag{B.27}$$

the elements σ_{s1}^2 and σ_{s2}^2 measure the amounts of random variation among the intercepts and slopes, respectively, and $-1 \leq \rho \leq 1$ is the correlation between these two. If ρ is positive, those sites which make a positive contribution to the intercept also on average make a positive contribution to the slope, while if ρ is negative, the opposite is true.

To discuss another way of making linear models hierarchical, we return to the multispecies model with $y_{ij} = L_{ij} + \varepsilon_{ij}$, where $L_{ij} = \sum_{l=1}^{k} x_{il}\beta_{lj}$. We have already connected the models among the species (indexed by j) by assuming that the residuals are correlated. Another way to connect the models among the species is to assume that the regression coefficients β_{lj} have a joint structure, e.g. that they are distributed according to the multivariate normal distribution. Thus, denoting by $\boldsymbol{\beta}_{\cdot j}$ the vector of all regression coefficients for species j, we may assume that $\boldsymbol{\beta}_{\cdot j} \sim N(\boldsymbol{\mu}, \boldsymbol{\Sigma})$. In this model, $\boldsymbol{\mu}$ and $\boldsymbol{\Sigma}$ are community-level parameters, whereas the regression coefficients $\boldsymbol{\beta}$ are species-level parameters. The vector $\boldsymbol{\mu}$ models the response of a typical species to environmental covariates, and the variance-covariance matrix $\boldsymbol{\Sigma}$ models the amounts of variation and co-variation in the species responses to the environmental covariates. As discussed in Section 4.5, the advantage of combining species-specific models with a community model is that it helps to parameterize the model especially for rare species with limited data, and that the community-level parameters $\boldsymbol{\mu}$ and $\boldsymbol{\Sigma}$ provide a compact summary of the community structure.

The models discussed in this section are of course only examples of the many ways in which models can be built hierarchically. More generally, hierarchical models can be built by modelling different parts of the main model by sub-models.

B.2 Model fitting with Bayesian inference

In this section, we discuss the basics of model fitting with Bayesian inference. Though we do not have any philosophical preference for either Bayesian or maximum likelihood (ML)-based inference, we focus here on the former. We do so because the Bayesian framework has turned out to be very flexible for fitting complex and hierarchical models (Gillies et al., 2006), which largely explains its increasing popularity in ecology and evolutionary biology (Cressie et al., 2009).

B.2.1 The concepts of likelihood, maximum likelihood, and parameter uncertainty

One of the most central concepts in model fitting is that of the likelihood. To illustrate what likelihood is, let us consider Bernoulli distributed data, i.e. data that consists of 0s or 1s, for example empty or occupied patches. The parameter of the Bernoulli distribution is the probability θ by which the observation is 1 (e.g. a patch is occupied), so that with probability $1 - \theta$ the observation is 0 (e.g. a patch is empty). Assume that we have $n = 5$ observations, out of which $y = 1$ are 1s and the remaining $n - y = 4$ are 0s. What can we say about the parameter θ? Intuitively, as 20% of the observations are 1s, $\theta = 0.2$ sounds like the best estimate. But the sample size is small, so also other values of θ could have plausibly produced the outcome of $y = 1$.

The range of 'plausible' values for θ can be quantified with the help of the likelihood of the data. Thus, we ask what would be the probability $p(y|\theta)$ of observing the data $y = 1$ out of the $n = 5$ trials, if the data are Bernoulli distributed with probability of success being θ. The notation $p(y|\theta)$ denotes the likelihood of observing the data y, given the parameter value θ. In this example, the likelihood is given by the binomial distribution,

$$p(y|\theta) = \frac{n!}{y!(n-y)!}\theta^y(1-\theta)^{n-y}. \tag{B.28}$$

The likelihood $p(y|\theta)$ of observing the data, given the parameter value θ, is a function of the parameter θ. As illustrated in Figure B.2A, the likelihood function peaks (obtains its highest value) at $\theta = 0.2$. This is why the estimate $\hat{\theta} = 0.2$ is called the maximum likelihood estimate. It can be considered as the best guess of the parameter value.

But how much uncertainty is there about the parameter value around our best guess of $\theta = 0.2$? Intuitively, this depends on how sharply the likelihood function declines around its maximal value. Figure B.2B shows the likelihood function with $y = 20$ ones out of $n = 100$ trials. The ML estimate is $\hat{\theta} = 0.2$ in this case too, but as there are more data, there is less uncertainty and thus the likelihood function declines more sharply around its peak. The amount of parameter uncertainty can be measured by the confidence interval (in the context of ML inference) or the credibility interval (in the context of Bayesian inference) of the parameter value. Both of these can be computed at different coverage levels: we may wish to construct e.g. the 50, 90, or 95% confidence or credibility interval.

The interpretation of the 95% Bayesian credibility interval is that the true parameter value belongs to the interval with 95% probability. The interpretation of the confidence interval of the ML estimate is, however, a bit different. This is because while in Bayesian inference the parameter is considered random, in ML estimation the parameter is considered fixed. If the parameter is random, it may belong to a given interval with e.g. 95% probability. But if it is fixed, it either belongs to an interval or not, so the probability is 0 or 100%. What is considered random with ML estimation is the confidence interval. Assume that one would repeat the experiment many times, and compute for each dataset the 95% confidence interval. Then the true parameter value θ would be contained within the 95% confidence interval in 95% of the cases.

B.2.2 Prior and posterior distributions, and the Bayes theorem

In Bayesian inference we are not interested in the likelihood of the data $p(y|\theta)$, but in $p(\theta|y)$, i.e. the probability (or probability density) that the parameter has the value θ, given the observed data y. While we call θ in this section as 'the parameter', we note that it can equally well be a vector of many parameters.

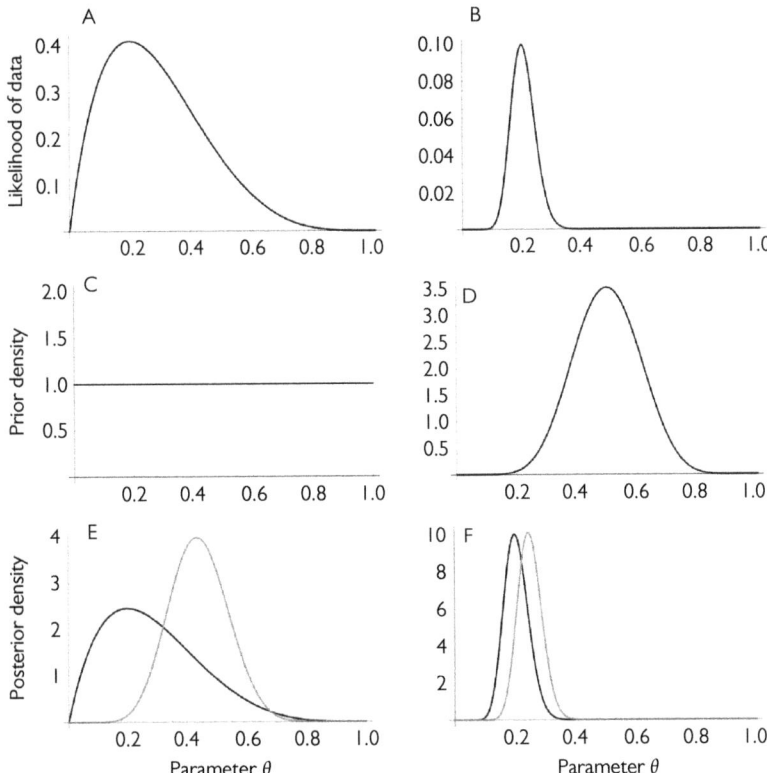

Figure B.2 Illustration of basic concepts of Bayesian inference. We consider the binomial model discussed in the text. (A, B) show the likelihood of the data $p(y|\theta)$ as a function of the parameter θ. In (A), the data consist of $n = 5$ observations, out of which $y = 1$ observation is 1 and the remaining $n - y = 4$ observations are 0s. In (B), $n = 100$ and $y = 20$. (C, D) show two alternative prior densities $p(\theta)$. In (C) we have assumed the uniform prior distribution Beta(1,1), whereas in (D) we have assumed the more informative prior distribution Beta(10,10). (E, F) show the posterior densities $p(\theta|y)$, (E) corresponding to the data in (A) and (F) corresponding to the data in (B). The black lines in (E, F) correspond to the prior of (C), whereas the grey lines correspond to the prior of (D).

To convert $p(y|\theta)$ into $p(\theta|y)$, we need the Bayes theorem, which states that

$$p(\theta|y) \propto p(\theta)p(y|\theta). \tag{B.29}$$

In this equation, the symbol '$\propto$' should be read as 'is proportional to'. In Eq. (B.19), $p(\theta)$ is the prior probability (or probability density) for θ, which quantifies what we knew of θ before seeing the data. The need for the prior distribution is both the strength and the weakness of the Bayesian approach. It is a strength because if prior information are available, it is possible to utilize it. But it is a weakness because one needs to define the prior even if there are no prior information. In such a case defining the prior may seem rather arbitrary, and thus one may worry on how much the choice of the prior influences the results. We will return to this issue later.

The Bayes theorem can be derived very simply from rules for conditional probabilities. If A and B are two random variables (see Appendix A.3), then the probability that A has the value a and that B has the value b can be written as $p(A = a$ and $B = b)$. We recall that this probability can be decomposed as $p(A = a$ and $B = b) = p(A = a)p(B = b|A = a)$. In the right-hand side, $p(A = a)$ is the probability that A has the value a, whereas $p(B = b|A = a)$ is the probability that B has the value b, conditional on A having the value a.

Reversing the roles of A and B, we obtain

$$p(A = a)p(B = b|A = a) = p(B = b)p(A = a|B = b). \tag{B.30}$$

Letting θ play the role of A and y play the role of B, we obtain $p(y)p(\theta|y) = p(\theta)p(y|\theta)$, which yields the Bayes theorem

$$p(\theta|y) = \frac{p(\theta)p(y|\theta)}{p(y)} \propto p(\theta)p(y|\theta). \tag{B.31}$$

Here $p(y)$ can be interpreted as the marginal probability of observing the data. It behaves as a normalizing constant which ensures that the posterior distribution is a probability distribution and thus integrates to 1 over all values of θ. The normalizing constant is generally not of interest as it does not depend on the parameter θ, and thus it can be ignored during parameter estimation. The main point of Bayes theorem is that it helps us to convert the prior distribution $p(\theta)$ to the posterior distribution $p(\theta|y)$. The prior distribution describes what we knew about the parameter value before seeing the data, whereas the posterior distribution describes our updated knowledge about the parameter after seeing the data.

To illustrate the Bayes theorem, we continue with the binomial example of Figure B.2. We have assumed two choices for the prior distribution $p(\theta)$. In Figure B.2C we have assumed that any value of θ within the range from 0 to 1 is equally likely, and thus assumed that θ follows a uniform distribution. In Figure B.2D we have used a more informative prior by assuming that the value of θ is close to 0.5. For example, assume that our experiment is about tossing a coin, so that θ is the probability of getting heads, and so $1 - \theta$ is the probability of getting tails. If there would be no reason to assume that the coin is biased, we would be quite certain that θ is close to 0.5.

Figures B.2E and F show the four posterior distributions of the parameter value θ that are obtained for these two prior distributions, and for the small ($n = 5$) and large ($n = 100$) data sets. In the case of the uninformative prior, the shapes of the posterior densities are identical with the likelihood profiles. With the informative prior, the posterior distributions are a compromise between what is suggested by the prior and what is suggested by the likelihood of the data. If there are not much data, the posterior distribution follows closely the prior (Figure B.2E). If there are lot of data, the posterior distribution follows closely the likelihood profile of the data (Figure B.2F). Thus, as the amount of the data increases, the role of the prior distribution becomes less important. This is intuitive: if there is only few data, there is only little information to change our prior belief about the parameter value. In such a case, the choice of the prior distribution will have a large influence on the results. But if there are much data, the posterior distribution is mostly determined by the data and thus less sensitive to the choice of the prior.

B.2.3 Methods for sampling the posterior distribution

Let us finally discuss the practicalities of the Bayesian approach: how can the posterior distribution be computed or sampled? Before answering this question, we make the somewhat provocative note that, for an ecologist, this is not actually very important to know. The reason here is that the posterior distribution is fully determined by the combination of the prior distribution and the description of the statistical model, out of which the latter determines the likelihood of the data. Thus, computing the posterior distribution is solely a technical step that does not require any ecological input. Yet, ecological literature that applies Bayesian inference often involves a lot of technical details on how the posterior distribution was sampled. This is because sampling the posterior distribution is often not an easy task.

In relatively simple and standard problems, it is possible to utilize general purpose software packages (e.g. R: Albert 2007; WinBUGS: Gimenez et al. 2009; STAN: Carpenter et al. 2015) to sample the posterior distribution. However, in more complex problems, a problem-specific implementation may be needed. Even if the ecologist may not need to program the computer algorithm but leave it for a collaborator, it is valuable to have some basic knowledge of the underlying techniques. For this reason, we will illustrate here two commonly used methods for sampling the posterior distribution. To do so in a very simple context, we will continue with the binomial distribution. So let us consider the case of little data and an uninformative prior distribution, for which the density function of the posterior distribution is shown by the black line in Figure B.2E. The question that we next try to answer is where does this line come from?

The first method that we describe is that of computing the posterior distribution analytically. Let us start by telling the end result: with uniform prior distribution, and with data with $y = 1$ out of the $n = 5$ trials, the posterior distribution for θ is the beta distribution with parameters Beta(2,5). The black line in Figure B.2E shows the probability density function of this distribution. The reason the posterior is Beta(2,5) involves some mathematical 'magic' related to the concept of conjugate distributions. To make the derivation, let us assume that the prior distribution is Beta(a, b), i.e. the beta distribution with parameters a and b. The probability density of the beta distribution is

$$p(\theta) = \frac{\theta^{a-1}(1-\theta)^{b-1}}{B(a,b)}, \tag{B.32}$$

where the beta function $B(a, b)$ is a normalization constant that ensures that the probability density integrates to 1 over its range, i.e. that $\int_{\theta=0}^{1} p(\theta)d\theta = 1$.

For binomially distributed data, the likelihood of the data $p(y|\theta)$ is given by Eq. (B.28). To apply the Bayes rule, $p(\theta|y) \propto p(\theta)p(y|\theta)$, we need to multiply Eq. (B.32) with Eq. (B.28). This gives

$$p(\theta|y) \propto \theta^{y}(1-\theta)^{n-y}\theta^{a-1}(1-\theta)^{b-1}, \tag{B.33}$$

which we have simplified by dropping all terms that do not depend on θ. This can be done as Eq. (B.33) is not yet the posterior density itself, just proportional to it. Thus, at this point we don't worry about the normalization constant, just on the dependency of the parameter θ. We next simplify the equation to

$$p(\theta|y) \propto \theta^{y+a-1}(1-\theta)^{n-y+b-1}. \tag{B.34}$$

The final step is to realize that Eq. (B.34) is proportional to the probability density of the beta distribution Beta($y + a, n - y + b$), as can be seen by comparing Eq. (B.34) to Eq. (B.32).

As the posterior density must integrate to 1, the normalizing constant must be that of the beta distribution, and thus $\theta|y \sim \text{Beta}(y + a, n - y + b)$.

After deriving the general expression for the posterior distribution, we just need to plug in the values of the parameters a, b, n, and y. The assumption of the uniform prior is equivalent with the prior being Beta(1, 1), as inserting the parameters $a = 1$ and $b = 1$ in Eq. (B.32) yields $p(\theta) = 1$. As our data were $y = 1$ and $n = 5$, the posterior distribution is Beta(2,5).

We have shown that the combination of a beta-distributed prior with binomially distributed data results into a beta-distributed posterior. This is why the beta distribution is called the conjugate prior distribution for binomially distributed data. If we had assumed any other prior distribution than the beta distribution, the mathematical magic would not have worked. Many standard probability distributions have conjugate prior distributions that pair with them in the sense that the posterior distribution follows the same distribution (but with different parameters) as the prior distribution, which is a practical reason why they are often utilized as prior distributions (e.g. Gelman et al., 2013).

Being able to compute the posterior distribution analytically is very convenient, but unfortunately it is possible only for specific cases. The second method that we will discuss is that of obtaining the posterior distribution numerically. This is actually not one method, but a large array of methods, and the need of finding faster and more generally applicable methods has generated an active area of statistical research. To illustrate numerical sampling with one specific method, we consider the Metropolis–Hastings algorithm, as it is probably the simplest and most widely applied Markov Chain Monte Carlo (MCMC) method. In this algorithm, one starts from any parameter $\theta_1 = \theta$. The next step is to propose a change in the parameter. For example, we may propose a new value $\hat{\theta}$ from the distribution $\hat{\theta} \sim N(\theta, \sigma^2)$, where we have set the expected value to the present value of the parameter, and the value of the variance σ^2 parameter determines how far from the present value we propose new values. The new value $\hat{\theta}$ is then compared to the present value θ in terms of the posterior density $p(\theta|y)$. If the proposed parameter is 'better' in the sense that it produces a higher posterior density, $p(\hat{\theta}|y) > p(\theta|y)$, it is always accepted. If it produces a lower posterior density, it is accepted with probability $p(\hat{\theta}|y)/p(\theta|y)$. Thus, parameter values that lower the posterior density only by a little are mostly accepted, whereas parameter values that lower the posterior density a lot are mostly rejected. If the proposed parameter $\hat{\theta}$ is accepted, we set the next iteration of the parameter to this value, so that $\theta_2 = \hat{\theta}$. If it is rejected, we keep the old value, and thus $\theta_2 = \theta_1$.

Mathematically, the values of the parameter $\theta_1, \theta_2, \ldots$ form a Markov chain (see Appendix A.4). The Markov chain converges to a stationary distribution, which corresponds exactly to the posterior distribution. Why this is the case involves again some mathematical magic, i.e. a mathematical proof showing that the stationary distribution of the Markov chain equals the posterior distribution. Such a proof, as well as many relevant extensions to this algorithm (asymmetric proposals, multivariate proposals, adaptive algorithms, etc.), can be found in textbooks on Bayesian methods, such as Casella (2006).

Running the Metropolis–Hasting algorithm for many enough iterations will generate a sample from the posterior distribution. This is illustrated in Figure B.3, where we have applied the Metropolis–Hastings algorithm for the case of the uninformative prior Beta(1,1) and the data $n = 5$ and $y = 1$. The left-hand panels show how the parameter value θ evolves in the Markov chain, whereas the right-hand panels compare thus obtained samples of the posterior distribution to the probability density of Beta(2,5) that we calculated earlier analytically. The three rows correspond to different values of the variance parameter σ^2. Let us first focus on the middle row of panels, which is the best scenario in the sense that the chain mixes the best.

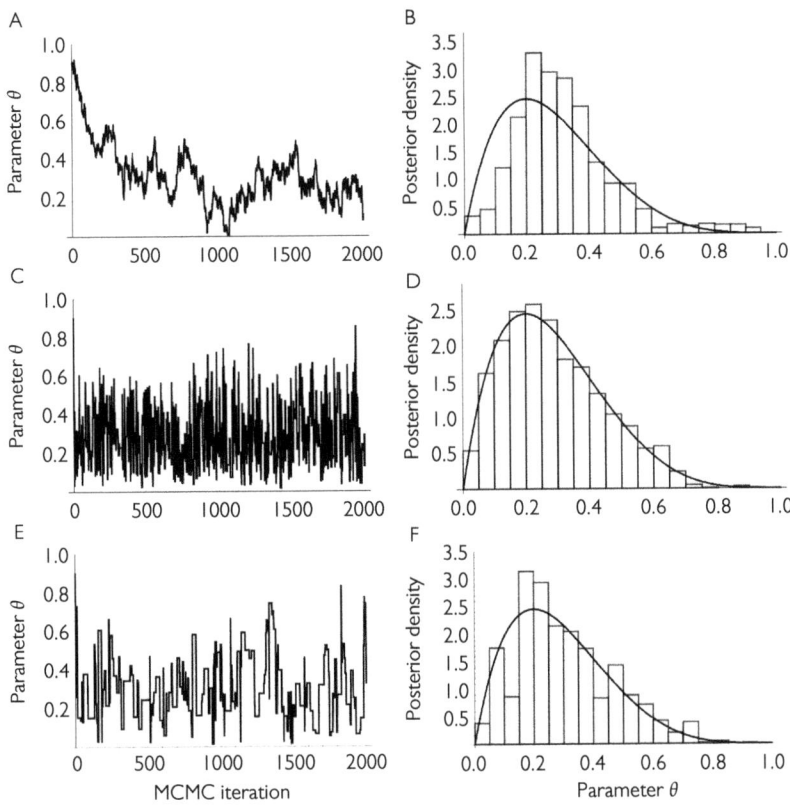

Figure B.3 Sampling the posterior distribution with the Metropolis–Hastings algorithm. (A, C, E) show the Markov Chain Monte Carlo (MCMC) iterations, and (B, D, F) show the corresponding histograms that approximate the posterior distributions. In (B, D, F), the black lines show the density of the exact posterior distribution Beta(2,5). The variance parameter of the proposal distribution is set to $\sigma^2 = 0.02^2$ (A, B), $\sigma^2 = 0.5^2$ (C, D), and $\sigma^2 = 2^2$ (E, F). The fraction of accepted proposals is 0.97 (A, B), 0.35 (C, D), and 0.10 (E, F).

Here the value of the parameter θ moves rapidly up and down, without much autocorrelation, and without a long transient in the beginning, even though we started the chain from $\theta = 0.9$ and thus far away from core of the posterior distribution. In this case, the posterior distribution is well approximated with a sample obtained by 2,000 iterations (Figure B.3D). In the upper panels, the variance parameter has a lower value. This does not lead to good mixing, because the parameter value changes only little at a time, and thus it takes a long time to move from a large value to a low value. This results in a long transient before the chain finds the core of the posterior distribution. As a consequence, the sample obtained by 2,000 iterations is not a very good approximation of the true posterior distribution (Figure B.3B). The lower panels involve an opposite problem: now the chain mixes slowly because the variance parameter is too high. In this case, the proposed values are often outside the plausible range, leading to their rejection, and thus the chain stays for a long time in one value before it moves on. As a result, also in this case the sample obtained by 2,000 iterations is not a very good approximation of the true posterior distribution (Figure B.3F).

Figure B.3 makes the point that while the Metropolis–Hastings algorithm will always converge to the posterior distribution, the details of the algorithm, such as the value of the variance parameter σ^2, will influence how fast this will happen. The same holds with basically all methods that can be used to sample the posterior distribution. When reading applications of Bayesian inference, the reader may encounter terms such as Gibbs sampling, particle filtering, grid sampling methods, adaptive MCMC, and reversible jump MCMC. From the ecological point of view it is not relevant which of these methods is applied, as they all will lead exactly to the same end result: the posterior distribution, which is uniquely determined by the prior and the statistical model. Thus, the sampling method matters just for how fast a representative sample of the posterior distribution can be achieved.

References

Abecasis, G.R., Cherny, S.S., Cookson, W.O. and Cardon, L.R. (2002). Merlin-rapid analysis of dense genetic maps using sparse gene flow trees. *Nat. Genet.* **30**, 97–101.

Abrams, P.A. (2005). 'Adaptive dynamics' vs. 'adaptive dynamics'. *J. Evol. Biol.* **18**, 1162–1165.

Abrams, P.A. and Ginzburg, L.R. (2000). The nature of predation: prey dependent, ratio dependent, or neither? *Trends Ecol. Evol.* **15**, 337–341.

Adriaensen, F., Chardon, J.P., De Blust, G., Swinnen, E., Villalba, S., Gulinck, H. and Matthysen, E. (2003). The application of 'least-cost' modelling as a functional landscape model. *Landsc Urban Plan* **64**, 233–247.

Agrawal, A.A., Ackerly, D.D., Adler, F., Arnold, A.E., Cáceres, C., Doak, D.F., Post, E., Hudson, P.J., Maron, J., Mooney, K.A., et al. (2007). Filling key gaps in population and community ecology. *Front. Ecol. Environ.* **5**, 145–152.

Albert, J. (2007). *Bayesian Computation with R*. New York: Springer.

Alcock, J. (1987). Leks and hilltopping in insects. *J. Nat. Hist.* **21**, 319–328.

Allesina, S., Alonso, D. and Pascual, M. (2008). A general model for food web structure. *Science* **320**, 658–661.

Almeida, P.J. a. L., Vieira, M. V., Kajin, M., Forero-Medina, G. and Cerqueira, R. (2010). Indices of movement behaviour: conceptual background, effects of scale and location errors. *Zoologia* **27**, 674–680.

Alt, W. (1990). Correlation analysis of two-dimensional locomotion paths. In: *Biological Motion* (ed. Alt, W. and Hoffmann, G.), pp. 254–268. Berlin: Springer.

Altermatt, F. and Ebert, D. (2010). Populations in small, ephemeral habitat patches may drive dynamics in a Daphnia magna metapopulation. *Ecology* **91**, 2975–2982.

Amarasekare, P. (2003). Competitive coexistence in spatially structured environments: a synthesis. *Ecol. Lett.* **6**, 1109–1122.

Amos, W. and Balmford, A. (2001). When does conservation genetics matter? *Heredity* **87**, 257–265.

Anderson, M.J., Crist, T.O., Chase, J.M., Vellend, M., Inouye, B.D., Freestone, A.L., Sanders, N.J., Cornell, H. V., Comita, L.S., Davies, K.F., et al. (2011). Navigating the multiple meanings of β diversity: a roadmap for the practicing ecologist. *Ecol. Lett.* **14**, 19–28.

Andreassen, H.P. and Ims, R.A. (2001). Dispersal in patchy vole populations: role of patch configuration, density dependence, and demography. *Ecology* **82**, 2911–2926.

Armstrong, R.A. and McGehee, R. (1980). Competitive exclusion. *Am. Nat.* **115**, 151–170.

Arnold, S.J., Bürger, R., Hohenlohe, P.A., Ajie, B.C. and Jones, A.G. (2008). Understanding the evolution and stability of the G-matrix. *Evolution* **62**, 2451–2461.

Avgar, T., Kuefler, D. and Fryxell, J.M. (2011). Linking rates of diffusion and consumption in relation to resources. *Am. Nat.* **178**, 182–190.

Baddeley, A. and Turner, R. (2005). Spatstat: an R package for analyzing spatial point patterns. *J. Stat. Softw.* 12, 1–42.
Barraquand, F., Ezard, T.H.G., Jørgensen, P.S., Zimmerman, N., Chamberlain, S., Salguero-Gómez, R., Curran, T.J. and Poisot, T. (2014). Lack of quantitative training among early-career ecologists: a survey of the problem and potential solutions. *PeerJ* 2, e285.
Bartumeus, F., da Luz, M.G.E., Viswanathan, G.M. and Catalan, J. (2005). Animal search strategies: a quantitative random-walk analysis. *Ecology* 86, 3078–3087.
Bartumeus, F., Catalan, J., Viswanathan, G.M., Raposo, E.P. and da Luz, M.G.E. (2008). The influence of turning angles on the success of non-oriented animal searches. *J. Theor. Biol.* 252, 43–55.
Bascompte, J. and Jordano, P. (2013). *Mutualistic Networks*. Princeton, NJ: Princeton University Press.
Bauer, S. and Klaassen, M. (2013). Mechanistic models of animal migration behaviour—their diversity, structure and use (ed Hays, G). *J. Anim. Ecol.* 82, 498–508.
Bee, J.N., Tanentzap, A.J., Lee, W.G., Lavers, R.B., Mark, A.F., Mills, J.A. and Coomes, D.A. (2009). The benefits of being in a bad neighbourhood: plant community composition influences red deer foraging decisions. *Oikos* 118, 18–24.
Begon, M., Harper, J. and Townsend, C. (1996). *Ecology: Individuals, Populations and Communities*, 3rd edn. Oxford: Blackwell Science.
Bélisle, M. (2005). Measuring landscape connectivity: the challenge of behavioral landscape ecology. *Ecology* 86, 1988–1995.
Bell, W.J. (1991). *Searching Behaviour—the Behavioural Ecology of Finding Resources*. London: Chapman and Hall.
Bell, G. (2012). Evolutionary rescue and the limits of adaptation. *Philos. Trans. R. Soc. Lond. B Biol. Sci.* 368, 20120080.
Bell, M.A. and Foster, S.A. (1994). *The Evolutionary Biology of the Threespine Stickleback*. Oxford: Oxford University Press.
Bell, G. and Gonzalez, A. (2009). Evolutionary rescue can prevent extinction following environmental change. *Ecol. Lett.* 12, 942–948.
Bell, G. and Gonzalez, A. (2011). Adaptation and evolutionary rescue in metapopulations experiencing environmental deterioration. *Science* 332, 1327–1330.
Bender, D.J., Tischendorf, L. and Fahrig, L. (2003). Using patch isolation metrics to predict animal movement in binary landscapes. *Landsc. Ecol.* 18, 17–39.
Benhamou, S. (2004). How to reliably estimate the tortuosity of an animal's path: straightness, sinuosity, or fractal dimension? *J. Theor. Biol.* 229, 209–220.
Benhamou, S. (2006). Detecting an orientation component in animal paths when the preferred direction is individual-dependent. *Ecology* 87, 518–528.
Benhamou, S. (2011). Dynamic approach to space and habitat use based on biased random bridges. *PLoS ONE* 6, e14592.
Benjamini, Y. and Hochberg, Y. (1995). Controlling the false discovery rate: a practical and powerful approach to multiple testing. *J. R. Stat. Soc. B* 57: 289–300.
Bergman, C.M., Schaefer, J.A. and Luttich, S.N. (2000). Caribou movement as a correlated random walk. *Oecologia* 123, 364–374.
Bergström, U., Englund, G. and Leonardsson, K. (2006). Plugging space into predator-prey models: an empirical approach. *Am. Nat.* 167, 246–259.
Bivand, R.S., Pebesma, E. and Gómez-Rubio, V. (2013). *Applied Spatial Data Analysis with R*. New York: Springer.

Black, A.J. and McKane, A.J. (2012). Stochastic formulation of ecological models and their applications. *Trends Ecol. Evol.* **27**, 337–345.

Blomqvist, D., Pauliny, A., Larsson, M. and Flodin, L.-A. (2010). Trapped in the extinction vortex? Strong genetic effects in a declining vertebrate population. *BMC Evol. Biol.* **10**, 33.

Bohonak, A.J. (2012). Dispersal, gene flow, and population structure. *Q. Rev. Biol.* **74**, 21–45.

Bolker, B.M. (2003). Combining endogenous and exogenous spatial variability in analytical population models. *Theor. Popul. Biol.* **64**, 255–270.

Bolker, B.M. (2007). *Ecological Models and Data in R*. Princeton, NJ: Princeton University Press.

Bolker, B.M., Brooks, M.E., Clark, C.J., Geange, S.W., Poulsen, J.R., Stevens, M.H.H. and White, J.-S.S. (2009). Generalized linear mixed models: a practical guide for ecology and evolution. *Trends Ecol. Evol.* **24**, 127–135.

Bolker, B.M. (2009). Evolution of dispersal scale and shape in heterogeneous environments: a correlation equation approach. In: *Spatial Ecology* (ed. Cosner, C., Cantrell, S. and Ruan, S.). London: Chapman and Hall/CRC.

Bolker, B.M. and Pacala, S.W. (1997). Using moment equations to understand stochastically driven spatial pattern formation in ecological systems. *Theor. Popul. Biol.* **52**, 179–197.

Bonin, M.C., Almany, G.R. and Jones, G.P. (2011). Contrasting effects of habitat loss and fragmentation on coral-associated reef fishes. *Ecology* **92**, 1503–1512.

Bonte, D., Van Dyck, H., Bullock, J.M., Coulon, A., Delgado, M., Gibbs, M., Lehouck, V., Matthysen, E., Mustin, K., Saastamoinen, M., et al. (2011). Costs of dispersal. *Biol. Rev.* **87**, 290–312.

Borcard, D., Gillet, F. and Legendre, P. (2011). *Numerical Ecology with R*. London: Springer.

Börger, L., Dalziel, B.D. and Fryxell, J.M. (2008). Are there general mechanisms of animal home range behaviour? A review and prospects for future research. *Ecol. Lett.* **11**, 637–650.

Bowler, D.E. and Benton, T.G. (2005). Causes and consequences of animal dispersal strategies: relating individual behaviour to spatial dynamics. *Biol. Rev.* **80**, 205–225.

Boyce, M.S., Pitt, J., Northrup, J.M., Morehouse, A.T., Knopff, K.H., Cristescu, B. and Stenhouse, G.B. (2010). Temporal autocorrelation functions for movement rates from global positioning system radiotelemetry data. *Philos. Trans. R. Soc. Lond. B Biol. Sci.* **365**, 2213–2219.

Bradshaw, H.D., Otto, K.G., Frewen, B.E., Mckay, J.K. and Schemske, D.W. (1998). Quantitative trait loci affecting differences in floral morphology between two species of monkeyflower (Mimulus). *Genet. Adapt.* **149**, 367–382.

Brännström, A. and Sumpter, D.J.T. (2005). Coupled map lattice approximations for spatially explicit individual-based models of ecology. *Bull. Math. Biol.* **67**, 663–682.

Brown, A.M., Warton, D.I., Andrew, N.R., Binns, M., Cassis, G. and Gibb, H. (2014). The fourth-corner solution—using predictive models to understand how species traits interact with the environment. *Methods Ecol. Evol.* **5**, 344–352.

Brown, J.H. and Kodric-Brown, A. (1977). Turnover rates in insular biogeography: effect of immigration on extinction. *Ecology* **58**, 445–449.

Buchmann, C.M., Schurr, F.M., Nathan, R. and Jeltsch, F. (2012). Movement upscaled—the importance of individual foraging movement for community response to habitat loss. *Ecography* **35**, 436–445.

Buchmann, C.M., Schurr, F.M., Nathan, R. and Jeltsch, F. (2013). Habitat loss and fragmentation affecting mammal and bird communities—The role of interspecific competition and individual space use. *Ecol. Inform.* **14**, 90–98.

Buckley, L.B., Urban, M.C., Angilletta, M.J., Crozier, L.G., Rissler, L.J. and Sears, M.W. (2010). Can mechanism inform species' distribution models? *Ecol. Lett.* **13**, 1041–1054.

Cadotte, M.W. (2007). Competition-colonization trade-offs and disturbance effects at multiple scales. *Ecology* **88**, 823–829.

Calabrese, J.M. and Fagan, W.F. (2004). A comparison-shopper's guide to connectivity metrics. *Front. Ecol. Environ.* **2**, 529–536.

Calvo, B. and Furness, R.W. (1992). A review of the use and the effects of marks and devices on birds. *Ringing and Migration* **13**, 129–151.

Cantrell, R.S., Cosner, C. and Lou, Y. (2009). Evolution of dispersal in heterogeneous landscapes. In: *Spatial Ecology* (ed. Cosner, C., Cantrell, S. and Ruan, S.), pp. 213–229. London: Chapman and Hall/CRC.

Cantrell, R.S. and Cosner, C. (2003). *Spatial Ecology via Reaction-Diffusion Equations*. New York: Wiley.

Capaldi, E.A., Smith, A.D., Osborne, J.L., Fahrbach, S.E., Farris, S.M., Reynolds, D.R., Edwards, A.S., Martin, A., Robinson, G.E., Poppy, G.M., et al. (2000). Ontogeny of orientation flight in the honeybee revealed by harmonic radar. *Nature* **403**, 537–540.

Cardon, L.R. and Palmer, L.J. (2003). Population stratification and spurious allelic association. *Lancet* **361**, 598–604.

Carlson, S.M., Cunningham, C.J. and Westley, P. a. H. (2014). Evolutionary rescue in a changing world. *Trends Ecol. Evol.* **29**, 521–530.

Carpenter, B., Gelman, A., Hoffman, M., Lee, D., Goodrich, B., Betancourt, M., Brubaker, M.A., Guo, J., Li, P. and Riddell, A. (2015). Stan: a probabilistic programming language. *J. Stat. Softw.*

Casella, G. (2006). *An Introduction to Bayesian Analysis*. New York: Springer.

Caswell, H. (2001). *Matrix Population Models: Construction, Analysis, and Interpretation*. Sunderland, MA: Sinauer Associates.

Cattin, M.-F., Bersier, L.-F., Banašek-Richter, C., Baltensperger, R. and Gabriel, J.-P. (2004). Phylogenetic constraints and adaptation explain food-web structure. *Nature* **427**, 835–839.

Charlesworth, D. and Charlesworth, B. (2012). Inbreeding depression and its evolutionary consequences. *Annu. Rev. Ecol. Syst.* **18**, 237–268.

Charnov, E.L. (1976). Optimal foraging, the marginal value theorem. *Theor. Popul. Biol.* **9**, 129–136.

Chase, J.M. (2014). Spatial scale resolves the niche versus neutral theory debate. *J. Veg. Sci.* **25**, 319–322.

Cheng, L., Connor, T.R., Sirén, J., Aanensen, D.M. and Corander, J. (2013). Hierarchical and spatially explicit clustering of DNA sequences with BAPS software. *Molec. Biol. Evol.* **30**, 1224–1228.

Chesson, P. (2000). Mechanisms of maintenance of species diversity. *Annu. Rev. Ecol. Syst.* **31**, 343–366.

Chiarucci, A., Bacaro, G. and Scheiner, S.M. (2011). Old and new challenges in using species diversity for assessing biodiversity. *Philos. Trans. R. Soc. Lond. B Biol. Sci.* **366**, 2426–2437.

Chipperfield, J.D., Holland, E.P., Dytham, C., Thomas, C.D. and Hovestadt, T. (2011). On the approximation of continuous dispersal kernels in discrete-space models. *Methods Ecol. Evol.* **2**, 668–681.

Chisholm, R.A. and Muller-Landau, H.C. (2011). A theoretical model linking interspecific variation in density dependence to species abundances. *Theor. Ecol.* **4**, 241–253.

Clobert, J., Danchin, E., Dhondt, A.A. and Nichols, J.D. (2001). *Dispersal*. Oxford: Oxford University Press.

Clutton-Brock, T. (1988). *Reproductive Success. Studies of Individual Variation in Contrasting Breeding Systems*. Chicago: University of Chicago Press.

Codling, E.A., Plank, M.J. and Benhamou, S. (2008). Random walk models in biology. *J. R. Soc. Interface* **5**, 813–834.

Codling, E.A., Bearon, R.N. and Thorn, G.J. (2010). Diffusion about the mean drift location in a biased random walk. *Ecology* **91**, 3106–3113.

Codling, E.A. and Hill, N.A. (2005). Sampling rate effects on measurements of correlated and biased random walks. *J. Theor. Biol.* **233**, 573–88.

Cohen, J.E. and Newman, C.M. (1985). A stochastic theory of community food webs: I. models and aggregated data. *Proc. R. Soc. B* **224**, 421–448.

Colin, S.P. and Dam, H.G. (2007). Comparison of the functional and numerical responses of resistant versus non-resistant populations of the copepod Acartia hudsonica fed the toxic dinoflagellate Alexandrium tamarense. *Harmful Algae* **6**, 875–882.

Colwell, R.K. and Rangel, T.F. (2009). Hutchinson's duality: the once and future niche. *Proc. Natl. Acad. Sci. USA* **106**, 19651–19658.

Comins, H.N., Hamilton, W.D. and May, R.M. (1980). Evolutionarily stable dispersal strategies. *J. Theor. Biol.* **82**, 205–230.

Conradt, L. and Roper, T.J. (2006). Nonrandom movement behavior at habitat boundaries in two butterfly species: implications for dispersal. *Ecology* **87**, 125–132.

Corder, G.W. and Foreman, D.I. (2009). *Nonparametric Statistics for Non-Statisticians: A Step-by-Step Approach*. New York: Wiley.

Corlatti, L., Hackländer, K. and Frey-Roos, F. (2009). Ability of wildlife overpasses to provide connectivity and prevent genetic isolation. *Conserv. Biol.* **23**, 548–556.

Cottenie, K., Michels, E., Nuytten, N. and De Meester, L. (2003). Zooplankton metacommunity structure: regional vs. local processes in highly interconnected ponds. *Ecology* **84**, 991–1000.

Cottenie, K. (2005). Integrating environmental and spatial processes in ecological community dynamics. *Ecol. Lett.* **8**, 1175–1182.

Courchamp, F., Clutton-Brock, T. and Grenfell, B. (1999). Inverse density dependence and the Allee effect. *Trends Ecol. Evol.* **14**, 405–410.

Crawley, M. (2007). *The R Book*. London: Wiley.

Cressie, N., Calder, C.A., Clark, J.S., Hoef, J.M. Ver and Wikle, C.K. (2009). Accounting for uncertainty in ecological analysis: the strengths and limitations of hierarchical statistical modeling. *Ecol. Appl.* **19**, 553–570.

Crone, E.E. and Schultz, C.B. (2008). Old models explain new observations of butterfly movement at patch edges. *Ecology* **89**, 2061–2067.

Curwen, V., Eyras, E., Andrews, T.D., Clarke, L., Mongin, E., Searle, S.M.J. and Clamp, M. (2004). The Ensembl automatic gene annotation system. *Genome Res.* **14**, 942–950.

Cushman, S.A., McRae, B., Adriaensen, F., Beier, P., Shirley, M. and Zeller, K. (2013). Biological corridors and connectivity. In: *Key Topics in Conserv. Biol.* (ed. Macdonald, D.W. and Willis, K.J.). London: Wiley.

Darroch, J.N. and Seneta, E. (1967). On quasi-stationary distributions in absorbing continuous-time finite markov chains. *J. Appl. Probab.* **4**, 192–196.

Darwin, C. (1859). *On the Origin of Species by Means of Natural Selection, or the Preservation of Favoured Races in the Struggle for Life*. London: John Murray.

Dayanandan, S., Dole, J., Bawa, K. and Kesseli, R. (1999). Population structure delineated with microsatellite markers in fragmented populations of a tropical tree, Carapa guianensis (Meliaceae). *Mol. Ecol.* **8**, 1585–1592.

Denno, R.F. and Peterson, M.A. (1995). Density-dependent dispersal and its consequences for population dynamics. In: *Population Dynamics* (ed. Cappuccino, N. and Price, P.), pp. 113–130. San Diego, CA: Academic Press.
Dercole, F. and Rinaldi, S. (2008). *Analysis of Evolutionary Processes: The Adaptive Dynamics Approach and Its Applications*. Princeton, NJ: Princeton University Press.
Dickinson, M.H., Farley, C.T., Full, R.J., Koehl, M.A., Kram, R. and Lehman, S. (2000). How animals move: an integrative view. *Science* **288**, 100–106.
Didham, R.K., Kapos, V. and Ewers, R.M. (2012). Rethinking the conceptual foundations of habitat fragmentation research. *Oikos* **121**, 161–170.
Dieckmann, U., O'Hara, B. and Weisser, W. (1999). The evolutionary ecology of dispersal. *Trends Ecol. Evol.* **14**, 88–90.
Dingle, H. (1996). *Migration: The Biology of Life on the Move*. Oxford: Oxford University Press.
Dingle, H. and Drake, V.A. (2007). What is migration? *BioScience* **57**, 113–121.
Dobzhansky, T. (1973). Nothing in biology makes sense except in the light of evolution. *Am. Biol. Teach.* **35**, 125–129.
Donalson, D.D. and Nisbet, R.M. (1999). Population dynamics and spatial scale: effects of system size on population persistence. *Ecology* **80**, 2492–2507.
Dormann, C.F., McPherson, J.M., Araújo, M.B., Bivand, R., Bolliger, J., Carl, G., Davies, R.G., Hirzel, A.H., Jetz, W., Kissling, W.D., et al. (2007). Methods to account for spatial autocorrelation in the analysis of species distributional data: a review. *Ecography* **30**, 609–628.
Dormann, C.F., Schymanski, S.J., Cabral, J., Chuine, I., Graham, C., Hartig, F., Kearney, M., Morin, X., Römermann, C., Schröder, B., et al. (2012). Correlation and process in species distribution models: bridging a dichotomy. *J. Biogeogr.* **39**, 2119–2131.
Douglas, A. (1994). *Symbiotic Interactions*. Oxford: Oxford University Press.
Edelaar, P., Burraco, P. and Gomez-Mestre, I. (2011). Comparisons between Q(ST) and F(ST)—how wrong have we been? *Mol. Ecol.* **20**, 4830–4839.
Edwards, A.W.F. (2011). Mathematizing Darwin. *Behav. Ecol. Sociobiol.* **65**, 421–430.
Ehrlich, P.R. and Hanski, I. (2004). *On the Wings of Checkerspots: A Model System for Population Biology*. Oxford: Oxford University Press.
Elith, J., Graham, H., Anderson, P., Dudik, M., Ferrier, S., Guisan, A., Hijmans, J., Huettmann, F., Leathwick, R., Lehmann, A., et al. (2006). Novel methods improve prediction of species' distributions from occurrence data. *Ecography* **29**, 129–151.
Elith, J. and Leathwick, J.R. (2009). Species distribution models: ecological explanation and prediction across space and time. *Annu. Rev. Ecol. Evol. System.* **40**, 677–697.
Ellis, T.H.N., Hofer, J.M.I., Timmerman-Vaughan, G.M., Coyne, C.J. and Hellens, R.P. (2011). Mendel, 150 years on. *Trends Plant Sci.* **16**, 590–596.
Ellner, S.P. (2001). Pair approximation for lattice models with multiple interaction scales. *J. Theor. Biol.* **210**, 435–447.
Ewers, R.M. and Didham, R.K. (2006). Confounding factors in the detection of species responses to habitat fragmentation. *Biol. Rev.* **81**, 117–142.
Fagan, W.F., Lewis, M.A., Auger-Méthé, M., Avgar, T., Benhamou, S., Breed, G., LaDage, L., Schlägel, U.E., Tang, W., Papastamatiou, Y.P., et al. (2013). Spatial memory and animal movement. *Ecol. Lett.* **16**, 1316–1329.
Fagan, W.F. and Holmes, E.E. (2006). Quantifying the extinction vortex. *Ecol. Lett.* **9**, 51–60.
Fahrig, L. (2007). Landscape heterogeneity and metapopulation dynamics. In: *Key Topics in Landscape Ecology* (ed. Wu, J. and Hobbs, R.J.), pp. 78–91. New York: Cambridge University Press.

Falconer, D.S. and Mackay, T.F.C. (1996). *Introduction to Quantitative Genetics*. London: Longman.
Fargione, J., Brown, C.S. and Tilman, D. (2003). Community assembly and invasion: an experimental test of neutral versus niche processes. *Proc. Natl. Acad. Sci. USA* **100**, 8916–8920.
Fauchald, P. and Tveraa, T. (2003). Using first-passage time in the analysis of area-restricted search and habitat selection. *Ecology* **84**, 282–288.
Ferrer, M., Newton, I. and Pandolfi, M. (2009). Small populations and offspring sex-ratio deviations in eagles. *Conserv. Biol.* **23**, 1017–1025.
Ferriere, R. and Legendre, S. (2013). Eco-evolutionary feedbacks, adaptive dynamics and evolutionary rescue theory. *Philos. Trans. R. Soc. Lond. B Biol. Sci.* **368**, 20120081.
Fisher, R.A. (1930). *The Genetical Theory of Natural Selection*. Oxford: Oxford University Press.
Fisher, R.A. (1932). The Evolutionary Modification of Genetic Phenomena. *Proceedings of the 6th International Congress of Genetics* **1**, 165–172.
Fisher, R.A. (1936). Has Mendel's work been rediscovered? *Ann. Sci.* **1**, 115–137.
Fisher, N.I. (1993). *Statistical Analysis of Circular Data*. Cambridge: Cambridge University Press.
Flicek, P. and Birney, E. (2009). Sense from sequence reads: methods for alignment and assembly. *Nat. Methods* **6**, S6–S12.
Fonseca, D.M. and Hart, D.D. (2014). Density-dependent dispersal of black fly neonates is mediated by flow. *Oikos* **75**, 49–58.
Forman, R. and Godron, M. (1986). *Landscape Ecology*. New York: Wiley.
Fortin, D., Beyer, H.L., Boyce, M.S., Smith, D.W., Duchesne, T. and Mao, J.S. (2005). Wolves influence elk movements: Behavior shapes a trophic cascade in Yellowstone National Park. *Ecology* **86**, 1320–1330.
Fortuna, M.A. and Bascompte, J. (2006). Habitat loss and the structure of plant-animal mutualistic networks. *Ecol. Lett.* **9**, 281–286.
Frair, J.L., Merrill, E.H., Visscher, D.R., Fortin, D., Beyer, H.L. and Morales, J.M. (2005). Scales of movement by elk (Cervus elaphus) in response to heterogeneity in forage resources and predation risk. *Landsc. Ecol.* **20**, 273–287.
Frair, J.L., Fieberg, J., Hebblewhite, M., Cagnacci, F., DeCesare, N.J. and Pedrotti, L. (2010). Resolving issues of imprecise and habitat-biased locations in ecological analyses using GPS telemetry data. *Philos. Trans. R. Soc. Lond. B Biol. Sci.* **365**, 2187–2200.
Fretwell, S.D. and Lucas, H.L. (1969). On territorial behavior and other factors influencing habitat distribution in birds. *Acta Biotheor.* **19**, 16–36.
Fronhofer, E.A., Hovestadt, T. and Poethke, H.-J. (2012). From random walks to informed movement. *Oikos* **122**, 857–866.
Fryxell, J.M., Wilmshurst, J.F. and Sinclair, A.R.E. (2004). Predictive models of movement by Serengeti grazers. *Ecology* **85**, 2429–2435.
Fryxell, J.M., Hazell, M., Börger, L., Dalziel, B.D., Haydon, D.T., Morales, J.M., McIntosh, T., Rosatte, R.C. and Juan, M. (2008). Multiple movement modes by large herbivores at multiple spatiotemporal scales. *Proc. Natl. Acad. Sci. USA* **105**, 19114–1919.
Fryxell, J.M. and Sinclair, A.R. (1988). Causes and consequences of migration by large herbivores. *Trends Ecol. Evol.* **3**, 237–241.
Gabriel, S.B., Schaffner, S.F., Nguyen, H., Moore, J.M., Roy, J., Blumenstiel, B., Higgins, J., DeFelice, M., Lochner, A., Faggart, M., et al. (2002). The structure of haplotype blocks in the human genome. *Science* **296**, 2225–2229.

Galliard, J. Le, Ferriere, R., Clobert, J. and Ranta, E. (2005). Juvenile growth and survival under dietary restriction: are males and females equal? *Oikos* **111**, 368–376.

Garzon-Lopez, C.X., Jansen, P.A., Bohlman, S.A., Ordonez, A. and Olff, H. (2014). Effects of sampling scale on patterns of habitat association in tropical trees. *J. Veg. Sci.* **25**, 349–362.

Gautestad, A.O. and Mysterud, I. (2010). Spatial memory, habitat auto-facilitation and the emergence of fractal home range patterns. *Ecol. Model.* **221**, 2741–2750.

Gelman, A., Carlin, J.B., Stern, H.S., Dunson, D.B., Vehtari, A. and Rubin, D.B. (2013). *Bayesian Data Analysis*, third edn. London: Chapman and Hall/CRC.

Geritz, S.A.H., Kisdi, É., Meszena, G. and Metz, J.A.J. (1998). Evolutionarily singular strategies and the adaptive growth and branching of the evolutionary tree. *Evol. Ecol.* **12**, 35–57.

Getz, W.M. and Saltz, D. (2008). A framework for generating and analyzing movement paths on ecological landscapes. *Proc. Natl. Acad. Sci. USA* **105**, 19066–19071.

Gibbs, M., Saastamoinen, M., Coulon, A. and Stevens, V.M. (2010). Organisms on the move: ecology and evolution of dispersal. *Biology Letters* **6**, 146–148.

Gilg, O., Sittler, B., Sabard, B., Hurstel, A., Sané, R., Delattre, P. and Hanski, I. (2006). Functional and numerical responses of four lemming predators in high arctic Greenland. *Oikos* **113**, 193–216.

Gillespie, D.T. (1977). Exact stochastic simulation of coupled chemical reactions. *J. Phys. Chem.* **81**, 2340–2361.

Gillies, C.S., Hebblewhite, M., Nielsen, S.E., Krawchuk, M.A., Aldridge, C.L., Frair, J.L., Saher, D.J., Stevens, C.E. and Jerde, C.L. (2006). Application of random effects to the study of resource selection by animals. *J. Anim. Ecol.* **75**, 887–898.

Gimenez, O., Bonner, S. and King, R. (2009). WinBUGS for population ecologists: Bayesian modeling using Markov Chain Monte Carlo methods. In: *Modeling Demographic Processes in Marked Populations* (ed. Thomson, D.L., Cooch, E.G. and Conroy, M.J.), pp. 883–915. Berlin: Springer.

Giuggioli, L., Potts, J.R. and Harris, S. (2012). Predicting oscillatory dynamics in the movement of territorial animals. *J. R. Soc. Interface* **9**, 1529–1543.

Goldberg, A.D., Allis, C.D. and Bernstein, E. (2007). Epigenetics: a landscape takes shape. *Cell* **128**, 635–638.

Gonda, A., Herczeg, G. and Merilä, J. (2009). Adaptive brain size divergence in nine-spined sticklebacks (Pungitius pungitius)? *J. Evol. Biol.* **22**, 1721–1726.

Gonzalez, A., Ronce, O., Ferriere, R. and Hochberg, M.E. (2013). Evolutionary rescue: an emerging focus at the intersection between ecology and evolution. *Philos. Trans. R. Soc. Lond. B Biol. Sci.* **368**, 20120404.

Gotelli, N.J. (1999). Ecology: how do communities come together? *Science* **286**, 1684–1685.

Götzenberger, L., de Bello, F., Bråthen, K.A., Davison, J., Dubuis, A., Guisan, A., Lepš, J., Lindborg, R., Moora, M., Pärtel, M., et al. (2012). Ecological assembly rules in plant communities-approaches, patterns and prospects. *Biol. Rev.* **87**, 111–127.

Grafen, A. and Hails, R. (2002). *Modern Statistics for the Life Sciences*. Oxford: Oxford University Press.

Grant, V. (1966). Block inheritance of viability genes in plant species. *Am. Nat.* **100**, 591–601.

Green, J.L. and Ostling, A. (2003). Endemics–area relationships: the influence of species dominance and spatial aggregation. *Ecology* **84**, 3090–3097.

Grimm, V. (1999). Ten years of individual-based modelling in ecology: what have we learned and what could we learn in the future? *Ecol. Model.* **115**, 129–148.

Grimm, V. and Railsback, S.F. (2013). *Individual-Based Modelling in Ecology*. Princeton, NJ: Princeton University Press.

Griswold, C.K., Logsdon, B. and Gomulkiewicz, R. (2007). Neutral evolution of multiple quantitative characters: a genealogical approach. *Genetics* **176**, 455–466.

Guisan, A., Edwards, T.C. and Hastie, T. (2002). Generalized linear and generalized additive models in studies of species distributions: setting the scene. *Ecol. Model.* **157**, 89–100.

Guisan, A. and Thuiller, W. (2005). Predicting species distribution: offering more than simple habitat models. *Ecol. Lett.* **8**, 993–1009.

Guisan, A. and Zimmermann, N.E. (2000). Predictive habitat distribution models in ecology. *Ecol. Model.* **135**, 147–186.

Gurarie, E., Andrews, R.D. and Laidre, K.L. (2009). A novel method for identifying behavioural changes in animal movement data. *Ecol. Lett.* **12**, 395–408.

Gurarie, E., Bracis, C., Delgado, M., Meckley, T.D., Kojola, I. and Wagner, C.M. (2016). What is the animal doing? Tools for exploring behavioural structure in animal movements. *J. Anim. Ecol.* **85**, 69–84.

Gurarie, E. and Ovaskainen, O. (2011). Characteristic spatial and temporal scales unify models of animal movement. *Am. Nat.* **178**, 113–123.

Gurarie, E. and Ovaskainen, O. (2013). Towards a general formalization of encounter rates in ecology. *Theor. Ecol.* **6**, 189–202.

Guttal, V. and Couzin, I.D. (2010). Social interactions, information use, and the evolution of collective migration. *Proc. Natl. Acad. Sci. USA* **107**, 16,172–16,177.

Hadfield, J.D. (2010). MCMC Methods for multi-response generalized linear mixed models: the MCMCglmm R package. *J. Stat. Softw.* **33**, 1–22.

Haegeman, B. and Loreau, M. (2011). A mathematical synthesis of niche and neutral theories in community ecology. *J. Theor. Biol.* **269**, 150–165.

Haldane, J.B.S. (1932a). Can Evolution be Explained in Terms of Known Genetical Facts? *Proceedings of the 6th International Congress of Genetics* **1**, 185–189.

Haldane, J.B.S. (1932b). *The Causes of Evolution*. London: Longmans, Green.

Hamilton, W.D. and May, R.M. (1977). Dispersal in stable habitats. *Nature* **269**, 578–581.

Hanski, I. (1991). Single-species metapopulation dynamics: concepts, models and observations. *Biol. J. Linn. Soc. Lond.* **42**, 17–38.

Hanski, I. (1994). A practical model of metapopulation dynamics. *J. Anim. Ecol.* **63**, 151–162.

Hanski, I., Kuussaari, M. and Nieminen, M. (1994). Metapopulation structure and migration in the butterfly Melitaea cinxia. *Ecology* **75**, 747–762.

Hanski, I. (1997). Predictive and practical metapopulation models: the incidence function approach. In: *Spatial Ecology* (ed. Tilman, D. and Kareiva, P.), pp. 21–45. Princeton, NJ: Princeton University Press.

Hanski, I. (1999). *Metapopulation Ecology*. Oxford: Oxford University Press.

Hanski, I. (2001). Spatially realistic theory of metapopulation ecology. *Naturwissenschaften* **88**, 372–381.

Hanski, I. (2011). Eco-evolutionary spatial dynamics in the Glanville fritillary butterfly. *Proc. Natl. Acad. Sci. USA* **108**, 14,397–14,404.

Hanski, I. and Mononen, T. (2011). Eco-evolutionary dynamics of dispersal in spatially heterogeneous environments. *Ecol. Lett.* **14**, 1025–1034.

Hanski, I., Mononen, T. and Ovaskainen, O. (2011). Eco-evolutionary metapopulation dynamics and the spatial scale of adaptation. *Am. Nat.* **177**, 29–43.

Hanski, I. and Ovaskainen, O. (2000). The metapopulation capacity of a fragmented landscape. *Nature* **404**, 755–758.

Hanski, I., Zurita, G. a, Bellocq, M.I. and Rybicki, J. (2013). Species-fragmented area relationship. *Proc. Natl. Acad. Sci. USA* **110**, 12,715–12,720.

Hardin, G. (1959). The competitive exclusion principle. *Science* **131**, 1292–1297.

Hardy, G.H. (1908). Mendelian proportions in a mixed population. *Science* **28**, 49–50.

Harrison, P.J., Hanski, I. and Ovaskainen, O. (2011). Bayesian state-space modeling of metapopulation dynamics in the Glanville fritillary butterfly. *Ecol. Monogr.* **81**, 581–598.

Hartl, D.L. (2014). *Essential Genetics: A Genomics Perspective.* Burlington: Jones and Bartlett.

Hartl, D.L. and Clark, A.G. (1997). *Principles of Population Genetics.* Sunderland, MA: Sinauer Associates.

Hastings, A. (1980). Disturbance, coexistence, history, and competition for space. *Theor. Popul. Biol.* **18**, 363–373.

Hastings, A. (1996). *Population Biology: Concepts and Models.* New York: Springer-Verlag, New York.

Hastings, A. (2010). Timescales, dynamics, and ecological understanding. *Ecology* **91**, 3471–3480.

He, H.-Q., Mao, W.-G., Pan, D., Zhou, J.-Y., Chen, P.-Y. and Fung, W.K. (2014). Detection of parent-of-origin effects for quantitative traits using general pedigree data. *J. Genet.* **93**, 339–347.

He, F. and Hubbell, S.P. (2011). Species-area relationships always overestimate extinction rates from habitat loss. *Nature* **473**, 368–371.

He, F. and Legendre, P. (2002). Species diversity patterns derived from species-area models. *Ecology* **83**, 1185–1198.

Hebblewhite, M. and Haydon, D.T. (2010). Distinguishing technology from biology: a critical review of the use of GPS telemetry data in ecology. *Philos. Trans. R. Soc. Lond. B Biol. Sci.* **365**, 2303–2312.

Helm, A., Hanski, I. and Pärtel, M. (2006). Slow response of plant species richness to habitat loss and fragmentation. *Ecol. Lett.* **9**, 72–77.

Henderson, C.R. (1950). Estimation of genetic parameters. *Biometrics* **6**, 186–187.

Henderson, C.R. (1976). A simple method for computing the inverse of a numerator relationship matrix used in prediction of breeding values. *Biometrics* **32**, 69–83.

Henig, R.M. (2001). *The Monk in the Garden: The Lost and Found Genius of Gregor Mendel, the Father of Modern Genetics.* Mariner Books.

Henle, K., Davies, K.F., Kleyer, M., Margules, C. and Settele, J. (2004). Predictors of species sensitivity to fragmentation. *Biodivers. Conserv.* **13**, 207–251.

Herczeg, G., Gonda, A. and Merilä, J. (2009a). Evolution of gigantism in nine-spined sticklebacks. *Evolution* **63**, 3190–3200.

Herczeg, G., Gonda, A. and Merilä, J. (2009b). Predation mediated population divergence in complex behaviour of nine-spined stickleback (*Pungitius pungitius*). *J. Evol. Biol.* **22**, 544–552.

Higgins, S.I. and Cain, M.L. (2002). Spatially realistic plant metapopulation models and the colonization-competition trade-off. *J. Ecol.* **90**, 616–626.

Hilborn, R. and Mangel, M. (1997). *The Ecological Detective: Confronting Models with Data.* Princeton, NJ: Princeton University Press.

Hirzel, A.H. and Le Lay, G. (2008). Habitat suitability modelling and niche theory. *J. Appl. Ecol.* **45**, 1372–1381.

Hjelm, J. and Persson, L. (2001). Size-dependent attack rate and handling capacity: intercohort competition in a zooplanktivorous fish. *Oikos* **95**, 520–532.

Hokkanen, P.J., Kouki, J. and Komonen, A. (2009). Nestedness, SLOSS and conservation networks of boreal herb-rich forests. *Appl. Veg. Sci.* **12**, 295–303.

Holden, C. (2006). Inching toward movement ecology. *Science* **313**, 779–782.

Holling, C.S. (1959). Some characteristics of simple types of predation and parasitism. *Canad. Entomol.* **91**, 385–398.

Holt, R.D., Lawton, J.H., Polis, G.A. and Martinez, N.D. (1999). Trophic rank and the species-area relationship. *Ecology* **80**, 1495–1504.

Holyoak, M., Casagrandi, R., Nathan, R., Revilla, E. and Spiegel, O. (2008). Trends and missing parts in the study of movement ecology. *Proc. Natl. Acad. Sci. USA* **105**, 19,114–19,119.

Hsu, S.B. (1980). A competition model for a seasonally fluctuating nutrient. *J. Math. Biol.* **9**, 115–132.

Hubbell, S.P. (2001). *The Unified Neutral Theory of Biodiversity and Biogeography*. Princeton, NJ: Princeton University Press.

Hui, F.K.C., Warton, D.I., Foster, S.D. and Dunstan, P.K. (2013). To mix or not to mix: comparing the predictive performance of mixture models vs. separate species distribution models. *Ecology* **94**, 1913–1919.

Hurford, A., Cownden, D. and Day, T. (2010). Next-generation tools for evolutionary invasion analyses. *J. R. Soc. Interface* **7**, 561–571.

Illian, J., Penttinen, A., Stoyan, H. and Stoyan, D. (2007). *Statistical Analysis and Modelling of Spatial Point Patterns*. London: Wiley.

Ives, A.R. and Helmus, M.R. (2010). Phylogenetic metrics of community similarity. *Am. Nat.* **176**, E128–E142.

Jabot, F. and Etienne, R.S. (2008). Reconciling neutral community models and environmental filtering: theory and an empirical test. *Oikos* **117**, 1308–1320.

Johnson, A.R., Wiens, J.A., Milne, B.T. and Crist, T.O. (1992). Animal movements and population dynamics in heterogeneous landscapes. *Landsc. Ecol.* **7**, 63–75.

Johnson, D.J., Bourg, N.A., Howe, R., McSHea, W.J. and Wolf, A. (2014). Conspecific negative density-dependent mortality and the structure of temperate forests. *Ecology* **95**, 2493–2503.

Johstron, S.E., McEwan, J.C., Pickering, N.K., Kijas, J.W., Beraldi, D., Pilkington, J.G., Pemberton, J.M. and Slate, J. (2011). Genome-wide association mapping identifies the genetic basis of discrete and quantitative variation in sexual weaponry in a wild sheep population. *Mol. Ecol.* **20**, 2555–2566.

Jones, A.G. and Ardren, W.R. (2003). Methods of parentage analysis in natural populations. *Mol. Ecol.* **12**, 2511–2523.

Jonsen, I.D., Basson, M., Bestley, S., Bravington, M. V., Patterson, T.A., Pedersen, M.W., Thomson, R., Thygesen, U.H. and Wotherspoon, S.J. (2013). State-space models for bio-loggers: A methodological road map. *Deep Sea Research II* **88–89**, 34–46.

Jost, L. (2007). Partitioning diversity into independent alpha and beta components. *Ecology* **88**, 2427–2439.

Kareiva, P., Mullen, A. and Southwood, R. (1990). Population dynamics in spatially complex environments: theory and data [and discussion]. *Philos. Trans. R. Soc. Lond. B Biol. Sci.* **330**, 175–190.

Kareiva, P.M. and Shigesada, N. (1983). Analyzing insect movement as a correlated random walk. *Oecologia* **56**, 234–238.

Karhunen, M., Merilä, J., Leinonen, T., Cano, J.M. and Ovaskainen, O. (2013). DRIFTSEL: an R package for detecting signals of natural selection in quantitative traits. *Mol. Ecol. Resources* **13**, 746–54.

Karhunen, M., Ovaskainen, O., Herczeg, G. and Merilä, J. (2014). Bringing habitat information into statistical tests of local adaptation in quantitative traits: a case study of nine-spined sticklebacks. *Evolution* **68**, 559–568.

Karhunen, M. and Ovaskainen, O. (2012). Estimating population-level coancestry coefficients by an admixture F model. *Genetics* **192**, 609–617.

Kass, R.E. and Wasserman, L. (1996). The selection of prior distributions by formal rules. *J. Am. Stat. Assoc.* **91**, 1343–1370.

Kearney, M., Simpson, S.J., Raubenheimer, D. and Helmuth, B. (2010). Modelling the ecological niche from functional traits. *Philos. Trans. R. Soc. Lond. B Biol. Sci.* **365**, 3469–3483.

Kearney, M., Simpson, S.J., Raubenheimer, D. and Kooijman, S.A.L.M. (2013). Balancing heat, water and nutrients under environmental change: a thermodynamic niche framework. *Funct. Ecol.* **27**, 950–966.

Kearney, M. and Porter, W.P. (2006). Ecologists have already started rebuilding community ecology from functional traits. *Trends Ecol. Evol.* **21**, 481–482.

Kearney, M. and Porter, W.P. (2009). Mechanistic niche modelling: combining physiological and spatial data to predict species' ranges. *Ecol. Lett.* **12**, 334–350.

Keeling, M.J., Wilson, H. B. and Pacala, S.W. (2002). Deterministic limits to stochastic spatial models of natural enemies. *Am. Nat.* **159**(1), 57–80.

Keller, L.F. and Waller, D.M. (2002). Inbreeding effects in wild populations. *Trends Ecol. Evol.* **17**, 230–241.

Kendall, B.E., Briggs, C.J., Murdoch, W.W., Turchin, P., Ellner, S.P., McCauley, E., Nisbet, R.M. and Wood, S.N. (1999). Why do population cycle? A synthesis of statistical and mechanistic modeling approaches. *Ecology* **80**, 1789–1805.

Kenward, R.E., Marcstrom, V. and Karlbom, M. (1999). Demographic estimates from radio-tagging: models of age-specific survival and breeding in the goshawk. *J. Anim. Ecol.* **68**, 1020–1033.

Keymer, J.E., Marquet, P.A., Velasco-herna, J.X., Levin, S.A., Velasco-Hernández, J.X. and Simon, A. (2000). Extinction thresholds and metapopulation persistence in dynamic landscapes. *Am. Nat.* **156**, 478–494.

Kie, J.G., Matthiopoulos, J., Fieberg, J., Powell, R.A., Cagnacci, F., Mitchell, M.S., Gaillard, J.-M. and Moorcroft, P.R. (2010). The home-range concept: are traditional estimators still relevant with modern telemetry technology? *Philos. Trans. R. Soc. Lond. B Biol. Sci.* **365**, 2221–2231.

Kim, S.-Y., Torres, R. and Drummond, H. (2009). Simultaneous positive and negative density-dependent dispersal in a colonial bird species. *Ecology* **90**, 230–239.

Kimura, M. and Ohta, T. (1969). The average number of generations until fixation of a mutant gene in a finite population. *Genetics* **61**, 763–771.

Kisdi, E. and Geritz, S.A.H. (2010). Adaptive dynamics: a framework to model evolution in the ecological theatre. *J. Math. Biol.* **61**, 165–169.

Klaassen, R.H.G., Nolet, B.A. and Bankert, D. (2006). Movement of foraging Tundra Swans explained by spatial pattern in cryptic food densities. *Ecology* **87**, 2244–2254.

Klingenberg, C.P. (2004). Integration and modularity of quantitative trait locus effects on geometric shape in the mouse mandible. *Genetics* **166**, 1909–1921.

de Knegt, H.J., Hengeveld, G.M., van Langevelde, F., de Boer, W.F. and Kirkman, K.P. (2007). Patch density determines movement patterns and foraging efficiency of large herbivores. *Behav. Ecol.* **18**, 1065–1072.

de Knegt, H.J., van Langevelde, F., Coughenour, M.B., Skidmore, A.K., de Boer, W.F., Heitkönig, I.M.A., Knox, N.M., Slotow, R., van der Waal, C. and Prins, H.H.T. (2010). Spatial autocorrelation and the scaling of species-environment relationships. *Ecology* **91**, 2455–2465.

Koehl, M.A.R. (1989). From individuals to populations. In: *Perspectives in Ecological Theory* (ed. Roughgarden, J., May, R. and Levin, S.), pp. 39–53. Princeton, NJ: Princeton University Press.

Kooijman, S.A.L.M. (2000). *Dynamic Energy and Mass Budgets in Biological Systems*. Cambridge: Cambridge University Press.

Kool, J.T., Moilanen, A. and Treml, E.A. (2012). Population connectivity: recent advances and new perspectives. *Landsc. Ecol.* **28**, 165–185.

Van de Koppel, J., Altieri, A.H., Silliman, B.R., Bruno, J.F. and Bertness, M.D. (2006). Scale-dependent interactions and community structure on cobble beaches. *Ecol. Lett.* **9**, 45–50.

Korpela, K., Helle, P., Henttonen, H., Korpimäki, E., Sundell, J., Koskela, E., Ovaskainen, O., Pietiäinen, H., Valkama, J. and Huitu, O. (2014). Predator – vole interactions in northern Europe: the role of small mustelids revised. *Proc. R. Soc. B* **281**, 20142119.

Korpimäki, E. and Norrdahl, K. (1989). Predation of Tengmalm's owls: numerical responses, functional responses and dampening impact on population fluctuations of microtines. *Oikos* **54**, 154–164.

Kraft, N.J.B., Adler, P.B., Godoy, O., James, E.C., Fuller, S. and Levine, J.M. (2015). Community assembly, coexistence and the environmental filtering metaphor. *Funct. Ecol.* **29**, 592–599.

Krause, J., Krause, S., Arlinghaus, R., Psorakis, I., Roberts, S. and Rutz, C. (2013). Reality mining of animal social systems. *Trends Ecol. Evol.* **28**, 541–551.

Krause, J. and Ruxton, G.D. (2002). *Living in Groups*. Oxford: Oxford University Press.

Kruuk, L.E.B. (2004). Estimating genetic parameters in natural populations using the 'animal model'. *Philos. Trans. R. Soc. Lond. B Biol. Sci.* **359**, 873–890.

Kuang, Y. (1993). *Delay Differential Equations: With Applications in Population Dynamics*. San Diego, CA: Academic Press.

Kuussaari, M., Nieminen, M. and Hanski, I. (1996). An experimental study of migration in the Glanville fritillary butterfly Melitaea cinxia. *J. Anim. Ecol.* **65**, 791–801.

Lachance, J. and Tishkoff, S.A. (2013). Population genomics of human adaptation. *Annu. Rev. Ecol. Evol. System.* **44**, 123–143.

Lambin, X., Krebs, C.J., Moss, R. and Yoccoz, N.G. (2002). Population cycles: inferences from experimental, modeling, and time series approaches. In: *Population Cycles: The Case for Trophic Interactions* (ed. Berryman, A.A.), pp. 155–176. Oxford: Oxford University Press.

Lande, R. (1987). Extinction thresholds in demographic models of territorial populations. *Am. Nat.* **130**, 624–635.

Lande, R. (1993). Risks of population extinction from demographic and environmental stochasticity and random catastrophes. *Am. Nat.* **142**, 911–927.

Lande, R. (1996). Statistics of partitioning species diversity, and similarity among multiple communities. *Oikos* **76**, 5–13.

Lande, R., Engen, S. and Saether, B.-E. (2003). *Stochastic Population Dynamics in Ecology and Conservation*. Oxford: Oxford University Press.

Lande, R. and Arnold, S.J. (1983). The measurement of selection on correlated characters. *Evolution* **37**, 1210–1226.

Law, R., Murrell, D.J. and Dieckmann, U. (2003). Population growth in space and time: spatial logistic equations. *Ecology* **84**, 252–262.

Lawson, D. and Jensen, H.J. (2006). The species-area relationship and evolution. *J. Theor. Biol.* **241**, 590–600.

Legendre, P. and Legendre, L. (2012). *Numerical Ecology*. London: Elsevier.

Leibold, M.A. (1995). The niche concept revisited: mechanistic models and community context. *Ecology* **76**, 1371–1382.

Leibold, M.A. (1998). Similarity and local co-existence of species in regional biotas. *Evol. Ecol.* **12**, 95–110.

Leibold, M.A., Holyoak, M., Mouquet, N., Amarasekare, P., Chase, J.M., Hoopes, M.F., Holt, R.D., Shurin, J.B., Law, R., Tilman, D., et al. (2004). The metacommunity concept: a framework for multi-scale community ecology. *Ecol. Lett.* **7**, 601–613.

Leibold, M.A. and Mcpeek, M.A. (2006). Coexistence of the niche and neutral perspectives in community ecology. *Ecology* **87**, 1399–1410.

Lele, S.R., Merrill, E.H., Keim, J.L. and Boyce, M.S. (2013). Selection, use, choice and occupancy: clarifying concepts in resource selection studies. *J. Anim. Ecol.* **82**, 1183–1191.

Levin, S.A., Muller-landau, H.C., Nathan, R., Chave, J. and Levin, A. (2003). The ecology and evolution of seed dispersal: a theoretical perspective. *Annu. Rev. Ecol. Evol. System.* **34**, 575–604.

Levine, J.M. and Rees, M. (2002). Coexistence and relative abundance in annual plant assemblages: the roles of competition and colonization. *Am. Nat.* **160**, 452–467.

Levins, R. (1969). Some demographic and genetic consequences of environmental heterogeneity for biological control. *Bull. Entomol. Soc. Am.* **15**, 237–240.

Levins, R. (1970). Extinction. In: *Some mathematical problems in Biology* (ed. Gerstenhaber, M.). Province, RI: American Mathematical Society.

Levins, R. and Culver, D. (1971). Regional coexistence of species and competition between rare species. *Proc. Natl. Acad. Sci. USA* **68**, 1246–1248.

Li, R., Zhu, H., Ruan, J., Qian, W., Fang, X., Shi, Z., Li, Y., Li, S., Shan, G., Kristiansen, K., et al. (2010). De novo assembly of human genomes with massively parallel short read sequencing. *Genome Res.* **20**, 265–272.

Li, J., Fenton, A., Kettley, L., Roberts, P. and Montagnes, D.J.S. (2013). Reconsidering the importance of the past in predator-prey models: both numerical and functional responses depend on delayed prey densities. *Proc. R. Soc. B* **280**, 20131389.

Liberal, I.M., Burrus, M., Suchet, C., Thébaud, C. and Vargas, P. (2014). The evolutionary history of Antirrhinum in the Pyrenees inferred from phylogeographic analyses. *BMC Evol. Biol.* **14**, 146.

Liebhold, A., Koenig, W.D. and Bjørnstad, O.N. (2004). Spatial Synchrony in Population Dynamics. *Annu. Rev. Ecol. Evol. System.* **35**, 467–490.

Lindenmayer, D.B. and Fischer, J. (2007). Tackling the habitat fragmentation panchreston. *Trends Ecol. Evol.* **22**, 127–132.

Loarie, S.R., Duffy, P.B., Hamilton, H., Asner, G.P., Field, C.B. and Ackerly, D.D. (2009). The velocity of climate change. *Nature* **462**, 1052–1055.

Logue, J.B., Mouquet, N., Peter, H. and Hillebrand, H. (2011). Empirical approaches to metacommunities: a review and comparison with theory. *Trends Ecol. Evol.* **26**, 482–491.

Loman, N.J., Misra, R. V., Dallman, T.J., Constantinidou, C., Gharbia, S.E., Wain, J. and Pallen, M.J. (2012). Performance comparison of benchtop high-throughput sequencing platforms. *Nat. Biotechnol.* **30**, 434–439.

Lotka, A.J. (1932). The growth of mixed populations: two species competing for a common food supply. *J. Washington Acad. Sci.* **22**, 461–469.

Lukacs, P.M. and Burnham, K.P. (2005). Review of capture-recapture methods applicable to noninvasive genetic sampling. *Mol. Ecol.* **14**, 3909–3919.

Lush, J.L. (1943). *Animal Breeding Plans*. Ames, IA: Iowa State College Press.

Lynch, M. and Walsh, B. (1998). *Genetics and Analysis of Quantitative Traits*. Sunderland, MA: Sinauer Associates.
MacArthur, R.H. and Levins, R. (1967). The limiting similarity, convergence, and divergence of coexisting species. *Am. Nat.* **101**, 377–387.
Mahalanobis, P.C. (1936). On the generalised distance in statistics. *Proc. Natl. Inst. Sci. India* **2**, 49–55.
Mangel, M. (2006). *The Theoretical Biologist's Toolbox: Quantitative Methods for Ecology and Evolutionary Biology*. New York: Cambridge University Press.
Manly, B.F.J., McDonald, L.L. and Thomas, D.L. (1993). *Resource Selection by Animals. Statistical Design and Analysis for Field Studies*. Lond: Chapman and Hall.
Martin, G., Chapuis, E. and Goudet, J. (2008). Multivariate QST-FST comparisons: a neutrality test for the evolution of the g matrix in structured populations. *Genetics* **180**, 2135–2149.
Martin, J., van Moorter, B., Revilla, E., Blanchard, P., Dray, S., Quenette, P.-Y., Allainé, D. and Swenson, J.E. (2013). Reciprocal modulation of internal and external factors determines individual movements. *J. Anim. Ecol.* **82**, 290–300.
Massot, M., Clobert, J., Pilorge, T., Lecomte, J. and Barbault, R. (2014). Density dependence in the common lizard: demographic consequences of a density manipulation. *Ecology* **73**, 1742–1756.
Mattila, A.L.K., Duplouy, A., Kirjokangas, M., Lehtonen, R., Rastas, P. and Hanski, I. (2012). High genetic load in an old isolated butterfly population. *Proc. Natl. Acad. Sci. USA* **109**, E2496–E2505.
May, R.M. and Hassell, M.P. (1981). The dynamics of multi- parasitoid host interactions. *Am. Nat.* **117**, 234–261.
McClintock, B.T., Johnson, D.S., Hooten, M.B., Ver Hoef, J.M. and Morales, J.M. (2014). When to be discrete: the importance of time formulation in understanding animal movement. *Mov. Ecol.* **2**, 21.
McClintock, B. and King, R. (2012). A general discrete-time modeling framework for animal movement using multistate random walks. *Ecol. Monogr.* **82**, 335–349.
McDonald, T.L. (2013). The point process use-availability or presence-only likelihood and comments on analysis. *J. Anim. Ecol.* **82**, 1174–1182.
McIntire, E.J.B. and Fajardo, A. (2009). Beyond description: the active and effective way to infer processes from spatial patterns. *Ecology* **90**, 46–56.
McMullen, M.D., Kresovich, S., Villeda, H.S., Bradbury, P., Li, H., Sun, Q., Flint-Garcia, S., Thornsberry, J., Acharya, C., Bottoms, C., et al. (2009). Genetic properties of the maize nested association mapping population. *Science* **325**, 737–740.
McNamara, J.M., Barta, Z., Klaassen, M. and Bauer, S. (2011). Cues and the optimal timing of activities under environmental changes. *Ecol. Lett.* **14**, 1183–1190.
McNamara, J.M. and Houston, A.I. (2008). Optimal annual routines: behaviour in the context of physiology and ecology. *Philos. Trans. R. Soc. Lond. B Biol. Sci.* **363**, 301–319.
Mcpeek, M.A. and Holt, R.D. (1992). The evolution of dispersal in spatially and temporally varying environments. *Am. Nat.* **140**, 1010–1027.
McRae, B.H. (2006). Isolation by resistance. *Evolution* **60**, 1551–1561.
McRae, B.H., Dickson, B.G., Keitt, T.H. and Shah, V.B. (2008). Using circuit theory to model connectivity in ecology, evolution, and conservation. *Ecology* **89**, 2712–2724.
Merilä, J. (2013). Nine-spined stickleback (Pungitius pungitius): An emerging model for evolutionary biology research. *Ann. N.Y. Acad. Sci.* **1289**, 18–35.
Metcalfe, A. V. and Cowpertwait, P. (2009). *Introductory Time Series with R*. New York: Springer.

Meyer, K. (1985). Maximum likelihood estimation of variance components for a multivariate mixed model with equal design matrices. *Biometrics* **41**, 153–165.

Millspaugh, J.J. and Marzluff, J.M. (2001). *Radio Tracking and Animal Populations*. San Diego, CA: Academic Press.

Minasny, B. and McBratney, A.B. (2005). The Matérn function as a general model for soil variograms. *Geoderma* **128**, 192–207.

Moilanen, A. (1999). Patch occupancy models of metapopulation dynamics: efficient parameter estimation using implicit statistical inference. *Ecology* **80**, 1031–1043.

Moilanen, A. (2011). On the limitations of graph-theoretic connectivity in spatial ecology and conservation. *J. Appl. Ecol.* **48**, 1543–1547.

Moilanen, A. and Hanski, I. (1995). Habitat destruction and coexistence of competitors in a spatially realistic metapopulation model. *J. Anim. Ecol.* **64**, 141–144.

Moilanen, A. and Nieminen, M. (2002). Simple connectivity measures in spatial ecology. *Ecology* **83**, 1131–1145.

Molofsky, J. and Ferdy, J.-B. (2005). Extinction dynamics in experimental metapopulations. *Proc. Natl. Acad. Sci. USA* **102**, 3726–3731.

Moorcroft, P.R., Lewis, M.A. and Crabtree, R.L. (1999). Home range analysis using a mechanistic home range model. *Ecology* **80**, 1656–1665.

Moorcroft, P.R., Lewis, M.A. and Crabtree, R.L. (2006). Mechanistic home range models capture spatial patterns and dynamics of coyote territories in Yellowstone. *Proc. R. Soc. B* **273**, 1651–1659.

Moorcroft, P.R. and Barnett, A.H. (2008). Mechanistic home range models and resource selection analysis: a reconciliation and unification. *Ecology* **89**, 1112–1119.

Moorcroft, P.R. and Lewis, M.A. (2006). *Mechanistic Home Range Analysis*. Princeton, NJ: Princeton University Press.

Morales, J.M. (2002). Behavior at habitat boundaries can produce leptokurtic movement distributions. *Am. Nat.* 160: 531–538.

Morales, J.M. and Carlo, T.A. (2006). The effects of plant distribution and frugivore density on the scale and shape of dispersal kernels. *Ecology* **87**, 1489–1496.

Morin, P.A., Luikart, G. and Wayne, R.K. (2004). SNPs in ecology, evolution and conservation. *Trends Ecol. Evol.* **19**, 208–216.

Morton, A.C. (1982). The effects of marking and capture on recapture frequencies of butterflies. *Oecologia* **53**, 105–110.

Mouquet, N. and Loreau, M. (2002). Coexistence in metacommunities: the regional similarity hypothesis. *Am. Nat.* **159**, 420–426.

Mouquet, N. and Loreau, M. (2003). Community patterns in source–sink metacommunities. *Am. Nat.* **162**, 544–557.

Mueller, T. and Fagan, W.F. (2008). Search and navigation in dynamic environments—from individual behaviors to population distributions. *Oikos* **117**, 654–664.

Murdoch, W.W., Briggs, C.J. and Nisbet, R.M. (2003). *Consumer–Resource Dynamics*. Princeton, NJ: Princeton University Press.

Murrell, D.J., Dieckmann, U. and Law, R. (2004). On moment closures for population dynamics in continuous space. *J. Theor. Biol.* **229**, 421–432.

Mustonen, V. and Lässig, M. (2009). From fitness landscapes to seascapes: non-equilibrium dynamics of selection and adaptation. *Trends Genet.* 25, 111–119.

Mutshinda, C.M., O'Hara, R.B. and Woiwod, I.P. (2009). What drives community dynamics? *Proc.R. Soc. B* **276**, 2923–2929.

Myles, H. and Wolfe, D.A. (1999). *Nonparametric Statistical Methods*. London: Wiley.

Nathan, R., Getz, W.M., Revilla, E., Holyoak, M., Kadmon, R., Saltz, D. and Smouse, P.E. (2008). A movement ecology paradigm for unifying organismal movement research. *Proc. Natl. Acad. Sci. USA* **105**, 19,052–19,059.

Nathan, R. and Giuggioli, L. (2013). A milestone for movement ecology research. *Mov. Ecol.* **1**, 1.

Nei, M. (1975). *Molecular Population Genetics and Evolution*. Dordrecht: North Holland.

Newman, T.J., Ferdy, J.-B. and Quince, C. (2004). Extinction times and moment closure in the stochastic logistic process. *Theor. Popul. Biol.* **65**, 115–126.

Newton, I. (1989). *Lifetime Reproduction in Birds*. San Diego, CA: Academic Press.

Norden, R. (1982). On the distribution of the time to extinction in the stochastic logistic population model. *Adv. Appl. Probab.* **14**, 687–708.

North, A., Cornell, S. and Ovaskainen, O. (2011). Evolutionary responses of dispersal distance to landscape structure and habitat loss. *Evolution* **65**, 1739–1751.

Nouvellet, P., Bacon, J.P. and Waxman, D. (2009). Fundamental insights into the random movement of animals from a single distance-related statistic. *Am. Nat.* **174**, 506–14.

O'Connell, A.F., Nichols, J.D. and Karanth, K.U. (2011). *Camera Traps in Animal Ecology: Methods and Analyses* (ed. O'Connell, A.F., Nichols, J.D. and Karanth, K.U.). New York: Springer.

Okubo, A. and Levin, S.A. (2001). *Diffusion and Ecological Problems: Modern Perspectives*. New York: Springer.

Otto, S.P. and Day, T. (2007). *A Biologist's Guide to Mathematical Modeling in Ecology and Evolution*. Princeton, NJ: Princeton University Press.

Ovaskainen, O. (2001). The quasistationary distribution of the stochastic logistic model. *J. Appl. Probab.* **38**, 898–907.

Ovaskainen, O. (2002). Long-term persistence of species and the SLOSS problem. *J. Theor. Biol.* **218**, 419–433.

Ovaskainen, O. (2004). Habitat-specific movement parameters estimated using mark-recapture data and a diffusion model. *Ecology* **85**, 242–257.

Ovaskainen, O. (2008). Analytical and numerical tools for diffusion-based movement models. *Theor. Popul. Biol.* **73**, 198–211.

Ovaskainen, O. and Cornell, S.J. (2003). Biased movement at a boundary and conditional occupancy times for diffusion processes. *J. Appl. Probab.* **40**, 557–580.

Ovaskainen, O. and Cornell, S.J. (2006a). Asymptotically exact analysis of stochastic metapopulation dynamics with explicit spatial structure. *Theor. Popul. Biol.* **69**, 13–33.

Ovaskainen, O. and Cornell, S.J. (2006b). Space and stochasticity in population dynamics. *Proc. Natl. Acad. Sci. USA* **103**, 12,781–12,786.

Ovaskainen, O. and Crone, E.E. (2009). Modeling animal movement with diffusion. In: *Spatial Ecology* (ed. Cosner, C., Cantrell, S. and Ruan, S.), pp. 63–83. London: Chapman and Hall/CRC.

Ovaskainen, O., Finkelshtein, D., Kutoviy, O., Cornell, S.J., Bolker, B.M. and Kondratiev, Y. (2014). A general mathematical framework for the analysis of spatiotemporal point processes. *Theor. Ecol.* **7**, 101–113.

Ovaskainen, O. and Hanski, I. (2001). Spatially structured metapopulation models: global and local assessment of metapopulation capacity. *Theor. Popul. Biol.* **60**, 281–302.

Ovaskainen, O. and Hanski, I. (2002). Transient dynamics in metapopulation response to perturbation. *Theor. Popul. Biol.* **61**, 285–295.

Ovaskainen, O. and Hanski, I. (2003a). Extinction threshold in metapopulation models. *Annu. Zool. Finnici* **40**, 81–97.

Ovaskainen, O. and Hanski, I. (2003b). How much does an individual habitat fragment contribute to metapopulation dynamics and persistence? *Theor. Popul. Biol.* **64**, 481–495.

Ovaskainen, O. and Hanski, I. (2004a). Metapopulation dynamics in highly fragmented landscapes. In: *Ecology, Genetics, and Evolution in Metapopulations* (ed. Hanski, I. and Gaggiotti, O.), pp. 73–103. San Diego, CA: Academic Press.

Ovaskainen, O. and Hanski, I. (2004b). From individual behavior to metapopulation dynamics: unifying the patchy population and classic metapopulation models. *Am. Nat.* **164**, 364–377.

Ovaskainen, O., Karhunen, M., Zheng, C., Arias, J.M.C. and Merilä, J. (2011). A new method to uncover signatures of divergent and stabilizing selection in quantitative traits. *Genetics* **189**, 621–632.

Ovaskainen, O., Luoto, M., Ikonen, I., Rekola, H., Meyke, E. and Kuussaari, M. (2008c). An empirical test of a diffusion model: predicting clouded apollo movements in a novel environment. *Am. Nat.* **171**, 610–619.

Ovaskainen, O. and Meerson, B. (2010). Stochastic models of population extinction. *Trends Ecol. Evol.* **25**, 643–652.

Ovaskainen, O., Rekola, H., Meyke, E. and Arjas, E. (2008b). Bayesian methods for analyzing movements in heterogeneous landscapes from mark-recapture data. *Ecology* **89**, 542–554.

Ovaskainen, O., Smith, A.D., Osborne, J.L., Reynolds, D.R., Carreck, N.L., Martin, A.P., Niitepõld, K. and Hanski, I. (2008a). Tracking butterfly movements with harmonic radar reveals an effect of population age on movement distance. *Proc. Natl. Acad. Sci. USA* **105**, 19,090–19,095.

Ovaskainen, O. and Soininen, J. (2011). Making more out of sparse data: hierarchical modeling of species communities. *Ecology* **92**, 289–295.

Painter, K.J. (2013). Multiscale models for movement in oriented environments and their application to hilltopping in butterflies. *Theor. Ecol.* **7**, 53–75.

Patlak, C.C.S. (1953a). Random walk with persistence and external bias. *Bull. Math. Biophys.* **15**, 311–338.

Patlak, C.S. (1953b). A mathematical contribution to the study of orientation of organisms. *Bull. Math. Biophys.* **15**, 431–476.

Patterson, T.A., Thomas, L., Wilcox, C., Ovaskainen, O. and Matthiopoulos, J. (2008). State-space models of individual animal movement. *Trends Ecol. Evol.* **23**, 87–94.

Pelletier, F., Garant, D. and Hendry, A.P. (2009). Eco-evolutionary dynamics. *Philos. Trans. R. Soc. Lond. B Biol. Sci.* **364**, 1483–1489.

Penteriani, V. and Delgado, M.M. (2011). Birthplace-dependent dispersal: are directions of natal dispersal determined a priori? *Ecography* **34**, 729–737.

Péron, G., Crochet, P.-A., Doherty, P.F. and Lebreton, J.-D. (2010). Studying dispersal at the landscape scale: efficient combination of population surveys and capture-recapture data. *Ecology* **91**, 3365–75.

van der Plas, F., Janzen, T., Ordonez, A., Fokkema, W., Reinders, J., Etienne, R.S. and Olff, H. (2015). A new modeling approach estimates the relative importance of different community assembly processes. *Ecology* **96**, 1502–1515.

Polansky, L., Wittemyer, G., Cross, P.C., Tambling, C.J. and Getz, W.M. (2010). From moonlight to movement and synchronized randomness: Fourier and wavelet analyses of animal location time series data. *Ecology* **91**, 1506–1518.

Pollock, L.J., Tingley, R., Morris, W.K., Golding, N., O'Hara, R.B., Parris, K.M., Vesk, P.A. and Mccarthy, M.A. (2014). Understanding co-occurrence by modelling species simultaneously with a Joint Species Distribution Model (JSDM). *Methods Ecol. Evol.* **5**, 397–406.

Poole, D. (2014). *Linear Algebra: A Modern Introduction*. Cengage Learning.

Post, D.M. and Palkovacs, E.P. (2009). Eco-evolutionary feedbacks in community and ecosystem ecology: interactions between the ecological theatre and the evolutionary play. *Philos. Trans. R. Soc. Lond. B Biol. Sci.* **364**, 1629–1640.

Postlethwaite, C.M., Brown, P. and Dennis, T.E. (2013). A new multi-scale measure for analysing animal movement data. *J. Theor. Biol.* **317**, 175–185.

Potts, J.R., Bastille-Rousseau, G., Murray, D.L., Schaefer, J.A. and Lewis, M.A. (2014). Predicting local and non-local effects of resources on animal space use using a mechanistic step selection model. *Methods Ecol. Evol.* **5**, 253–262.

Powell, T.M. and Steele, J.H. (1995). *Ecological Time Series*. Berlin: Springer.

Provine, W.B. (1971). *The Origins of Theoretical Population Genetics*. Chicago: University of Chicago Press.

Pulliam, H.R. (1988). Sources, sinks, and population regulation. *Am. Nat.* **132**, 652–661.

Queller, D.C. (1992). Quantitative genetics, inclusive fitness, and group selection. *Am. Nat.* **139**, 540–558.

Ramenofsky, M. and Wingfield, J.C. (2007). Regulation of migration. *BioScience* **57**, 135–143.

Ranta, E.S.A., Kaitala, V. and Lindstrom, J.A.N. (1995). Synchrony in population dynamics. *Proc. R. Soc. B* **262**, 113–118.

Ranta, E., Lundberg, P. and Kaitala, V. (2006). *Ecology of Populations*. Cambridge: Cambridge University Press.

Rastas, P., Paulin, L., Hanski, I., Lehtonen, R. and Auvinen, P. (2013). Lep-MAP: fast and accurate linkage map construction for large SNP datasets. *Bioinformatics* **29**, 3128–3134.

Rayfield, B., Fortin, M.-J. and Fall, A. (2011). Connectivity for conservation: a framework to classify network measures. *Ecology* **92**, 847–858.

Renner, I.W., Elith, J., Baddeley, A., Fithian, W., Hastie, T., Phillips, S.J., Popovic, G. and Warton, D.I. (2015). Point process models for presence-only analysis. *Methods Ecol. Evol.* **6**, 366–379.

Ricker, W.E. (1958). *Handbook of Computations for Biological Statistics of Fish Populations*. Vancouver: Fisheries Research Board of Canada.

Roff, D.A. (1975). Population stability and the evolution of dispersal in a heterogeneous environment. *Oecologia* **19**, 217–237.

Roff, D.A. (1997). *Evolutionary Quantitative Genetics*. London: Chapman and Hall.

Roff, D.A. (2000). The evolution of the G matrix: selection or drift? *Heredity* **84**, 135–142.

Roland, J., Keyghobadi, N. and Fownes, S. (2000). Alpine parnassius butterfly dispersal: effects of landscape and population size. *Ecology* **81**, 1642–1653.

Ronce, O. (2007). How does it feel to be like a rolling stone? Ten questions about dispersal evolution. *Annu. Rev. Ecol. Evol. System.* **38**, 231–253.

De Roos, A.M. and Persson, L. (2001). Physiologically structured models: from versatile technique to ecological theory. *Oikos* **94**, 51–71.

Root, R.B. and Kareiva, P.M. (1984). The search for resources by cabbage butterflies (*Pieris rapae*): ecological consequences and adaptive significance of markovian movements in a patchy environment. *Ecology* **65**, 147–165.

Rosindell, J., Hubbell, S.P. and Etienne, R.S. (2011). The unified neutral theory of biodiversity and biogeography at age ten. *Trends Ecol. Evol.* **26**, 340–348.

Rosindell, J., Hubbell, S.P., He, F., Harmon, L.J. and Etienne, R.S. (2012). The case for ecological neutral theory. *Trends Ecol. Evol.* **27**, 203–208.

Rosindell, J. and Cornell, S.J. (2009). Species-area curves, neutral models, and long-distance dispersal. *Ecology* **90**, 1743–1750.

Rowcliffe, J.M., Carbone, C., Kays, R., Kranstauber, B. and Jansen, P.A. (2012). Bias in estimating animal travel distance: the effect of sampling frequency. *Methods Ecol. Evol.* **3**, 653–662.

Royall, R.M. (1986). Model robust confidence intervals using maximum likelihood estimators. *Int. Stat. Rev.* **54**, 221–226.

Rue, H., Martino, S. and Chopin, N. (2009). Approximate Bayesian inference for latent Gaussian models by using integrated nested Laplace approximations. *J. R. Stat. Soc. B* **71**, 319–392.

Sabeti, P.C., Reich, D.E., Higgins, J.M., Levine, H.Z.P., Richter, D.J., Schaffner, S.F., Gabriel, S.B., Platko, J. V., Patterson, N.J., McDonald, G.J., et al. (2002). Detecting recent positive selection in the human genome from haplotype structure. *Nature* **419**, 832–837.

Saccheri, I., Kuussaari, M., Kankare, M., Vikman, P., Fortelius, W. and Hanski, I. (1998). Inbreeding and extinction in a butterfly metapopulation. *Nature* **392**, 491–494.

Sachidanandam, R., Weissman, D., Schmidt, S.C., Kakol, J.M., Stein, L.D., Marth, G., Sherry, S., Mullikin, J.C., Mortimore, B.J., Willey, D.L., et al. (2001). A map of human genome sequence variation containing 1.42 million single nucleotide polymorphisms. *Nature* **409**, 928–933.

Schick, R.S., Loarie, S.R., Colchero, F., Best, B.D., Boustany, A., Conde, D.A., Halpin, P.N., Joppa, L.N., McClellan, C.M. and Clark, J.S. (2008). Understanding movement data and movement processes: current and emerging directions. *Ecol. Lett.* **11**, 1338–1350.

Schmitt, T. and Seitz, A. (2002). Influence of habitat fragmentation on the genetic structure of Polyommatus coridon (Lepidoptera: Lycaenidae): implications for conservation. *Biol. Conserv.* **107**, 291–297.

Schoener, T.W. (1973). Population growth limited by intraspecific competition for energy of time: some simple representations. *Theor. Popul. Biol.* **4**, 56–84.

Schoener, T.W. (1976). Alternatives to Lotka-Volterra competition models of intermediate complexity. *Theor. Popul. Biol.* **10**, 309–333.

Schoville, S.D., Bonin, A., François, O., Lobreaux, S., Melodelima, C. and Manel, S. (2012). Adaptive genetic variation on the landscape: methods and cases. *Annu. Rev. Ecol. Evol. System.* **43**, 23–43.

Schultz, C.B., Franco, A.M.A. and Crone, E.E. (2012). Response of butterflies to structural and resource boundaries. *J. Anim. Ecol.* **81**, 724–734.

Schultz, C.B. and Crone, E.E. (2001). Edge-mediated dispersal behavior in a prairie butterfly. *Ecology* **82**, 1879–1892.

Sebastián-González, E., Sánchez-Zapata, J.A., Botella, F. and Ovaskainen, O. (2010). Testing the heterospecific attraction hypothesis with time-series data on species co-occurrence. *Proc. R. Soc. B* **277**, 2983–2990.

Senft, R.L., Coughenour, M.B., Bailey, D.W., Rittenhouse, L.R., Sala, O.E. and Swift, D.M. (1987). Large herbivore foraging and ecological hierarchies. *BioScience* **37**, 789–799.

Serrano, D., Forero, M.G., Donazar, J.A. and Tella, J.L. (2004). Dispersal and social attraction affect colony selection. *Ecology* **85**, 3438–3447.

Shaw, R.G. (1987). Maximum-likelihood approaches applied to quantitative genetics of natural populations. *Evolution* **41**, 812–826.

Shikano, T., Shimada, Y., Herczeg, G. and Merilä, J. (2010). History vs. habitat type: Explaining the genetic structure of European nine-spined stickleback (*Pungitius pungitius*) populations. *Mol. Ecol.* **19**, 1147–1161.

Shimatani, I.K., Yoda, K., Katsumata, N. and Sato, K. (2012). Toward the quantification of a conceptual framework for movement ecology using circular statistical modeling. *PLoS ONE* **7**, e50309.

Shmida, A. and Wilson, M. V. (1985). Biological determinants of diversity. *J. Biogeogr.* **12**, 1–20.
Sibly, R. M. and Hone, J. (2002). Population growth rate and its determinants: an overview. *Philos. Trans. R. Soc. Lond. B Biol. Sci.* **357**(1425), 1153–1170.
Skelsey, P., With, K.A. and Garrett, K.A. (2012). Why dispersal should be maximized at intermediate scales of heterogeneity. *Theor. Ecol.* **6**, 203–211.
Slatkin, M. (1985). Gene flow in natural populations. *Annu. Rev. Ecol. Syst.* **16**, 393–430.
Smouse, P.E., Focardi, S., Moorcroft, P.R., Kie, J.G., Forester, J.D. and Morales, J.M. (2010). Stochastic modelling of animal movement. *Philos. Trans. R. Soc. Lond. B Biol. Sci.* **365**, 2201–2211.
Solomon, M.E. (1949). The natural control of animal population. *J. Anim. Ecol.* **18**, 1–35.
Steppan, S.J., Phillips, P.C. and Houle, D. (2002). Comparative quantitative genetics: evolution of the G matrix. *Trends Ecol. Evol.* **17**, 320–327.
Stevens, V.M. and Baguette, M. (2008). Importance of habitat quality and landscape connectivity for the persistence of endangered natterjack toads. *Conserv. Biol.* **22**, 1194–204.
Stoltzfus, A. and Yampolsky, L.Y. (2009). Climbing mount probable: mutation as a cause of nonrandomness in evolution. *J. Heredity* **100**, 637–647.
Sugden, A. (2006). When to go, where to stop. *Science* **313**, 775–775.
Sutherland, W.J., Freckleton, R.P., Godfray, H.C.J., Beissinger, S.R., Benton, T., Cameron, D.D., Carmel, Y., David, A., Coulson, T., Emmerson, M.C., et al. (2013). Identification of 100 fundamental ecological questions. *J. Ecol.* **101**, 58–67.
Swokowski, E.W. (2000). *Calculus: The Classic Edition*. Cengage Advance Books.
Takagi, H., Sato, M.J., Yanagida, T. and Ueda, M. (2008). Functional analysis of spontaneous cell movement under different physiological conditions. *PLoS ONE* **3**, e2648.
Taylor, P.D. (1996). Inclusive fitness arguments in genetic models of behaviour. *J. Math. Biol.* **34**, 654–674.
Tétard-Jones, C., Kertesz, M.A. and Preziosi, R.F. (2011). Quantitative trait loci mapping of phenotypic plasticity and genotype-environment interactions in plant and insect performance. *Philos. Trans. R. Soc. Lond. B Biol. Sci.* **366**, 1368–1379.
Thomson, F.J., Moles, A.T., Auld, T.D. and Kingsford, R.T. (2011). Seed dispersal distance is more strongly correlated with plant height than with seed mass. *J. Ecol.* **99**, 1299–1307.
Thuiller, W., Münkemüller, T., Lavergne, S., Mouillot, D., Mouquet, N., Schiffers, K. and Gravel, D. (2013). A road map for integrating eco-evolutionary processes into biodiversity models. *Ecol. Lett.* **16**(Suppl 1), 94–105.
Thurfjell, H., Ciuti, S. and Boyce, M.S. (2014). Applications of step-selection functions in ecology and conservation. *Mov. Ecol.* **2**, 4.
Tilman, D., May, R.M., Lehman, C.L. and Nowak, M.A. (1994). Habitat destruction and the extinction debt. *Nature* **371**, 65–66.
Tilman, D. (1994). Competition and biodiversity in spatially structured habitats. *Ecology* **75**, 2–16.
Tilman, D., Wedin, D. and Knops, J. (1996). Productivity and sustainability influenced by biodiversity in grassland ecosystems. *Nature* **379**, 718–720.
Tjur, T. (2009). Coefficients of determination in logistic regression models—a new proposal: the coefficient of discrimination. *Am. Stat.* **63**, 366–372.
Tobias, J.A., Cornwallis, C.K., Derryberry, E.P., Claramunt, S., Brumfield, R.T. and Seddon, N. (2014). Species coexistence and the dynamics of phenotypic evolution in adaptive radiation. *Nature* **506**, 359–363.

Tremblay, Y., Robinson, P.W. and Costa, D.P. (2009). A parsimonious approach to modeling animal movement data. *PLoS ONE* **4**, e4711.
Tripathi, N., Hoffmann, M., Willing, E.-M., Lanz, C., Weigel, D. and Dreyer, C. (2009). Genetic linkage map of the guppy, Poecilia reticulata, and quantitative trait loci analysis of male size and colour variation. *Proc. R. Soc. B* **276**, 2195–2208.
Trivers, R.L. and Willard, D.E. (1973). Natural selection of parental ability to vary the sex ratio of offspring. *Science* **179**, 90–92.
Tsoar, A., Nathan, R., Bartan, Y., Vyssotski, A., Dell, G., Ulanovsky, N. and Dell'Omo, G. (2011). Large-scale navigational map in a mammal. *Proc. Natl. Acad. Sci. USA* **108**, E718–E724.
Tuomisto, H., Ruokolainen, K. and Yli-halla, M. (2003). Floristic variation of western Amazonian forests. *Science* **299**, 241–245.
Tuomisto, H. (2010a). A diversity of beta diversities: straightening up a concept gone awry. Part 1. Defining beta diversity as a function of alpha and gamma diversity. *Ecography* **33**, 2–22.
Tuomisto, H. (2010b). A diversity of beta diversities: straightening up a concept gone awry. Part 2. Quantifying beta diversity and related phenomena. *Ecography* **33**, 23–45.
Tuomisto, H. (2010c). A consistent terminology for quantifying species diversity? Yes, it does exist. *Oecologia* **164**, 853–860.
Turchin, P. (1991). Translating foraging movements in heterogeneous environments into the spatial distribution of foragers. *Ecology* **72**, 1253–1266.
Turchin, P. (1998). *Quantitative Analysis of Movement: Measuring and Modeling Population Redistribution in Animals and Plants*. Sunderland, MA: Sinauer Associates.
Turchin, P. (2003). *Complex Population Dynamics: A Theoretical/empirical Synthesis*. Princeton, NJ: Princeton University Press.
Urban, D. and Keitt, T.H. (2001). Landscape connectivity: a graph-theoretic perspective. *Ecology* **82**, 1205–1218.
Vahl, W.K., Boiteau, G., de Heij, M.E., MacKinley, P.D. and Kokko, H. (2013). Female fertilization: effects of sex-specific density and sex ratio determined experimentally for Colorado potato beetles and Drosophila fruit flies. *PloS one* **8**, e60381.
Välimäki, K., Herczeg, G. and Merilä, J. (2012). Morphological anti-predator defences in the nine-spined stickleback: Constitutive, induced or both? *Biol. J. Linn. Soc. Lond.* **107**, 854–866.
Vellend, M. (2010). Conceptual synthesis in community ecology. *Q. Rev. Biol.* **85**, 183–206.
Vincent, T. (1988). The Evolution of ESS theory. *Annu. Rev. Ecol. Syst.* **19**, 423–443.
Viswanathan, G.M., Afanasyev, V., Buldyrev, S. V., Murphy, E.J., Prince, P.A. and Stanley, H.E. (1996). Lévy flight search patterns of wandering albatrosses. *Nature* **381**, 413–415.
Viswanathan, G.M., Buldyrev, S. V., Havlin, S., Da Luz, M.G.E., Raposo, E.P. and Stanley, H.E. (1999). Optimizing the success of random searches. *Nature* **401**, 911–914.
Viswanathan, G.M., Afanasyev, V., Buldyrev, S. V., Havlin, S., Da Luz, M.G.E., Raposo, E.P. and Stanley, H.E. (2000). Lévy flights in random searches. *Physica A* **282**, 1–12.
Viswanathan, G.M., da Luz, M.G.E., Raposo, E.P. and Stanley, H.E. (2011). *The Physics of Foraging: An Introduction to Random Searches and Biological Encounters*. New York: Cambridge University Press.
Volkov, I., Banavar, J.R., He, F., Hubbell, S.P. and Maritan, A. (2005). Density dependence explains tree species abundance and diversity in tropical forests. *Nature* **438**, 658–661.
Volterra, V. (1926). Variazioni e fluttuazioni del numero d'individui in specie animali conviventi. *Mem. Acad. Lincei Roma* **2**, 31–113.

de Vries, H. (1914). The principles of the theory of mutation. *Science* **40**, 77–84.
Wang, H., Nagy, J.D., Gilg, O. and Kuang, Y. (2009). The roles of predator maturation delay and functional response in determining the periodicity of predator-prey cycles. *Math. Biosci.* **221**, 1–10.
Watson, J.D.M. and Crick, F.H.C. (1953). Molecular structure of nucleic acids. *Nature* **171**, 737–738.
Waxman, D. and Gavrilets, S. (2005). 20 questions on adaptive dynamics. *J. Evol. Biol.* **18**, 1139–1154.
Wedin, D. and Tilman, D. (1993). Competition among grasses along a nitrogen gradient: Initial conditions and mechanisms of competition. *Ecol. Monogr.* **63**, 199–229.
Weinberg, W. (1908). Über den Nachweis der Vererbung beim Menschen. *Jahreshefte des Vereins für vaterländische Naturkunde in Württemberg* **64**, 368–382.
Weir, B.S. and Cockerham, C.C. (1984). Estimating F-statistics for the analysis of population structure. *Evolution* **38**, 1358–1370.
Wennekes, P.L., Rosindell, J. and Etienne, R.S. (2012). The neutral-niche debate: a philosophical perspective. *Acta Biotheor.* **60**, 257–271.
van der Werf, J., Graser, H.-U., Frankham, R. and Gondro, C. (2009). *Adaptation and Fitness in Animal Populations*. Berlin: Springer.
White, G.C. and Garrott, R.A. (1990). *Analysis of Wildlife Radio-Tracking Data*. San Diego, CA: Academic Press.
Whitlock, M.C. (1995). Variance-induced peak shifts. *Evolution* **49**, 252–259.
Whitlock, M.C. and Guillaume, F. (2009). Testing for spatially divergent selection: comparing QST to FST. *Genetics* **183**, 1055–1063.
Whitlock, M.C. and McCauley, D.E. (1999). Indirect measures of gene flow and migration: FST not equal to $1/(4Nm + 1)$. *Heredity* **82**, 117–125.
Whittaker, R.H. (1960). Vegetation of the Siskiyou Mountains, Oregon and California. *Ecol. Monogr.* **30**, 279–338.
Whittemore, A.S. and Halpern, J. (1994). Probability of gene identity by descent: computation and applications. *Biometrics* **50**, 109–117.
Wicklund, C. and Fagerstrom, T. (1977). Why do males emerge before females? *Oecologia* **31**, 153–158.
Wiens, J.A. (1976). Population responses to patchy environments. *Annu. Rev. Ecol. Syst.* **7**, 81–120.
Wilcox, B.A. and Murphy, D.D. (1985). Conservation strategy: the effects of fragmentation on extinction. *Am. Nat.* **125**, 879–887.
Willi, Y., Van Buskirk, J. and Hoffmann, A.A. (2006). Limits to the adaptive potential of small populations. *Annu. Rev. Ecol. Evol. System.* **37**, 433–458.
Williams, R.J. and Martinez, N.D. (2000). Simple rules yield complex food webs. *Nature* **404**, 180–183.
Wilson, H.B., Hassell, M.P. and Holt, R.D. (1998). Persistence and area effects in a stochastic tritrophic model. *Am. Nat.* **151**, 587–595.
Wisz, M.S., Pottier, J., Kissling, W.D., Pellissier, L., Lenoir, J., Damgaard, C.F., Dormann, C.F., Forchhammer, M.C., Grytnes, J.-A., Guisan, A., et al. (2013). The role of biotic interactions in shaping distributions and realised assemblages of species: implications for species distribution modelling. *Biol. Rev.* **88**, 15–30.
Wittemyer, G., Polansky, L., Douglas-Hamilton, I. and Getz, W.M. (2008). Disentangling the effects of forage, social rank, and risk on movement autocorrelation of elephants using Fourier and wavelet analyses. *Proc. Natl. Acad. Sci. USA* **105**, 19108–19113.

World Resources Institute. (2005). *Millennium Ecosystem Assessment. Ecosystems and Human Well-Being: Biodiversity Synthesis.* Washington, DC.
Wright, S. (1931). Evolution in Mendelian populations. *Genetics* **16**, 97–159.
Wright, S. (1932). The roles of mutation, inbreeding, crossbreeding, and selection in evolution. *Proceedings of the 6th International Congress of Genetics* **1**, 356–366.
Wright, S. (1937). The distribution of gene frequencies in populations. *Proc. Natl. Acad. Sci. USA* **23**, 307–320.
Wright, S. (1942). Statistical genetics and evolution. *Bull. Am. Math. Soc.* **48**, 223–246.
Wright, S. (1948). On the roles of directed and random changes in gene frequency in the genetics of populations. *Evolution* **2**, 279–294.
Wright, S.J. (2002). Plant diversity in tropical forests: a review of mechanisms of species coexistence. *Oecologia* **130**, 1–14.
Wu, C., Orozco, C., Boyer, J., Leglise, M., Goodale, J., Batalov, S., Hodge, C.L., Haase, J., Janes, J., Huss, J.W., et al. (2009). BioGPS: an extensible and customizable portal for querying and organizing gene annotation resources. *Genome Biol.* **10**, R130.
Yi, N. and Xu, S. (2008). Bayesian LASSO for quantitative trait loci mapping. *Genetics* **179**, 1045–1055.
Yu, J., Pressoir, G., Briggs, W.H., Vroh Bi, I., Yamasaki, M., Doebley, J.F., McMullen, M.D., Gaut, B.S., Nielsen, D.M., Holland, J.B., et al. (2006). A unified mixed-model method for association mapping that accounts for multiple levels of relatedness. *Nat. Genet.* **38**, 203–208.
Zemek, R. and Nachman, G. (1999). Interactions in a tritrophic acarine predator-prey metapopulation system: prey location and distance moved by *Phytoseiulus persimilis* (Acari: Phytoseiidae). *Exp. Appl. Acar.* **23**, 21–40.
Zheng, C., Ovaskainen, O. and Hanski, I. (2009a). Modelling single nucleotide effects in phosphoglucose isomerase on dispersal in the Glanville fritillary butterfly: coupling of ecological and evolutionary dynamics. *Philos. Trans. R. Soc. Lond. B Biol. Sci.* **364**, 1519–1532.
Zheng, C., Pennanen, J. and Ovaskainen, O. (2009b). Modelling dispersal with diffusion and habitat selection: Analytical results for highly fragmented landscapes. *Ecol. Model.* **220**, 1495–1505.
Zimmerman, E., Palsson, A. and Gibson, G. (2000). Quantitative Trait Loci Affecting Components of Wing Shape in Drosophila melanogaster. *Genetics* **155**, 671–683.
Zurell, D., Berger, U., Cabral, J.S., Jeltsch, F., Meynard, C.N., Münkemüller, T., Nehrbass, N., Pagel, J., Reineking, B., Schröder, B., et al. (2010). The virtual ecologist approach: simulating data and observers. *Oikos* **119**, 622–635.
Zuur, A., Ieno, E. N. and Smith, G. M. (2007). *Analysing Ecological Data.* New York: Springer.
Zuur, A.F., Ieno, E.N., Walker, N.J., Saveliev, A.A., Smith, G.M. and Park, W. (2009). *Mixed Effects Modelling and Extensions in Ecology with R.* New York: Springer.
Zuur, A.F., Ieno, E.N. and Elphick, C.S. (2010). A protocol for data exploration to avoid common statistical problems. *Methods Ecol. Evol.* **1**, 3–14.

Index

A

abiotic environment 122–3, 125, 135, 166
absorbing boundary condition
  (*see* boundary condition)
absorbing state 79, 232
adaptive
  dynamics 194–8, 210
  radiation 126
additive genetic
  effect 177, 181, 240, 242
  relationship 177 (*see also* coancestry coefficient)
  value (*see* breeding value)
  variation 179–81, 183, 184 (*see also* genetic variation, G-matrix)
advection 25, 30–5 (*see also* diffusion)
aggregation (*see* spatial aggregation)
Åland Islands 15–7, 43–4
Allee effect 70
allele 169–71, 175–6, 180, 228
  dominant 170
  recessive 170
allele frequency (*see* gene frequency)
allelic
  effect 170 (*see also* genetic effect)
  variant 170, 176, 183, 228
ancestral population (*see* population)
anisotropic diffusion (*see* diffusion)
animal
  breeding 171, 181
  model 180, 198, 209–10, 212, 240–3
ansatz 220
antagonistic interaction 124–5 (*see also* biotic interaction)
artificial selection (*see* selection)
association analysis 204
asymmetric competition (*see* competition)
autocorrelation function (ACF) 58–9, 67, 103–4, 120–1 (*see also* spatial autocorrelation, temporal autocorrelation, velocity autocorrelation function)

autoregressive model (AR) 102–5, 150–5
  first order 102–5, 150–1
  second order 152–4

B

Bayes
  rule 249
  theorem 246–8
Bayesian statistics 5, 61–2, 105–6, 109–11, 245–9
behavioural switching 64
Bernoulli distribution (*see* distribution)
Bessel function 23, 45
beta distribution (*see* distribution)
biased random walk (*see* random walk)
binomial distribution (*see* distribution)
biomechanical paradigm (*see* paradigm)
biotic
  environment 13, 123–5, 135
  interaction 123–5, 129, 140
birth rate 76, 81, 105, 119, 135–8
boundary condition 39–40
  absorbing 30
  reflecting 30, 32
breeder's equation 188–90
  multivariate 188, 190
breeding value 176, 180–90, 204, 228, 240
Brownian motion 17 (*see also* random walk)
butterfly metapopulation model 72, 89–93, 100–2, 115–18, 120

C

capture-mark-recapture data (CMR) 2, 16–17, 35, 40, 51–3
capture probability 61–3
carrying capacity 69, 75, 78, 81, 83–6, 90–1, 111
categorical variable (*see* explanatory variable)
Cauchy distribution (*see* distribution)
character displacement 126
characteristic scale 59
chromosome 169–70
circuit theory 66

clonal reproduction (*see* reproduction)
Clouded Apollo butterfly 40–2
coancestry 177–80
  coefficient 179, 182, 193, 202
  matrix 202, 208, 241
coexistence 126, 130–3, 138, 165–6
cognitive paradigm (*see* paradigm)
colonization 17, 43, 69, 71, 73–4, 115–17, 165
  -competition trade-off 126, 141
  -extinction dynamics (*see* metapopulation dynamics)
  rate 71, 94–6
commensalism 124 (*see also* biotic interaction)
community
  assembly 122
  -level model 158–9
  local 126
  similarity 160–1
competition 69–70, 76–8, 122–5, 129–34, 141–5 (*see also* kin competition)
  among kin 200, 211
  asymmetric 130
  interference 70, 135
  kernel (*see* kernel)
  sensitivity to 149–50
competitive
  ability 72, 125–6, 165
  community 144, 158–9
  exclusion 125, 129 (*see also* coexistence)
  influence 149
  interaction 129–30, 145 (*see also* biotic interaction)
complete spatial randomness 75–7
confidence interval (CI) 55, 57, 109, 246
conjugate distribution (*see* distribution)
connectivity 35–6, 66, 73–4, 116–17
  functional 35–8, 40–3,
  population dynamical 116
  structural 36
consumer-resource interaction (*see* resource-consumer interaction)
consumption rate (*see* resource consumption rate)
continuous explanatory variable (*see* explanatory variable)
continuous-time model (*see* model)
convolution 81–3, 119, 217–19, 227
co-occurrence 156–7, 162–4
correlated random walk (*see* random walk)
correlation 225 (*see also* autocorrelation function)
correspondence analysis (CA) 161
corridor 36, 40–3
covariance 225

covariate (*see* explanatory variable)
credibility interval 110, 114, 246
cross-validation 3
cumulative distribution function (CDF) (*see* distribution)
cyclic dynamics (*see* oscillatory dynamics)

D

dam 175, 177–9, 184, 240 (*see also* sire)
death rate 22, 30, 46, 69–70, 76–8, 105, 130, 141
degree of freedom 234
delayed density dependence (*see* density dependence)
demographic heterogeneity 118
  stochasticity 79–80, 85–6, 102–5
  structure 68–9, 71–2
density dependence 89–90, 118, 135, 138–40, 151–2
  to conspecifics 89–90, 118, 135
  delayed 152–4
  direct 70
  to heterospecifics 125, 135
  inverse 70
deoxyribonucleic acid (DNA) 168–71, 211
dependent variable (*see* response variable)
derivative 210, 217–22
design matrix 237–41
deterministic model (*see* model)
developmental instability 176
difference equation 7, 222
differential equation 7, 39–40, 130, 136, 218–22
  autonomous 220
  linear 108, 220
  non-linear 130, 136
  ordinary 221–2
  partial (PDE) 20–1, 119, 221–2
  system of 220–1, 232
diffusion
  -advection model 25–6
  -advection-reaction model 29–30, 39, 44
  anisotropic 30
  approximation 56, 64
  coefficient 20–1, 23–4, 30, 34–5, 55–9, 63–4
  isotropic 30
  model 20, 22, 25, 34, 47, 60–5
  rate 30–5, 38, 44–5, 57
diffusive movement 64
diploid 169–70, 175
direct density dependence (*see* density dependence)

directional
  bias 25
  persistence 23
  selection (*see* selection)
discrete-time model (*see* model)
dispersal 13–14
  active 12, 27, 70, 120
  distance 199–201, 212
  evolution 195, 211
  kernel (*see* kernel)
  limitation 125, 167
  passive 12, 65, 70, 120
  of propagules 12–13, 21, 70, 76–7, 81, 86–8, 195–8
displacement 2, 18–21, 54
  mean squared (MSD) 52, 56, 58
distribution (*see also* random variable)
  Bernoulli 79, 235–6, 243, 246
  beta 247, 249–51
  binomial 246–9
  bivariate normal 75, 187, 192, 199, 225
  Cauchy 28, 54–5, 57
  conjugate 249–50
  cumulative distribution function (CDF) 112, 155, 224–5, 236
  exponential 37, 50, 54, 231
  geometric 223–4
  joint 111, 225
  joint posterior 109, 111
  log-normal 54, 92, 143
  marginal 62–3, 109–11
  multivariate normal 180, 225, 237, 242, 245
  normal 85, 92–3, 112, 155, 223–5, 235–6
  Poisson 89–90, 236
  posterior 5, 62–3, 109–11, 246–52
  prior 110–11, 246–52
  probability density function (PDF) of 223–7
  quasi-stationary 36–7, 79–80, 95–6, 232
  stationary 26, 79, 135, 200, 232
  uniform 19, 112, 143, 201, 227, 248
  Weibull 28, 54–7
  wrapped Cauchy 28, 54–5, 57
diversifying selection (*see* selection)
DNA (*see* deoxyribonucleic acid)
dominance 170, 178 (*see also* allele)
drift (*see* ecological drift, genetic drift)

E

eco-evolutionary model 198, 209–11
ecological
  community 122–4, 127
  drift 123, 127

interaction (*see* biotic interaction)
niche (*see* niche)
niche model (ENM) 112, 165
succession (*see* succession)
edge-mediated behaviour 15, 27–9, 34–5, 52–3, 63, 65
eigenvalue (*see* matrix)
eigenvector (*see* matrix)
emigration 12, 14, 69–71, 120, 169
  probability 44–6, 50
  rate 3, 43, 71
endemics-area relationship 146
endogenous factor 69
environmental
  filtering 122, 140, 166
  fluctuations (*see* environmental stochasticity)
  heterogeneity 25, 27, 83, 87, 119–20, 166
  stochasticity 13, 70, 84–6, 92–3, 101–5, 118–19 (*see also* regional stochasticity)
epigenetic effect 211
epistasis 170, 176
equilibrium state 75, 94–5, 131–40 (*see also* stationary state)
error distribution 234
establishment probability 81, 87, 138, 141–3
Eulerian approach 18–22
Euler's formula 221
evolution 170–1 (*see also* selection, genetic drift, mutation)
  convergent 173
evolutionary
  evolutionary dynamics 168, 172, 194
  invasion analysis (*see* adaptive dynamics)
  rescue effect 71, 172
  stable strategy (ESS) 197
exogenous factor 69
explanatory variable (predictor) 105, 155–7, 233–44 (*see also* fixed effect, random effect)
  categorical factor 234, 238–9
  continuous covariate 114–15, 155–61, 233–4, 238, 244–5
  interaction 234
exponential distribution (*see* distribution)
exponential growth 219, 222
extinction
  debt 146, 148
  local 17, 71, 92, 100, 115, 119
  mean time to 86, 95–6
  rate 94–6
  risk of 71, 86, 92, 98, 100, 119, 166
  threshold 71, 73, 80, 94, 101–2
  vortex 173, 194
extrapolation 2

F

factor (*see* explanatory variable)
fat-tailed distribution (*see* leptokurtic distribution)
fecundity rate 70, 76, 84–5, 87–8, 130, 141–4
first passage time (FPT) 52, 59–60
fitness 172–3, 175, 188
    function 185–9
    inclusive 198
fixed effect 238–44 (*see also* random effect, explanatory variable)
foraging 12–4, 64
forward
    approach 2–5
    persistence (*see* directional persistence)
fragmentation (*see* habitat loss and fragmentation)
F_{ST} 193, 206–7, 213
F_{ST}-Q_{ST} test 213
functional connectivity (*see* connectivity)
functional response 124–5
fundamental niche (*see* niche)

G

gamete 170–1
gene 169–70, 210–11, 228
    annotation 213
    flow 169–72, 184–6, 194, 202–3
    frequency 169–71, 184
    modifier 170
    regulation 211
generalist species 143 (*see also* specialist species)
generalized linear model (*see* linear model)
genetic
    architecture 175
    code 168
    composition 170–1
    covariance 179
    differentiation 170–1
    divergence (*see* population divergence)
    drift 169–71, 183–6, 190–1, 194, 207–10
    effect 177, 181, 203
    linkage 175, 213
    marker 201–2, 208, 212–13
    population structure 193, 201–2
    segregation 175
    variation 179–84, 188
genome 169, 204
    assembly 213
    -wide association analysis 204
genotype 169–71, 176–7, 185–8, 202
genotype-environment interaction 177
genotype-phenotype map 177, 210
geometric distribution (*see* distribution)
Gillespie algorithm 106, 231
Glanville fritillary butterfly 15–17, 43–4, 73–4, 89
Global Positioning System (GPS) data 50–4, 66
G-matrix 189–90, 210
    evolution of 210
graph-theoretic analysis 66
growth rate 70, 74–5, 80–1, 90, 103, 111, 151, 154, 195–7

H

habitat
    area 97–8, 146–8
    association 167, 211
    loss and fragmentation 71, 97, 120, 145, 194
    patch 16–17, 27, 43–4, 73–4, 89, 94–100, 201
    patch network 7, 16, 43–4, 47, 73–4
    preference 34–5, 41, 44–9
    quality 12–13, 30, 34, 97
    selection 13, 27–9, 63, 67
    selection analysis 53
    selection ratio 52–3, 61
    suitability model (HSM) 112 (*see also* species distribution model)
Hamilton-May model 194, 211
haploid 169, 175
haplotype 170, 213
Hardy-Weinberg equilibrium 170
heritability 168–71, 181, 187–9, 203–4
    broad-sense 181
    narrow-sense 181
heterogeneous-space model 27, 83–4, 86, 89, 140 (*see also* homogeneous-space model)
hierarchical model 244–5
highly fragmented landscape 27, 43, 71, 115, 120 (*see also* habitat loss and fragmentation)
hill-topping behaviour 33
hitting probability 36–40, 50, 62
home-range model 25–6, 210
homogeneous-space model 17, 73–4, 129, 162–3 (*see also* heterogeneous-space model)
homoscedasticity (*see* residual)
host-parasite interaction 124, 129, 135 (*see also* biotic interaction)
H-test 208–9 (*see also* S-test)
hypothesis 1, 3, 8, 50

I

ideal free distribution 65
identity by descent 179, 228
identity matrix (*see* matrix)

immigration 14, 44–6, 69, 71
  probability 45–6
inbreeding 12, 98, 172–3, 179, 194
  depression 173
inclusive fitness (*see* fitness)
independent assortment 170, 175
independent variable (*see* explanatory variable)
individual-based model (IBM) 6–7, 17, 75–8, 86–8, 105–6, 131, 135, 137, 198–9, 209–10
infinitesimal model 175, 185
initial condition 28, 131, 139–40, 218–22
integral 22, 36–9, 83, 110, 217–18
integro-differential equation 83, 119 (*see also* differential equation)
intercept 103, 155, 233–4, 244–5
interference competition (*see* competition)
internal state 11
interspecific interaction 69, 122–5 (*see also* intraspecific interaction, biotic interaction)
intraspecific interaction 69, 130, 133 (*see also* interspecific interaction, biotic interaction)
invasion fitness 197 (*see also* fitness, adaptive dynamics)
inverse approach 2–5
inverse density dependence (*see* density dependence)
isotropic diffusion (*see* diffusion)

J
joint species distribution model (JSDM) 155, 167, 241 (*see also* species distribution model)

K
kernel
  competition 76, 78, 81–3, 130, 142–4
  dispersal 21–3, 76, 96, 218
  radially symmetric 22, 76
kin competition (*see* competition)
Kronecker product 181, 243, 245
Kronecker's delta 180, 228, 243

L
Lagrangian approach 18–21
Lande's metapopulation model 101–2
leptokurtic distribution 23
Levins metapopulation model 93, 100, 119 (*see also* Lande's metapopulation model)
Lévy walk 64
likelihood 61–2, 107–11, 246–9
limit cycle 139–40
lineage 191–3, 202–3

linear
  predictor 155, 160, 234–44
  regression (*see* regression)
linear algebra (*see* matrix algebra)
linear model 103–4, 153–4, 233
  generalized 112, 233–6
  mixed 233, 238
  multivariate 241
  with random effects (*see* random effect)
linkage (*see* genetic linkage)
link function 112, 234
  log 236
  logit 235
  probit 235–6
local adaptation 190–2, 206–9
locus 169–70, 176, 179–80, 185, 205–7
log link function (*see* link function)
log-normal distribution (*see* distribution)
logistic model 74–83, 102–9
  deterministic 74–5, 80–2
  heterogeneous space 86–8
  individual-based 76–80, 86–8, 105–6
  mean-field 74–5, 105–7
  spatial 81–3
  stochastic 78
logistic regression (*see* regression)
logit link function (*see* link function)

M
marginal distribution (*see* distribution)
marker (*see* genetic marker)
Markov
  chain 7, 229–31
  process 7, 50, 65, 78, 94–6, 105–6, 231–2
Markov chain Monte Carlo (MCMC) 250–2
mass-effect paradigm (*see* paradigm)
matching condition 34–5
Matérn covariance structure 113–14
mathematical model (*see* model)
matrix
  algebra 47, 215–17, 229–32
  correlation 156, 244
  dimension of 216, 237
  eigenvalue 97, 216, 220–1, 231–2
  eigenvector 216, 220–1, 231–2
  identity 75, 181, 192, 232
  inverse of 189, 216, 220
  multiplication 215–16, 230, 237
  positive definite 42–4
  symmetric 156–8, 179, 238, 242–4
  transpose 216
  unsuitable 17, 27–8, 40–4, 52, 115, 145–6, 200

matrix (*continued*)
  variance-covariance 75, 158, 180–1, 192, 199, 225, 237–45
maximum likelihood (ML) 108, 238, 246 (*see also* likelihood)
mean
  -field assumption 75
  -field model (MFM) 84–8, 95, 106–7, 129–42
Melitaea cinxia (*see* Glanville fritillary butterfly)
Mendel's laws 140, 168, 170, 175
metacommunity 126, 164
metapopulation
  capacity 73–4, 98, 102
  dynamics 17, 73–4, 91–5, 115, 211
  model 72, 74, 89, 93, 100, 115
  spatially realistic model 96, 102, 115
Metropolis-Hastings algorithm 109–10, 117, 250–2
migration 13–14, 43–6, 69–71, 116–17 (*see also* dispersal)
mixed model (*see* linear model)
model
  continuous-time 7, 65, 76, 222, 230–1
  definition 8
  deterministic 6, 80–4, 94–6, 119, 140, 222
  discrete-time 7, 18, 28, 89, 222, 229–31
  fitting (parameterization) 4–5, 60, 105, 113, 153–5, 245
  individual-based (*see* individual-based model)
  mathematical 1–2, 6–8
  mean-field (*see* mean-field model)
  null (*see* null model)
  prediction 3–5, 32, 35, 94, 113, 233
  statistical 1–2, 6–8, 233
  stochastic 6–7, 17, 229
  uncertainty 3–5 (*see also* parameter uncertainty)
  validation 5
mortality rate (*see* death rate)
motion capacity 11–12
movement
  corridor (*see* corridor)
  directional 23, 25
  distance (*see* displacement)
  probability 40–4, 47–50
  spatial scale of 59
  speed 12–13, 17, 25, 58–9
  temporal scale of 59
  trajectory 19, 24, 26, 54, 131
  types of 13
multiple testing 204–5
mutant 197–8
mutation 169–72, 185–6, 197–200
mutualistic interaction 124, 129, 145 (*see also* biotic interaction)

N

natural selection (*see* selection)
navigation capacity 11, 13
neutral
  genetic drift (*see* genetic drift)
  marker (*see* genetic marker)
  paradigm (*see* paradigm)
  theory 123, 167
niche 112, 120–1, 123, 125–7, 148–50, 165–7
  differentiation 126, 166
  fundamental 71, 87–8, 120–1, 125–6, 141, 165
  partitioning 166
  realized 88, 125–6, 141
  separation 125, 150
non-linear dynamics 92, 152, 221
nucleobase 168
nucleotide 168, 202
null
  expectation 164, 208
  model 18, 74, 127, 167, 184
numerical response 124

O

observation
  model 61, 66, 106
  process 50–1, 61–2
occupancy 36–9, 73–4, 91–3, 112–7
  time 36–9, 44, 46–50
  time density 36–9
optimality paradigm (*see* paradigm)
ordinary differential equation (*see* differential equation)
ordination 8, 161–2
oscillatory dynamics 90, 105, 138
over-depletion of resources 13

P

pairwise invisibility plot (PIP) 196–7 (*see also* adaptive dynamics)
paradigm 14–15, 126, 164
  biomechanical 14
  cognitive 15
  mass-effect 126, 165
  neutral 126–7, 165
  optimality 14, 65
  patch-dynamics 126
  species sorting 165
parameter uncertainty 4, 110–11, 208, 246

parasitic interaction 124 (*see also* biotic interaction)
parent-offspring regression 203–4, 212
partial autocorrelation function (PACF) 120–1 (*see also* autocorrelation function)
partial derivative 20, 222 (*see also* derivative)
partial differential equation (*see* differential equation)
particle-sampling 27–8
patch (*see* habitat patch)
  -dynamics paradigm (*see* paradigm)
  network (*see* habitat patch network)
  occupancy (*see* occupancy)
patch-dynamics paradigm (*see* paradigm)
pedigree 177–81
phase-space plot 130–3, 137–8
phenotype 169–71, 175–7
phenotypic value 176, 181, 185–90, 203
phylogenetic relationship 123–4, 167
plant
  breeding 171, 181
  community model 142–7, 155–61
  population model 86, 98, 112–13, 198, 211
point pattern analysis 162
point process 7, 67, 76, 162
Poisson distribution (*see* distribution)
Poisson regression (*see* regression)
polygenetic trait 170, 175
polynucleotide 168
population
  ancestral 178–84, 190–1
  dispersion 68
  divergence 173, 182–4, 190, 206–9
  dynamical connectivity (*see* connectivity)
  dynamics 68, 81, 83
  growth rate (*see* growth rate)
  local 40, 43, 69, 71, 90–2, 99–100, 182, 190–3
  size 68, 150
positional error 66
positive definite matrix (*see* matrix)
posterior distribution (*see* distribution)
power-law 64, 86, 148
predator-prey interaction 124, 129, 138–40, 150–5, 162–3 (*see also* biotic interaction)
prediction (*see* model prediction)
predictor (*see* explanatory variable)
pre-emption 13
presence-absence data 88, 112–13, 142, 157, 234
principal component analysis (PCA) 161
prior distribution (*see* distribution)
probability density function (*see* distribution)

probit
  link function (*see* link function)
  regression (*see* regression)
process model 61, 106
proximate cause 12, 14

Q

Q_{ST} 213
Quantitative
  genetics 175, 198, 228
  trait loci (QTL) 204–7
  trait loci mapping 204–6
quasi-stationary distribution (*see* distribution)

R

random effect 115, 238, 244 (*see also* fixed effect, explanatory variable)
random variable 79, 92–3, 210, 211–12, 222, 229 (*see also* distribution)
  continuous-valued 223
  discrete-valued 222
  expectation of 223, 226–7
  independence of 225–7
  joint distribution of 225
  sum of 226
  variance of 223
random walk 15, 17–29, 64
  biased 25
  correlated 23–5, 54–6
  heterogeneous-space 27–9
realized niche (*see* niche)
recolonization (*see* colonization)
recombination 169–71, 185–6, 209–10
redundancy analysis (RDA) 161
reflecting boundary condition (*see* boundary condition)
regional stochasticity 89
regression 128, 203–6, 212 (*see also* linear model)
  coefficient 155–8, 233–4, 244
  logistic 115, 235–6, 243–5
  Poisson 236
  probit 112–13, 236, 243
reproduction
  clonal 170, 175, 186–9
  sexual 72, 169–71, 175
rescue effect 71, 172
resident population 197–8
residual 233–45
  homoscedasticity 233, 237
  independence of 234–9
  spatially correlated 113–15, 121

resource
    availability 10–14, 69–71, 97, 124–9, 135–40
    -consumer interaction 129, 135, 150, 162–3 (see also biotic interaction)
    consumption rate 135–6
    conversion efficiency 135–6
    renewal rate 136–8
response variable 233–44 (see also state variable)
ribonucleic acid (RNA) 168
Ricker model 89–90,
R-package
    driftsel 208
    MCMCglmm 212
    RAFM 202
    spatstat 164

S

sampling interval 50, 54–7
    irregular 50, 66
second-order spatial moment (see spatial moment)
segregation (see genetic segregation, spatial segregation)
selection
    artificial 171
    differential 186
    directional 188–9, 210
    diversifying 208–9, 213
    natural 167–71, 185–6, 194
    negative 171
    positive 171
    ratio 53
    stabilizing 208–13
sexual reproduction (see reproduction)
single nucleotide polymorphism (SNP) 202, 204–5, 207
sire 175, 177–9, 184, 204, 240–1 (see also dam)
source-sink dynamics 71, 88, 121, 125–6, 172
spatial
    aggregation 162–3
    autocorrelation 114–15, 162, 237–8
    heterogeneity 27, 86, 89
    logistic model (see logistic model)
    moment 75–7, 162
    randomness 75–7, 135
    scale 73, 96, 114 (see also characteristic scale)
    segregation 162
spatially correlated environmental stochasticity (see regional stochasticity)
spatially realistic metapopulation model (see metapopulation model)
spatiotemporal point process (see point process)
specialist species 70, 143, 150, 161

species
    -abundance relationship 127, 167
    -area relationship (SAR) 127, 145, 166–7
    composition 122, 159
    distribution model (SDM) 88, 112, 234, 243 (see also joint species distribution model)
    richness 123, 128, 143–4, 159–60
    sorting paradigm (see paradigm)
stabilizing selection (see selection)
stable equilibrium 131–40, 221
state-space model 61, 105–6
state variable 6, 18, 20 (see also response variable)
stationary
    distribution (see distribution)
    state 32, 76, 79–80, 230–2
statistical
    independence (see residual)
    model (see model)
step length
    distribution 23–5, 29, 54–7, 64
    mean of 55
    squared mean of 55
    variability index 59
step selection
    analysis 53–4
    function 67
S-test 208–9, 213 (see also H-test)
stickleback 173–4, 190–1
stochastic
    logistic model (see logistic model)
    model (see model)
    patch occupancy model (SPOM) 7, 115–17 (see also metapopulation model)
    process 18–19, 167, 222, 229
stochasticity
    demographic, see demographic stochasticity
    environmental, see environmental stochasticity
    regional, see regional stochasticity
stopping rate 2–24
structural connectivity (see connectivity)
succession 124, 126
symbiosis 124 (see also biotic interaction)
symmetric matrix (see matrix)
system of differential equations (see differential equation)
system of linear equations (see matrix algebra)

T

temporal
    autocorrelation 84–6, 105
test-statistic 8

time-series analysis 102, 105, 150, 238
　multivariate 167
Tjur's R^2 113–14
tortuosity 67
total response 124
tracking data 16, 50
trait 10–11, 123, 149–50, 168–213, 241–3
trajectory (*see* movement trajectory)
transition
　matrix 95, 229–32
　rate 78, 231–2
transpose (*see* matrix)
turning angle distribution 16, 18–19, 23–9, 54–9
　mean cosine of 55–9

U

ultimate cause 12, 14
unbiased estimate 62, 66

uniform distribution (*see* distribution)
unstable equilibrium 131, 133 (*see also* stable equilibrium)
unsuitable matrix (*see* matrix)

V

validation (*see* model validation)
variable (*see* explanatory variable, random variable, response variable)
variance-covariance matrix (*see* matrix)
vector 215
velocity autocorrelation function (VAF) 58–9 (*see also* autocorrelation function)
vital rate 83

W

Weibull distribution (*see* distribution)

The manufacturer's authorised representative in the EU for product safety is
Oxford University Press España S.A. of el Parque Empresarial San Fernando de
Henares, Avenida de Castilla, 2 – 28830 Madrid (www.oup.es/en or product.
safety@oup.com). OUP España S.A. also acts as importer into Spain of products
made by the manufacturer.

www.ingramcontent.com/pod-product-compliance
Ingram Content Group UK Ltd.
Pitfield, Milton Keynes, MK11 3LW, UK
UKHW022241230426
12048UKWH00018BA/1403